STATISTICS

STATISTICS

The Essentials for Research

HENRY E. KLUGH

ALMA COLLEGE

John Wiley & Sons, Inc.

New York · London · Sydney · Toronto

Library of Congress Catalogue Card Number: 72-104188

SBN 471 49370 8

Printed in the United States of America

10 9 8 7 6 5 4 3 2 1

Preface to the Instructor

This textbook was written for the one-semester course in statistics normally required of students majoring in the behavioral sciences, particularly in psychology or education. Most of these students have had little college level mathematics, no previous statistics, and they probably will not have any additional course work in statistics unless it is required in graduate work. For a majority of these students the course in statistics is both introductory and terminal. Any text for such a course should accomplish several purposes that are, unfortunately, almost mutually exclusive. First, the text must be easy to understand, must require relatively little mathematical sophistication and, where necessary, must sacrifice rigor for clarity and teachability. Second, the text must provide the student with enough statistical sophistication so that he can read research literature with understanding and can conduct experiments of his own which are complex enough to be interesting. Finally, the text should also provide a foundation for additional course work in graduate school. However, if the first two purposes have not been met, if the student is lost in a labyrinth of rigor or plods through material irrelevant for research, he is unlikely to be very enthusiastic about graduate study which requires additional work in statistics.

It should be possible to cover most, if not all, of this text in a four credit one-semester course, even if it is taught to students who have had no formal mathematics instruction beyond a good high school program. However, this text includes some rather difficult material. It reflects the changing trends in the kinds of statistical analyses *used* by investigators in the behavioral sciences. Evidence is available that the introductory semester of statistics must now go beyond the t test, the correlation coefficient and chi square, if it is to help undergraduates understand very much of contemporary research literature. Edgington[1] has noted a particularly interesting trend in the use of statistical analyses by investigators pub-

[1] Edgington, E. S. "A Tabulation of Inferential Statistics Used in Psychology Journals." *American Psychologist*, 1964, **19**, 202–203.

lishing in the American Psychological Association Journals from 1948 to 1962. He found that the use of the analysis of variance increased from 11 percent to 55 percent of all articles, while use of the t test and correlational techniques decreased from about one-half to approximately 20 percent. Chi square remained stable at 10 percent, and the incidence of nonparametric statistics rose from 3 percent to 20 percent, with the Mann-Whitney U test the most common.

This trend is reflected in the coverage of this text. Analysis of variance occupies almost one-third of the book and most of the major nonparametric statistics are explained with one or more example problems. The emphasis on analysis of variance in an introductory course has required economies in some more traditional areas. Little space is devoted to confidence intervals, and the chapter on correlation and regression is quite brief, although it should be adequate for most purposes. Obviously, some rigor is also sacrificed; hopefully, an intuitive understanding will substitute for rigor at this stage of the student's preparation.

In summary, the aim of this text is to present clearly and concisely the logic and methods of calculation for statistical techniques most commonly used in the behavioral sciences, particularly in psychology and education, within the restrictions of a one-semester introductory course for mathematically unsophisticated students.

H. E. Klugh

Acknowledgments

I gratefully acknowledge the assistance of my wife, Barbara, whose encouragement and typing skills expedited the prepublication revisions of the text and thus contributed substantially to what clarity the manuscript may possess. I also thank the several classes of Alma College students who labored through prepublication versions of the text, discovering errors and making suggestions for improvement. Also I thank the editorial staff of the publisher for assistance and encouragement.

I am particularly indebted to the Literary Executor of the late Sir Ronald A. Fisher, F.R.S. Cambridge, and to Oliver and Boyd Ltd., Edinburgh for permission to reprint Tables III and IV from their book *Statistical Methods for Research Workers*, and for their permission, with that of Dr. Frank Yates, F.R.S. Rothampsted, to reprint Table VI from their book *Statistical Tables for Biological, Agricultural and Medical Research*. Acknowledgements to the authors and original publishers of these and other tables in the Appendix appear following each table.

H.E.K.

Contents

7 PROBABILITY

8 THE BINOMIAL DISTRIBUTION AND ITS NORMAL APPROXIMATION

9 CHI SQUARE

10 THE t DISTRIBUTION

11 ANALYSIS OF VARIANCE I

12 ANALYSIS OF VARIANCE II

STATISTICS

CHAPTER 1

Introduction

ADVICE TO THE STUDENT

Students in the behavioral sciences often undertake their first course in statistics wishing they had majored in some other discipline—any other discipline not requiring them to study statistics. They have convinced themselves that their "aptitude for math" is low, and they are prepared to find statistics difficult. Statistics is difficult for many students, but not, as a rule, because they have a low aptitude for mathematics. These students, who usually have little or no college work in the field, suffer from a serious misconception about the rate at which they should be able to read mathematical material. Accustomed to reading a fifty-page assignment for most courses in two or three hours, they discover that they have spent that much time before they understand a half-dozen pages of statistics. When this happens, they may become absolutely convinced that they have no aptitude for mathematics, so they hurry to drop the course and change their majors!

Not even mathematicians read mathematics as rapidly as they read other material. Of course, this does depend on the material and on the individual, but even professional mathematicians read unfamiliar mathematics at a considerably slower pace than they read anything else. When students require a great deal of time to understand a page of mathematics it *does not* reflect unfavorably on their intelligence or mathematical aptitude. Students should *expect* to read this material much more slowly than they read other textbooks, and they should expect to reread some sections a number of times before the relationships discussed become clear.

One reason for slower reading in a statistics course is that complex ideas are communicated by the use of unfamiliar symbols. In most courses, the student already knows the meaning of the words by which new ideas are communicated. His problem is to understand a novel thought communicated by a new arrangement of familiar symbols. In statistics, however, the symbols and the concepts are both unfamiliar. The student must begin by learning a new and fairly complex vocabulary of symbols before he can understand the concepts communicated by that vocabulary. For this reason he should make sure he understands the meaning of each new symbol or term before he reads beyond the section in which it is introduced.

The student can check his comprehension by answering the questions at the end of the chapters, and by reviewing the adequacy of his answer in the answer section before going on to the next question. Answers for most of the problems are supplied, and the procedures by which certain answers are obtained have also been included in the answer section. If the student cannot answer one of the questions he should go back and reread the appropriate section of that chapter. Above all, he must study the material regularly, but preferably not for more than a few hours at a time. Finding himself a chapter behind on the day before the test is not a position from which he can recover by an all-night study session. If the student is willing to exert consistent effort, he will probably finish the course with much more respect for his "mathematical aptitude" than he had when he began.

WHY STATISTICS?

All sciences, including psychology, try to describe and ultimately understand relationships between the empirical events in their disciplines. In some areas of science (notably physics and chemistry), but also in some sections of psychology, the relationships between these events may be clear cut and easy to demonstrate. For example, the length of time it will take a 1-cubic-inch marble to fall 4 feet can be determined with a fairly high degree of accuracy. If air density is kept constant, and our instruments are in order, we can probably obtain almost exactly the same result with all marbles of similar dimensions. In this example from physics one must consider the density of the medium and the shape of the marble, but for all practical purposes that ends the list.

On the other hand, we might wish to know the speed with which a rat will traverse a 4-foot alley for food reward. We can set up instruments for measuring elapsed time which are just as sophisticated as those used in the physics experiment, but it is quite unlikely that the psychologist's rats will produce the consistent speeds produced by the physicist's marbles. The behavioral scientist has a great many more variables to control. Of course, the rats to be compared should all be equally deprived of food, all of the same sex, age, and weight, all receive the same amount and type of food reward on earlier trials, and all be housed under identical conditions. If we carefully observe *all* of these controls, and then compare the running times of two rats chosen by lot, we shall almost certainly find the times to be different; not quite as different as they would have been without the controls, but different nevertheless.

In Table 1.1, in the theoretical column, we have recorded the running times one might expect for a group of "identical" rats, if these were obtainable and, in the observed column, the running time of real rats as they might be recorded in a real experiment.

Even if we have exerted every effort to hold constant the unwanted influences on running time, it is still quite safe to assume that we have not controlled all of them. Some rats may have been handled a bit more roughly than others; some may have had a fight with their cage mate just before running the alley; one may have noticed an attractive (or repulsive) odor left by the previous occupant of the start box and adjusted his time accordingly. In short, the study of behavior often involves a host of variables, not all of which can be controlled, that act to disguise the relationships between the variables under investigation.

All scientific observations, even those of physicists, contain a true component and an error component. The true component is equivalent to the theoretical running times of Table 1.1, and the error component is the sum of all the chance, or randomly operating, uncontrolled variables that, when added to the true component, give rise to each entry in the observed column. This error component often tends to disguise relationships between events just as static tends to disguise intelligible sound from a radio.

There are two ways to reduce the effect of error: experimentally, by careful laboratory procedures, and statistically, by increasing the numbers of observations and manipulating the data so that the relationships will be apparent in spite of the random or chance error. Statistics, as a

TABLE 1.1 Time to Traverse an Alley Maze

Rat	Theoretical Rats	Observed Rats
1	4 sec.	6 sec.
2	4 "	8 "
3	4 "	4 "
4	4 "	3 "
5	4 "	2 "

discipline, is concerned with this latter process. In its applied form, as we shall study it in this text, it is concerned with describing and drawing inferences from many observations; these observations are ordinarily translated into measurements or counts.

The study of statistics may be divided into two broad areas. They overlap to the extent that some authors prefer not to differentiate them. One of these areas is called inferential statistics because it deals with *inferences* about the true nature of the relationships between variables, even though the data include chance or random errors. Most of this book is devoted to inferential statistics. Before we can infer anything from observations, however, they must be described in a systematic fashion. This branch of statistics is called descriptive statistics. Descriptive statistics show us efficient ways to describe and summarize data, and consequently how to present it in the most usable form. It is this aspect of statistics that we now discuss.

CHAPTER 2

Graphing Distributions

In this chapter we shall discuss the graphical presentation of data, but first we shall comment briefly on the nature of the data with which the scientist works.

OPERATIONAL DEFINITIONS

An experiment, in its simplest form, is designed to investigate the effect of one variable upon another. Scientific convention uses the term "independent variable" to designate any variable presumed to exert the effect, and the term "dependent variable" to designate the variable presumably affected. If we were to investigate the effect of hunger on activity, hunger would be the independent variable and activity would be the dependent variable.

If an investigator were to keep his subjects away from food and were to observe any systematic changes in their tendency to be active, he would have some of the elements of an experiment. The investigator might take notes on the behavior of his subjects, and then summarize his observations in a written description of their behavior. Unfortunately, if another investigator conducted the same experiment he might write a different report, not necessarily because of differences in the behavior of the animals but, perhaps, because of differences between the *observers* regarding the kinds of behavior they considered to be indicative of "hunger" and "activity."

We can increase the objectivity and reliability of such an experiment if we define hunger and activity in a way that permits their measurement. If we define hunger by specifying the operations used to produce or measure it, the definition is called an operational definition. We can define the degree of hunger operationally in terms of the number of hours since food was last available to the animal. Thus, by definition, an animal deprived of food for 24 hours is "hungrier" than one which has been deprived of food for 6 hours. Notice that such a definition does *not* describe the internal stimuli produced by the absence of food, nor does it describe the sensations presumably endured by a hungry animal. In fact, there are a number of ways in which such a definition might be deficient, but it *is* an operational definition; it specifies the operations by which "hunger" is produced. If we accept this operational definition of the independent variable we can form different "hunger" subgroups by depriving some animals of food for 6 hours, some for 12 hours, and some for 24 hours.

Similarly, activity can be operationally defined as the number of rotations of an activity wheel made by the animal during a 5-minute test period. Each subject in the different "hunger" subgroups can then be given an activity score, and average activity scores would be compared among the hunger subgroups. If different experimenters repeat this experiment using the same operational definitions of the variables under investigation, and the same procedures and apparatus, they should arrive at essentially the same description of the relationship between these variables. The operational definition is thus an extremely valuable scientific tool; it helps investigators to know if they are really discussing the same phenomena.

SCALES OF MEASUREMENT

When variables are operationally defined, the methods by which they are measured must be specified. Consequently, the process of measurement and the various kinds of measurement scales are of considerable importance to the behavioral scientist.

There are four principal types of measurement scales, and we shall discuss each one briefly. The *nominal* scale consists simply of the specification of attributes so that the variable in question can be divided into mutually exclusive categories. For example, political parties constitute a nominal scale. To use such a scale we specify the membership requirements of Republican, Democratic, Socialist, and other parties, so that an individual belonging to one of these groups can be identified and counted. The categories composing the elements of a nominal scale must be exhaustive and mutually exclusive; every individual must fall into one, but only one category. Examples of other nominal scales are marital status, college major, and sex. Nominal scales are sometimes referred to as nonorderable countables. In a nominal scale, such as "marital status," we can only count or enumerate the individuals in each of the following categories.

Marital Status	*Frequency*
1. Single, never married	65
2. Presently married	330
3. Divorced—not remarried	29
4. Widowed—not remarried	10
Total	434

Although we can assign the numerals 1 to 4 to assist in the designation of the various marital statuses, it is clear that the categories are not intrinsically orderable; the numerals have no quantitative significance. Four single individuals are not equivalent to one widowed person simply because the numeral "1" designates single and "4" designates widowed. Thus, in a nominal scale, where numbers may be used to designate non-orderable categories, the numbers are only used in place of names; they have no other significance. Standard arithmetical operations with these numbers would yield quite meaningless results.

The *ordinal* scale is a somewhat more sophisticated measuring device. The categories in an ordinal scale do imply order. For example, if we wished to measure "friendliness," we could develop a scale consisting of the categories: extremely friendly, very friendly, friendly, slightly friendly, and acquainted. While such a scale has a variety of shortcomings, it does have a definite order so that the categories, from left to right, represent decreasing amounts of friendliness. Since numbers can also represent decreasing amounts of a quantity, we can use numbers to represent this property of the scale. Thus we might have: (5) extremely friendly, (4) very friendly, (3) friendly, (2) slightly friendly, and (1) acquainted. Notice that in the nominal scale, which we discussed previously, numbers were used only as numerals, as *names* for categories. In the ordinal scale we also let the order of the numbers represent the *order* in the categories.

A difficulty now arises because when *numbers* represent the categories of an ordinal scale, the numbers have other properties *not* possessed by the scale itself. For example, the intervals between the numbers 1, 2, 3, etc. are all the same. There is exactly one unit between each number. However, we do not know the size of the "friendliness" intervals. We do not know if it requires as much "friendliness" to move from "acquainted" to "slightly friendly" as it does to move from "friendly" to "very friendly." The intervals between the *numbers* designating these scale categories are the same of course, but that does not mean that there are equal intervals between the *actual scale categories* represented by the chosen numbers. An ordinal scale implies only order. It does not imply that there are equal intervals between the scale categories, or that the scale numbering begins with a true zero point. Since the scale intervals are not equal we cannot meaningfully add or subtract the numbers representing ordinal scale categories unless we arbitrarily *assume* that the intervals between them are equal.

Psychologists make frequent use of ordinal scales. For example, any rank ordering of individuals or objects produces an ordinal scale. A girl ranked first in a beauty contest is prettier than the girl ranked second, and she in turn is prettier than the girl ranked third. But such a scale cannot tell us if the *difference* in beauty between the first and second ranked girls is the same as the difference in beauty between the second and third ranked girls. The second ranked girl could be very close to the girl ranked first, or very close to the girl ranked third; either way she receives the rank of second.

Equal intervals between scale values first emerge in the *interval* scale. Temperature measured by degrees Centigrade or Fahrenheit is measured by an interval scale. The difference in temperature between 18 and 19° F is the same as the difference between 180 and 181° F. Similarly, the difference between 1 and 2° C is exactly the same as the difference between any other two consecutive degree marks on a centigrade thermometer.

Psychologists have not been entirely successful with their efforts to develop interval scales. The earliest attempts were made in the field of psychophysics by using the concept of the "just noticeable difference" (jnd) to provide a constant unit of sensation. More recent attempts have been made to develop interval scales in the measurement of attitudes and intelligence.

Since, by definition, the interval scale has equal intervals, the numbers that represent the scale divisions can be added and subtracted. With the centigrade scale, it will follow that since the numbers $2 + 1 = 3$, then $2° C + 1° C$ will equal $3° C$.

While an interval scale permits the operations of addition and subtraction, it does not permit the formation of ratios. This is because the interval scale does not have a true zero point. For example, 4° C does not represent four times as much temperature as is represented by 1° C. This is because 0° C does not represent the absence of temperature. The *absence* of temperature occurs at −273° C. This means that while 4° C is three degrees more temperature than 1° C it is not four times as much temperature, even though the number 4 is four times as large as the number 1. Notice that 1° C is not 1° of *temperature* but $273° + 1°$ of temperature, similarly 4° C is $273° + 4°$ of temperature. Therefore, 4° C is 277/274 as much *temperature* as 1° C. Perhaps another example will clarify this point.

Suppose we have three sticks of different and unknown lengths, and arrange them all so that they stand on end on some flat surface. We might find that stick S is the shortest stick, stick A is equal to stick S plus 1 inch, and stick B is equal to stick S plus 4 inches. Accordingly, we may let stick $A = S + 1$ in. and $B = S + 4$ in. A is clearly one unit longer than standard, and B is four units longer than standard. However, B is not four times as long as A unless the standard is zero. Note that $4(S + 1 \text{ in.}) = S + 4$ in. *only* when $S = 0$.

When measurements begin at a true zero point and the scale also has equal intervals we have a *ratio* scale. Length, mass, and time are measured ordinarily with ratio scales, but temperature can also be measured with a ratio scale when we record it in degrees Kelvin. Zero degrees Kelvin represents the absence of temperature, and this point is just $-273°$ C. Meaningful temperature ratios can be formed with the Kelvin scale because it has a true zero.

Such concepts as time, distance, and mass are measured by ratio scales. Since these concepts enter the operational definitions of a number of independent and dependent variables, the behavioral scientist will be using ratio scales with reasonable frequency. For example, at the beginning of this chapter we described a hypothetical experiment to investigate the effect of hunger drive on activity. In that illustration we operationally defined hunger drive as hours since feeding, and activity as activity-wheel revolutions per five minutes. Both of these are ratio scales.

Some of the measurement scales we have just discussed should only be subjected to limited forms of statistical analyses. For example, finding the average of a nominal scale would result in a totally meaningless figure because the individual numbers are only used to stand for names; they have no quantitative referent. In fact, averaging ordinally scaled data is not really appropriate either. It would be much like averaging the lengths of a group of objects measured with a ruler having "inches" of varying and unknown size. Averaging data can only result in meaningful numbers when we have interval or ratio scales.

However, in spite of these logical restrictions, it is quite common to average psychological test scores, which are ordinarily no more than ordinal scales. In such instances we simply *assume* that no serious errors will be incurred and, in most cases, the assumption is probably safe. The ubiquitous grade point average, or honor point ratio, with which most college students are familiar and which decides probation or graduation with honor, is the result of averaging course grades which were probably achieved by the use of ordinal scales. Similar situations may be found in the literature when ordinal scaling techniques are used to measure personality characteristics and other psychological dimensions.

FREQUENCY DISTRIBUTIONS

Regardless of the scale of measurement used, the data from an experiment should be presented in an orderly fashion. Suppose we wish to compare the effectiveness of two different methods of instruction. We may have test scores from one group of students taught by the lecture method and another group taught by the discussion method, and we may wish to compare the two sets of scores. The data may be compared more easily if we first tabulate the scores into two frequency distributions. A frequency distribution is a listing of all the different score values in order of magnitude with a tally or count of the number of scores at each value. Table 2.1 shows two frequency distributions which might result if our data were presented in this form.

With the scores pictured as they are in Table 2.1 we can see some differences between the distributions. The lecture method seems to produce higher achievement, but the range of scores is about the same. We can also observe that the scores of the "lecture students" tend to be concentrated toward the top of the distribution, while the scores of the discussion group seem to be more symmetrically distributed about a

Table 2.1 Frequency Distributions of Examination Scores for Students Taught by Lecture and by Discussion Methods

Lecture Method			Discussion Method		
Score	Tally	Frequency	Score	Tally	Frequency
54	//	2	54		0
53	////	4	53		0
52	卌 ///	8	52	/	1
51	卌 卌 卌	15	51	//	2
50	卌 卌 卌 //	17	50	///	3
49	卌 卌 卌 卌 //	22	49	卌	5
48	卌 卌 卌 ///	18	48	卌 //	7
47	卌 卌	10	47	卌 卌 卌	15
46	卌 ////	9	46	卌 卌 卌 ///	18
45	卌 //	7	45	卌 卌 卌 卌 卌	25
44	卌	5	44	卌 卌 卌	15
43	///	3	43	卌 卌 ///	13
42	///	3	42	卌 ///	8
41	//	2	41	卌	5
40	/	1	40	////	4
39	//	2	39	//	2

central value. These differences, rather easily seen in Table 2.1, would be completely obscured if the scores were simply presented in haphazard order.

FREQUENCY POLYGONS

The same data that are presented in Table 2.1 as a frequency distribution can also be presented as a graph. Two kinds of graphs are commonly used to illustrate data derived from continuous measurement scales. One of these graphs is called a frequency polygon and the other is called a histogram. When data are based on counts rather than measurements, the preferred graph is called a bar chart. The bar chart is mandatory if the data consist of nonorderable countables. Thus we might represent intelligence test scores by a frequency polygon or a histogram, but if we have

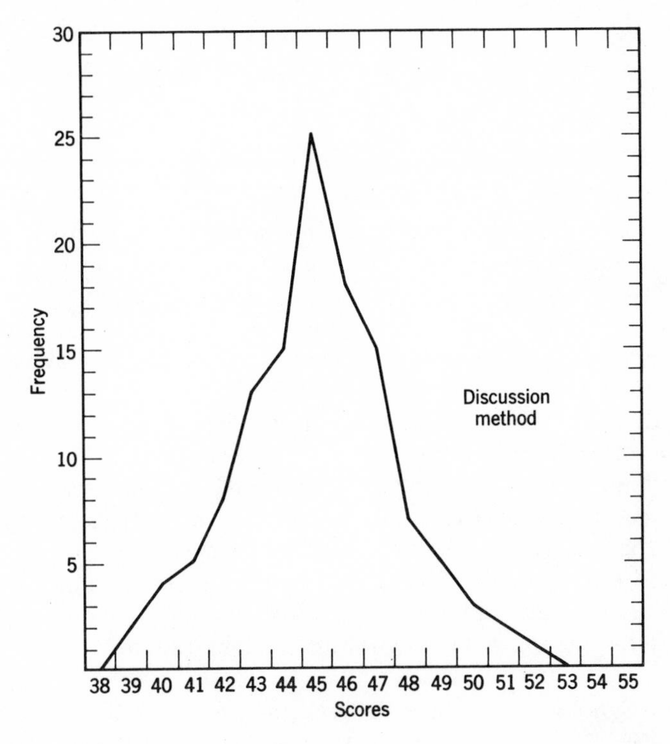

FIGURE 2.1 Frequency polygon of the examination scores tallied in Table 2.1.

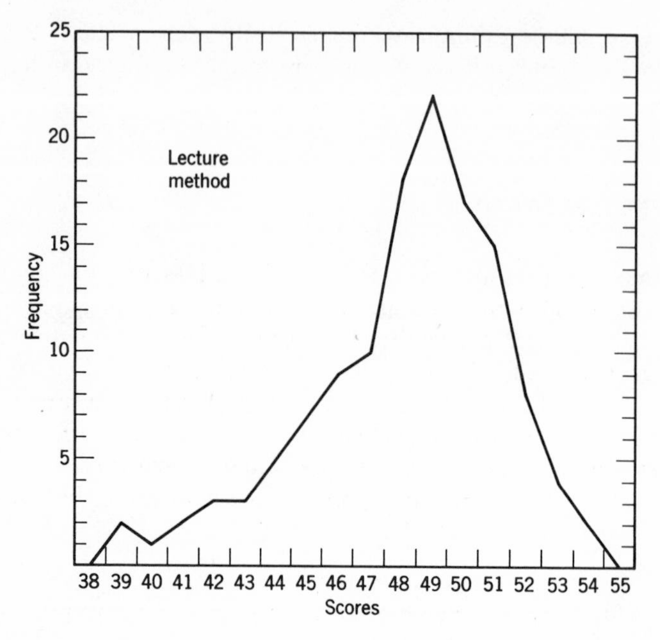

FIGURE 2.1a Frequency polygon of the examination scores tallied in Table 2.1.

tallied membership in several different political parties the data should be presented in the form of a bar chart. Of course test scores are not really continuous, since we normally count the number of right answers. Nevertheless, they are usually *assumed* to be, and are *treated* as if they were continuous; so the usual procedure is to depict distributions of test scores with frequency polygons or histograms. We shall use the data in Table 2.1 to illustrate such graphs.

In the frequency polygon, frequency is graphed as a function[1] of score. The value of each score is recorded on the horizontal axis, or *abscissa*, and the frequency of these values is recorded on the vertical axis, or *ordinate*. The points on the graphs are plotted directly above the midpoints of the intervals on the horizontal axis which represent the scores. You should also remember that the polygon is a closed figure; the ends meet the abscissa one full score unit above the highest tabulated score and one full score unit below the lowest tabulated score.

[1] If one variable is said to be a "function" of another, as when frequency is a function of score magnitude, a relationship is implied between the variables. The graph of the function in Figure 2.1 is simply a pictorial presentation of this relationship.

HISTOGRAMS

Figure 2.2 and 2.2a illustrate the histogram. The histogram consists of a series of adjacent bars whose heights represent the number of subjects obtaining a score and whose location on the abscissa represents the value of the score. Notice that the vertical lines marking off the bars do not originate from the center of the score interval but from its edges. The edges of the individual bars mark the theoretical limits of the score intervals along the abscissa.

Sometimes frequency polygons, or histograms of two different distributions, will both be plotted on the same set of coordinates. If the differences between the distributions are subtle, this procedure may highlight them.

BAR CHARTS

Bar charts are the preferred graphs when data are discrete, that is, when they result from the process of counting. This convention is some-

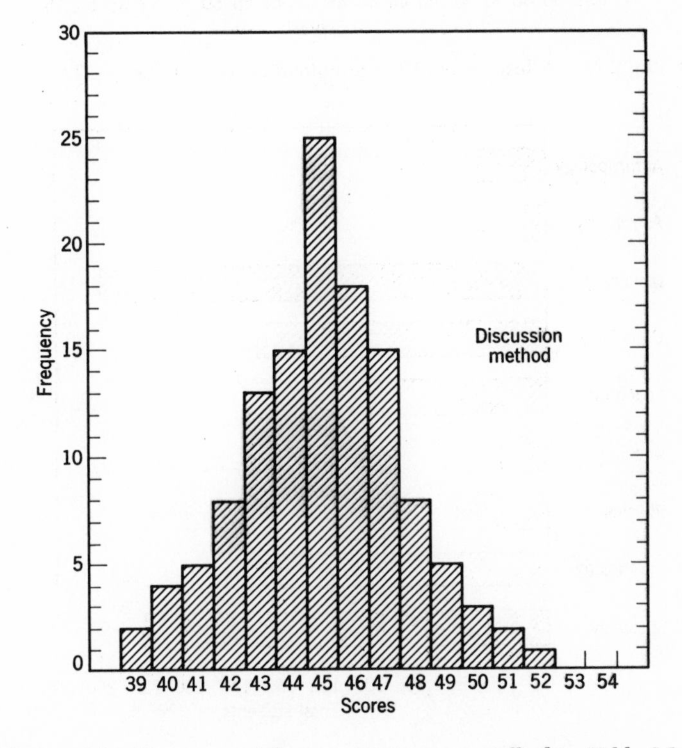

FIGURE 2.2 Histogram of the examination scores tallied in Table 2.1.

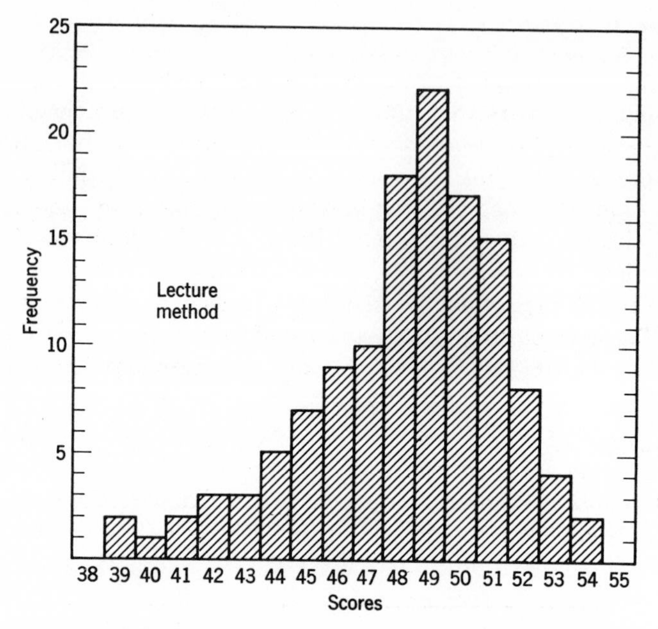

FIGURE 2.2a A histogram of the examination scores tallied in Table 2.1.

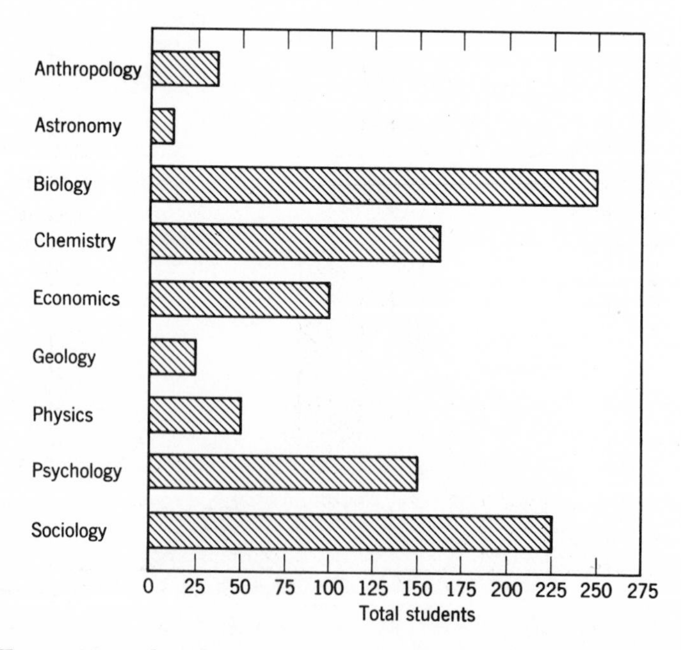

FIGURE 2.3 A bar chart of enrollment in introductory science courses.

what fluid in psychology, where ordinal scales are concerned, but it should be followed without exception for nominally scaled data. The bar chart is very much like the histogram except that spaces are left between the bars in the bar chart. It is also common with bar charts to use the vertical axis to represent categories and the horizontal axis to represent frequency of occurrence. Study the bar chart in Figure 2.3 where we have graphed the enrollment in introductory courses for science departments at a typical college.

GROUPED FREQUENCY DISTRIBUTIONS

We will now consider a more complex kind of frequency distribution called a grouped frequency distribution. We must first call attention to the approximate nature of all measurements. A length may be measured to the nearest inch, the nearest tenth inch, or the nearest hundredth inch. Each of these measurements is considered to be accurate only within certain limits. The limits usually are one half of the unit of measure above and below the unit in question.

If we are measuring with a ruler on which an inch is the smallest subdivision, a measurement will be recorded to the nearest inch. If an object is actually 7.25 inches long and another object 6.57 inches long, they will both be classified as 7 inches long because they are seen as closer to 7 than to either 6 or 8. That is the most accurate measurement we can make with such a ruler. Therefore, the measurements labeled "7 inches" really include all those from 6.5 to 7.5 inches. We may say that a measurement of 7 inches could really *extend* from 6.5 to (but not including) 7.5 inches, and that 6.5 and 7.5 are the theoretical limits of that unit, or measurement interval. This is the situation when the unit of measurement is an inch; the device is accurate only to within a half unit above and below the unit of measurement.

Similarly if the ruler is marked off in tenths of inches, a measurement of 6.8 inches would include measurements from 6.75 to 6.85 inches. If we can read hundredths of inches on our ruler, a measurement of 6.88 will include all measures from 6.875 to 6.885. Even when we can measure to the hundredth of an inch, our accuracy is within plus or minus half the unit of measurement. In each case we think of the measurement as extending one-half unit above and below the recorded value, from one theoretical limit to the next.

This situation is assumed to exist whether we are measuring inches, seconds, or achievement. A score of 176 on an achievement test is assumed to have an accuracy extending one half unit to either side of the obtained score, from 175.5 to 176.5. A score of 176 is thought of occupying an interval from 175.5 to 176.5 just as a measure of 8 inches occupies the interval from 7.5 to 8.5 inches. Scores, then, are thought of as occupying intervals, and these intervals have theoretical limits. A score of 77 occupies the interval 76.5 to 77.5, and 76.5 is the lower theoretical limit of the interval while 77.5 is the upper theoretical limit of the interval. A score of 78 has a lower theoretical limit of 77.5 (the same as the upper limit of the next lower score interval) and an upper theoretical limit of 78.5. The midpoint of the score interval is the recorded value of the score itself. When we constructed a frequency polygon we made use of these ideas. Notice that the scale on the abscissa in Figure 2.1 is continuous, the upper theoretical limit of one interval coinciding with the lower theoretical limit of the next. The points forming the outline of the polygon were plotted above the midpoints of the intervals.

Let's return for a moment to Table 2.1 and apply these ideas to the data in just one interval of the frequency distribution for the lecture class. We shall assume that the 22 students in the lecture class who had scores of 49, might have had, with a more accurate measuring device, scores spread over the interval from 48.5 to 49.5. We will, therefore, assume that a score of 49 encompasses that distance, that it includes those hypothetically finer measurements from 48.5 to 49.5. We have assumed that the achievement measured is continuous even though our measuring device (the number of right answers) was not.

One should notice that the size of the unit of measurement is quite arbitrary and depends on the device at hand. For example, we could transform the data of Table 2.1 by using a coarser grouping; this would be analogous to the use of a less accurate scale of measurement. We could group together scores of 39, 40, and 41; scores of 42, 43, and 44; scores of 45, 46, and 47, etc., throughout the distribution. This would be a grouped frequency distribution. If we were to group together scores of 39, 40, and 41 into a new expanded interval, the new interval would extend from 38.5 to 41.5, and the midpoint of the interval would be 40.

We shall now distinguish between the score limits of a grouped interval and the theoretical limits of a grouped interval. The score limits of an interval are the highest and lowest scores within it which can actually be obtained. In the interval consisting of scores of 39, 40, and 41, the score limits are 39 and 41. The theoretical limits of the interval extend a half unit of measurement above the upper score limit and below the lower score limit. Thus, in the interval consisting of scores of 39, 40, and 41, the theoretical limits of the interval are 38.5 and 41.5. In the interval formed by grouping together scores of 42, 43, and 44, the score limits are 42 and 44; the theoretical limits are 41.5 and 44.5. If the lower *theoretical* limit is subtracted from the upper *theoretical* limit of any interval the grouping interval size (i) is obtained, which in this example is three. Three original intervals of one unit each are used to form the grouped intervals.

The utility of a grouped frequency distribution is not apparent from the data in Table 2.1. In Table 2.2 are all of the scores from a real examination in introductory psychology. Suppose we had to make up a frequency distribution for these scores. Notice that the range of scores covers 50 *different* score magnitudes. An ungrouped frequency distribution would require tallies for 50 different intervals and, obviously, would be rather clumsy. We simplify our task considerably if we construct a grouped frequency distribution. The first decision involves a choice of size for the grouping interval. How many of the single unit intervals should be grouped together to form each new grouped interval?

There are two considerations when making this decision. First, one must decide how many grouped intervals will be required to reduce work and at the same time to preserve the essential configurations of the distribution. Statisticians normally use between 10 and 20 intervals when constructing frequency distributions, and i, the size of the group interval,

TABLE 2.2 Test Scores from a Final Examination in Introductory Psychology

86	74	66	63	58	54	51	45
85	74	66	62	58	54	51	45
84	73	66	62	57	53	50	45
84	73	66	62	57	53	50	45
84	73	66	62	57	53	50	45
84	72	66	61	57	53	49	43
83	72	66	61	56	53	49	43
82	72	65	61	56	53	49	43
82	71	65	61	56	53	49	43
80	71	65	61	56	53	49	42
79	71	64	61	56	53	48	41
79	70	64	61	55	53	48	41
78	70	64	61	55	53	48	41
78	70	64	60	55	53	47	41
78	67	64	60	55	52	47	39
77	67	64	60	55	52	47	38
76	67	64	58	55	52	47	38
76	67	64	58	54	52	46	37
76	67	63	58	54	52	46	
75	67	63	58	54	52	46	
75	67	63	58	54	51	46	
75	67	63	58	54	51	46	

should be chosen so as to obtain about that number. If we have 50 original intervals, grouping by threes or by fours ($i = 3$ or $i = 4$) would result in 13 or 18 of the new expanded intervals. It would seem that either of these values for i would do, but a second consideration is to use an interval size (i) that would yield an interval midpoint that will be an easy number to work with later. If our interval size is always an odd number, our interval midpoint will be a whole number. If our interval size is an even number, our interval midpoint will end with an additional .5. For example, the midpoint of the interval for the following group of six scores ($i = 6$) 41, 42, 43, 44, 45, 46 is 43.5. If we had a group of five scores, 41, 42, 43, 44, 45, the mid-point of the interval would be 43. Therefore, to avoid the dangling half, it is usually more convenient to have i equal some odd number. This will be convenient if we need to use the interval midpoints in later calculations. Since our choice for i in this distribution is between 3 and 4, we will select 3.

We should also begin grouping with a number that will make the grouped interval midpoints an even multiple of i. This is not necessarily the lowest score in the original distribution. In Table 2.3 we have selected 35–37 as the lowest interval, even though we have no scores of 35 or 36. This was done so that the grouped interval midpoints would be even multiples of i. In Table 2.3 the midpoints of the grouping intervals are 36, 39, 42, etc.; each is an even multiple of $i = 3$.

Table 2.3, which illustrates a grouped frequency distribution, is a much more manageable table than would have been produced if we had tallied scores in each of the 50 original intervals. It is certainly a clearer picture of the performance of the class than Table 2.2. Notice that Table 2.3 has 18 grouped intervals; this is within the 10–20 range we have suggested. With $i = 3$, the interval midpoints are whole numbers, and we began the grouping so that the values of the interval midpoints would be even multiples of i. Generally, if there are more than 25 original intervals, it is wise to use a grouped frequency distribution, but this is largely a matter of judgment. Also, note that adjacent grouping intervals *cannot overlap*. Intervals of 5–10, 10–15, and 15–20 cannot be used, because the interval into which scores of 10, 15, etc., are to be tallied would be ambiguous. Our grouping intervals must be mutually exclusive. Intervals must be so constructed that a score can be tallied in only one interval.

CUMULATIVE DISTRIBUTIONS

Psychological data are sometimes usefully presented in the form of cumulative frequency and cumulative proportion graphs. Table 2.3 con-

Table 2.3 Grouped Cumulative Frequency and Cumulative Proportion Distribution Based on Data of Table 2.2

Upper Theoretical Limits	Score Limits ($i = 3$)	Tally	Frequency	Cumulative Proportion	Cumulative Frequency
88.5	86–88	/	1	1.000	172
85.5	83–85	𝍸 /	6	.994	171
82.5	80–82	///	3	.959	165
79.5	77–79	𝍸 /	6	.942	162
76.5	74–76	𝍸 ///	8	.907	156
73.5	71–73	𝍸 ////	9	.860	148
70.5	68–70	///	3	.808	139
67.5	65–67	𝍸 𝍸 𝍸 ///	18	.791	136
64.5	62–64	𝍸 𝍸 𝍸 //	17	.686	118
61.5	59–61	𝍸 𝍸 /	11	.587	101
58.5	56–58	𝍸 𝍸 𝍸 //	17	.523	90
55.5	53–55	𝍸 𝍸 𝍸 𝍸 𝍸	25	.424	73
52.5	50–52	𝍸 𝍸 ///	13	.279	48
49.5	47–49	𝍸 𝍸 //	12	.203	35
46.5	44–46	𝍸 𝍸	10	.134	23
43.5	41–43	𝍸 ////	9	.076	13
40.5	38–40	///	3	.023	4
37.5	35–37	/	1	.006	1

tains columns headed "Cumulative Frequency" and "Cumulative Proportion." The cumulative frequency column lists, opposite the upper theoretical limit of each interval, the cumulative frequency, or total of all scores below that point. Notice that opposite the interval whose score limits are 47–49, and whose upper theoretical limit is 49.5, we have recorded a cumulative frequency of 35. The 35 was obtained by adding (cumulating) the frequency of all measurements below this upper theoretical limit, i.e. 12, 10, 9, 3, and 1. At the top of the cumulative frequency column we will always find the total number of cases in the distribution (N). This has to be the case because we must find exactly all of the measurements below the upper theoretical limit of the highest interval.

The entries in the column headed cumulative proportion are obtained by dividing each cumulative frequency entry by N (the total number of measures) and listing this figure as a decimal. Thus the highest entry is 172/172 or 1.000. The entry opposite the upper theoretical limit of the interval whose score limits are 68–70 is .808. This means that the proportion of scores below 70.5 is .808, or about 81 percent. This proportion was obtained by adding the frequencies in all the intervals below this *theoretical* limit and then dividing the sum by N.

A cumulative frequency and cumulative proportion graph is shown in Figure 2.4. These are constructed by plotting the cumulative frequency and cumulative proportion opposite the *upper theoretical limit* of the

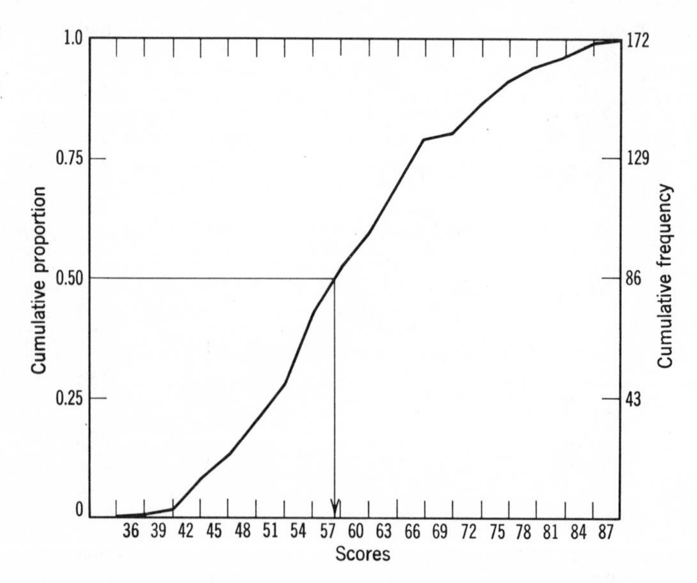

FIGURE 2.4 A cumulative proportion and frequency graph of the data in Table 2.3.

intervals listed along the abscissa. Note this subtle but important difference between noncumulative and cumulative graphs; the plot of points for noncumulative graphs are always above the midpoints of intervals, for cumulative graphs they are always above the upper theoretical limits of the intervals.

Once the cumulative proportion graph has been drawn we can determine the score values which separate different proportions of the distribution. For example, if we want to obtain the score point which just separates the distribution into two equal halves, we shall locate the .50 point on the ordinate, construct a horizontal line until it intersects the graph, then drop a vertical line and read the score point from the abscissa. These lines have been drawn in Figure 2.4 and the score point is approximately 58. To check this answer, count up 172/2 (or 86 scores) from the lowest score recorded in Table 2.2; the value should be 58.

This discussion has not included enough material to make you an expert on tabulating and graphing data, but it should enable you to present a group of measurements in the form of a grouped frequency distribution, frequency polygon, histogram, cumulative frequency or cumulative proportion graph.

REVIEW

When a fairly large number of observations have been recorded, the characteristics of the data are often made clearer if they are cast into a frequency distribution (Tables 2.1 and 2.3). In this form the different values are arranged from highest to lowest, and a tally (or count) of each value is recorded. Once this has been accomplished, graphs can be constructed. Graphs are typically constructed with the abscissa or horizontal axis, representing the values of the scores. The histogram (Figure 2.2) is a type of graph constructed of adjacent vertical bars; the height of the bar represents the frequency with which the score occurred. The frequency polygon is a closed figure drawn by connecting points plotted above the midpoints of the score intervals which are located along the abscissa. The distance of the points above the abscissa represents the frequency of the score's occurrence.

A measurement is assumed to occupy an interval and this interval extends one-half unit of measurement above and below the score. These are called the theoretical limits of the score interval. When many different measurements are obtained, it is often convenient to group several adjacent score intervals together. Such grouping intervals have score limits determined by the highest and lowest scores in them, and theoretical limits determined by the theoretical limits of these same scores. The grouped intervals must be mutually exclusive. The preference is to use an odd number of original score intervals to form the new grouped intervals, and to choose a value for this interval size (i) which will result in from 10 to 20 new grouped intervals. Grouping should proceed so that the interval midpoints are even multiples of i. Such a grouping of score units into intervals, and the tally of measurements within these separate intervals, yields a grouped frequency distribution. Data may also be graphed in the form of a cumulative frequency curve or a cumulative proportion curve (Figure 2.4). These curves are obtained by having the cumulative frequency of measurements below the upper theoretical limit of each interval represented on the ordinate, and the intervals represented on the abscissa. The maximum value on the ordinate of a cumulative frequency distribution will always equal N, the total number of observations. A cumulative proportion graph is similar to a cumulative frequency graph, except that each cumulative frequency is divided by N to obtain a proportion. These cumulative proportions are then plotted opposite the upper theoretical limit of the appropriate intervals.

Exercises

1. Draw cumulative proportion graphs of the data in Table 2.1, and by means of a graphical solution find the score value at the midpoint of each distribution.
2. Using $i = .003$, and the conventions for grouped frequency distributions, construct a grouped frequency distribution for the data below.

1.699	1.695	1.693	1.686	1.678
1.699	1.694	1.692	1.685	1.678
1.697	1.694	1.692	1.683	1.677
1.697	1.694	1.689	1.681	1.675
1.696	1.694	1.687	1.679	1.674
1.695	1.693	1.687	1.679	1.671

3. What are the theoretical limits of each of the two lowest intervals in Exercise 2?
4. What scale of measurement is represented by the following?
   (a) Numbers on football players' jerseys.
   (b) Rank of naval officers.
   (c) Jail sentences given by a criminal court.
   (d) Scores on a statistics test.
   (e) Typing errors.

CHAPTER 3

Measures of Central Tendency

Up to this point our concern has been with the use of graphs that picture the characteristics of a distribution of measurements. We can also use more exact methods and actually measure the characteristics of a distribution just as a score within a distribution measures some characteristic of an individual subject.

One characteristic of a distribution is its size. This is symbolized by N, the number of measurements that make up the distribution. If we have recorded the IQ scores of 100 students, $N = 100$. Figure 3.1 shows two distributions based on the intelligence test scores of students majoring in fields A and B. The two distributions pictured are based on equal numbers of subjects; therefore, $N_A = N_B$. This equality is reflected in the equal areas under the two curves. The area enclosed by a frequency polygon or histogram is directly proportional to the number of cases on which

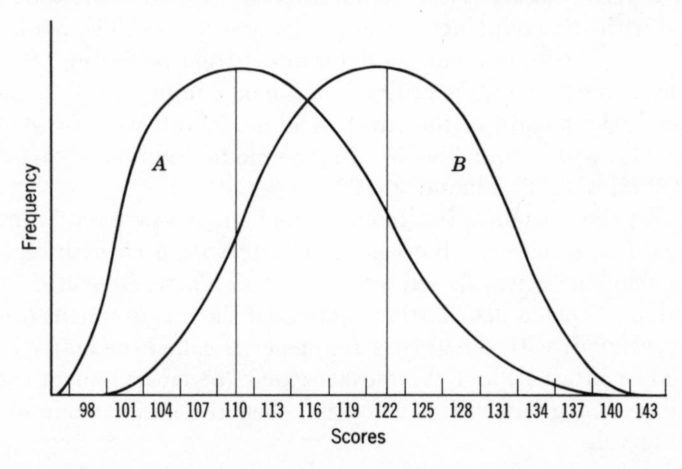

FIGURE 3.1 Distributions of intelligence test scores of students majoring in Departments A and B.

it is based. When frequency polygons and histograms are constructed, each case contributes an increase in the height of the curve or line opposite some score interval on the abscissa. Consequently, each measurement that goes into the distribution adds an increment to the area encompassed by the curve or line.

While the N's are equal and the areas are therefore the same, the distributions in Figure 3.1 differ in several ways. Although the scores overlap, there seems to be a tendency for students majoring in A to have lower IQ scores than students majoring in B. This difference between the distributions is reflected in their relative location on the abscissa or score continuum. Notice that the majority of scores in each distribution cluster, or "pile up," in a particular region. The location of this clustering, called the central tendency of the distribution, can be determined in several ways, and it provides a useful method of comparing distributions.

MODE

One measure of central tendency is called the mode. The mode is the most frequent score in a distribution. In Table 2.1 the mode for the "Lecture Method" is 49, since that score occurs more often than any other. Similarly, the mode for the "Discussion Method" is 45. Notice that the mode is the *value* of the score with the highest frequency; it is not the frequency of that score.

When measures have been grouped into intervals, we call the interval with the highest frequency the modal interval. In Figure 3.1 the abscissa is labeled with the midpoints of the grouping intervals. The modal intervals for each distribution can be determined from inspection. Since frequency is represented on the ordinate, the modal interval will fall directly below the highest point of the curve. The modal interval of distribution A is 109–111, with a midpoint of 110; the modal interval of distribution B is 121–123, with a midpoint of 122.

Some distributions have two modes, which are not necessarily intervals with equal frequencies (f), but separated intervals, each with higher frequencies than the intervals adjacent to them. These are called bimodal distributions. When a distribution is bimodal the measurement (or measurement interval) with the greater frequency is called the major mode or major modal interval; and the measurement (or measurement interval) with the second greatest frequency is called the minor mode or minor modal interval.

Bimodal distributions usually result from including two different kinds of subjects in the same distribution. If we make up a single frequency

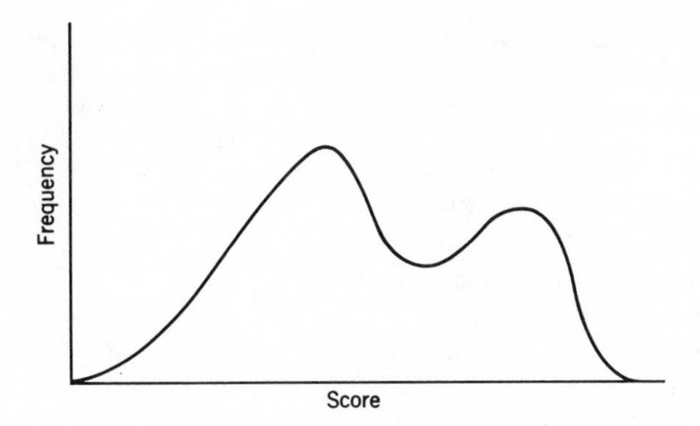

FIGURE 3.2 A bimodal distribution.

distribution from the intelligence test scores of freshmen and first year medical students at a university, we will probably obtain a bimodal distribution such as that in Figure 3.2. The major mode, contributed by the large number of freshmen, is lower on the score continuum than the minor mode contributed by the smaller number of generally brighter medical students.

RULES OF SUMMATION

There are four different rules of summation that are applied when we sum, or add, a series of terms. These rules all follow directly from elementary algebra, but they involve a new symbol for addition: the summation sign (Σ), which is a capital Greek sigma. This sign directs us to sum all of the measures $X_1, X_2, X_3, X_4 \cdots X_N$. Each of these measures is called a variate, and the dimension measured is called a variable. If X stands for the variable IQ, and if we have five variates, the IQ scores of five individuals so that $X_1 = 100$, $X_2 = 100$, $X_3 = 105$, $X_4 = 110$, and $X_5 = 120$, then $\Sigma X = X_1 + X_2 + X_3 + X_4 + X_5 = 535$. If Y represents running time in a maze for N rats, then $\Sigma Y = Y_1 + Y_2 + Y_3 + \cdots + Y_N$.

There are occasions when we may wish to sum only a part of a series, for example, the first 10 members. If this is the case, we can place limits on the summation sign thus $\sum_{i=1}^{i=10} X_i$. This means that we are to sum the first ten values of X, from $X_{i=1}$ to $X_{i=10}$. If we wished to be very formal about it we should show the limits of summation even when all N members of a series are to be summed. Thus, $\sum_{i=1}^{i=N} X_i$ instructs us to sum all N members of the series from $X_{i=1}$ to $X_{i=N}$. In this text we will ordinarily require the summation of all members of any series, so it will be understood that when ΣX or ΣY appear, they require the summation of all instances of the variable, unless otherwise noted.

The first rule of summation we shall discuss concerns the summation of a constant. A constant, by definition, can have only one value. In the formula for the area of a circle, $A = \pi r^2$, A and r symbolize variables while π symbolizes a constant. If we have a constant, C, repeated N times, and we wish to sum these N instances of the constant, we could find the sum either by adding the N instances of C, or by multiplying N times C. If these equivalent methods are stated in mathematical terms we have $\Sigma C = NC$.

The second rule of summation is used when we sum a series in which each term consists of a variate multiplied by a constant. The series might consist of the terms $X_1C + X_2C + X_3C + \cdots + X_NC$. The summation of this series could be symbolized by ΣXC or ΣCX. It would be possible to simplify this expression if we factored the constant term from the series and rewrote the expression as $C(X_1 + X_2 + X_3 + \cdots + X_N)$. Then the terms within parenthesis could be replaced by ΣX, and the entire expression could be written as $C\Sigma X$. Thus, $\Sigma CX = C\Sigma X$.

The third rule applies to the situation in which the series is composed of terms which are themselves the sum of scores on two or more variables. The series might consist of the terms $(X_1 + Y_1) + (X_2 + Y_2) + (X_3 + Y_3) + \cdots + (X_N + Y_N)$. This series of terms could be regrouped so that it would appear as $(X_1 + X_2 + X_3 + \cdots + X_N) + (Y_1 + Y_2 + Y_3 + \cdots + Y_N)$. Since the summed series within the first parenthesis may be symbolized by ΣX, and the summed series within the second parenthesis by ΣY, the sum of these two series may be given as $\Sigma X + \Sigma Y$. Therefore, $\Sigma(X + Y) = \Sigma X + \Sigma Y$.

The fourth rule concerns the summation of a series in which each term is composed of a variate plus a constant. Such a series would consist of $(X_1 + C) + (X_2 + C) + (X_3 + C) + \cdots + (X_N + C)$. We can regroup these additions in the form $(X_1 + X_2 + X_3 + \cdots + X_N) + (C + C + C + \cdots + C_N)$. The terms in the first parenthesis are given by ΣX and those in the second as ΣC, or NC. Consequently, the sum of the entire series may be written as $\Sigma X + NC$.

Here are two sample problems that are applications of these rules of summation.

1. We shall show that

$$\Sigma\left[\left(X + \frac{C}{N}\right) - \left(Y - \frac{C}{N}\right)\right] = \Sigma X - \Sigma Y + 2C$$

$$\left.\begin{aligned}&\Sigma\left[X + \frac{C}{N} - Y + \frac{C}{N}\right]\\&\Sigma\left[X - Y + \frac{2C}{N}\right]\end{aligned}\right\}\ \text{Collect terms with the usual rules of elementary algebra.}$$

$$\Sigma X - \Sigma Y + N\frac{2C}{N} \qquad \text{Sum over terms.}$$

$$\Sigma X - \Sigma Y + 2C = \Sigma X - \Sigma Y + 2C$$

2. We shall expand and simplify the expression $\Sigma(X + C)^2$.

$$\Sigma(X + C)^2$$

$$\Sigma(X^2 + 2CX + C^2) \qquad \text{Expansion of } (X + C)^2.$$

$$\Sigma X^2 + 2C\Sigma X + NC^2 \qquad \text{Sum over terms.}$$

THE MEAN

With these rules of summation we shall discuss another very widely used measure of central tendency, the mean. The mean is determined by summing all of the measures in a distribution and then dividing this sum by the number of measures.

If, in a distribution of IQ scores, we let X symbolize IQ, the formula for the mean becomes:

$$M = \frac{\Sigma X}{N} \qquad \text{FORMULA 3.1 The mean}^{1}$$

This is the general formula, and it will give us the mean of a distribution regardless of the kinds of measures represented by the X's. If we calcu-

[1] When statisticians refer to the mean of a population they ordinarily designate it by M or μ. Sample means, however, are usually symbolized by $\bar{X}$. The methods we will describe in this chapter and the next for determining central tendency and variability apply to *populations*. In Chapter 6 we will distinguish between samples and populations, and we will discuss the estimation of population parameters.

late the mean of the distribution 8, 7, 6, and 5, the answer is 6.5, that is: $\Sigma X = 26$, $N = 4$, and $M = 26/4 = 6.5$.

If data are in the form of a frequency distribution, an equivalent, useful formula for the mean is $\Sigma fX/N$, where X is the magnitude of each score and f is the *number of times* scores of that magnitude occur. Since

TABLE 3.1 Calculation of the Mean from Grouped and from Ungrouped Data

Ungrouped data	
64	62
64	62
64	62
64	62
64	62
64	62
64	62
64	62
63	61
63	61
63	61
63	61
63	61
63	61
63	61
63	61
63	61
63	60
63	60
63	60
63	60
63	60
63	60
63	60
63	59
63	59
62	59
62	59
62	59
62	59

$\Sigma X = 3713$

$$M = \frac{\Sigma X}{N} = \frac{3713}{60} = 61.88$$

Score (X)	Grouped data frequency (f)	fX
64	8	512
63	18	1134
62	12	744
61	9	549
60	7	420
59	6	354
		$\Sigma fX = 3713$

$$M = \frac{\Sigma fX}{N} = \frac{3713}{60} = 61.88$$

Grouped data coded by subtraction

X	f	$X' = X - 62$	fX'
64	8	2	16
63	18	1	18
62	12	0	0
61	9	−1	−9
60	7	−2	−14
59	6	−3	−18
			$\Sigma fX' = -7$

$$M = \frac{\Sigma fX'}{N} + C = \frac{-7}{60} + 62 = 61.88$$

multiplying the frequency of occurrence by the magnitude is the same as adding together all scores of a particular magnitude, the two formulas are equivalent. When data have already been cast into a frequency distribution we shall find $M = (\Sigma fX)/N$, the preferred formula. Remember, however, that ΣfX is *exactly* equivalent to ΣX. The X's of ΣfX refer to *different* score magnitudes. When these different score magnitudes are each multiplied by their frequency of occurrence and summed, the result will be equivalent to summing the total array of scores. By using ΣfX we are simply replacing a great deal of addition by a little multiplication.

In Table 3.1 we have presented the same data in "raw," and the frequency distribution forms to illustrate these two equivalent methods of calculating the mean. We shall consider this table in some detail.

To find the mean of the ungrouped data we must add all 60 double-digit numbers. There is an easier way: if we wish to find the mean we can multiply each score value (X column) by the frequency of that score (f column) and list these products in the column headed fX. The sum of the numbers in this column is ΣX (or ΣfX). Thus, by tallying the same scores together and multiplying each score by its frequency, the calculation consists of only 6 multiplications and the addition of the resulting 6 products. This is a much easier process than adding a column of 60 double digit numbers. Since ΣX (or ΣfX) $= 3713$, the mean for this distribution is $(\Sigma X)/N = 3713/60 = 61.88$.

MEAN OF COMBINED GROUPS

Suppose we have two distributions, the first with a mean of 28 based on 60 cases, and the second with a mean of 32 based on 90 cases. We wish to determine the mean for the combined group of 150 cases and, obviously, we would prefer to do this without re-adding all 150 scores. Unfortunately, since the subgroups have unequal N's, we cannot simply sum the means of the two subgroups and divide the result by 2.

The mean of the combined group is given by the sum of scores in both distributions divided by the total number of cases in both distributions. Thus, $M_C = (\Sigma X_1 + \Sigma X_2)/N_1 + N_2$. We know $N_1 + N_2 = 150$, therefore, all we need is ΣX_1 plus ΣX_2. If $M_1 = \Sigma X_1/N_1$, then by simple algebra $N_1M_1 = \Sigma X_1$, and $N_2M_2 = \Sigma X_2$. Therefore, the formula $M_C = (\Sigma X_1 + \Sigma X_2)/(N_1 + N_2)$ can be rewritten $M_C = (N_1M_1 + N_2M_2)/(N_1 + N_2)$. Since we know the N's and means of both distributions, the mean of the combined distribution can be obtained. $M_C = (28 \times 60 + 32 \times 90)/(60 + 90) = 30.4$. The formula for M_C can be expanded to cover the combined mean of any number of groups.

CODING BY SUBTRACTION

The rules of summation and some elementary algebra allow us to simplify the calculation of the mean by a process called coding. Coding simplifies computations by allowing the substitution of smaller numbers for larger ones. The rules for coding may seem complex at first, but if you do not have an automatic calculator, coding can save a great deal of very tedious arithmetic.

Suppose we begin by subtracting a constant (C) from each score (X), and let X' symbolize the result, which we shall call a coded score.

$X - C = X'$	Defines X', the coded score.
$X = X' + C$	Solve for X, the original scores (add C to both sides).
$\Sigma X = \Sigma(X' + C)$	Sum over all terms.
$\Sigma X = \Sigma X' + NC$	Simplify terms (rule 4).
$\dfrac{\Sigma X}{N} = \dfrac{\Sigma X'}{N} + C$	Divide by N to obtain the mean.

We have shown that if we subtract a constant (C) from each score (X) to produce a coded score (X'), and find the mean of these coded scores $\left(M' = \dfrac{\Sigma fX'}{N}\right)$, we can then find the mean of the original scores $\left(M = \dfrac{\Sigma fX}{N}\right)$ by adding the constant (C) to the mean of the coded scores. Symbolically, $M = \dfrac{\Sigma fX'}{N} + C$. This may sound complex, but the following example may clarify the issue. The mean of 101, 102, 103, and 104 is the same as the mean of 1, 2, 3, and 4 plus 100. In this example $C = 100$, which, when subtracted from 101, 102, 103, and 104 yields X' values of 1, 2, 3, and 4. M' is then obtained by $(\Sigma X')/N = 10/4 = 2.5$. M, the mean of the original distribution is given by $M' + C$, or $2.5 + 100$. The mean of 101, 102, 103 and 104 is thus 102.5.

We shall apply this coding procedure to the data of Table 3.1. Subtracting $C = 62$ from each score yields the numbers in the X' column. When we multiply these X' values by the frequencies in the f column we have the entries in the fX' column. The sum of these entries is $\Sigma fX' = -7$. M' is $(\Sigma fX')/N = (-7)/60 = -.12$. We can now determine M since $M = M' + C$; thus $M = -.12 + 62 = 61.88$. Coding by subtraction gives a considerable advantage with these data because the fX' products are easier to obtain than the fX products.

In this example we deliberately set out to make $\Sigma fX'$, and consequently

M', negative. The sign of $\Sigma fX'$ is entirely dependent on the size of C, the constant chosen for subtraction. Had we subtracted 61 instead of 62, we would have had a positive value for $\Sigma fX'$. Of course, the sign of $\Sigma fX'$ can be either positive or negative without altering our procedure in any way.

Coding by subtraction is even more advantageous when the original scores are very large. For example, to illustrate this advantage, find the mean of the following distribution of scores, using uncoded and coded procedures!

X	f
11,985	17
11,984	39
11,983	41
11,982	29
11,981	19

Let us consider how we might find the mean of scores which have been cast into a grouped frequency distribution. To determine the mean of a *grouped* frequency distribution we assume that the midpoint of each grouping interval represents the mean of all the scores tallied in that interval. For example, the grouped frequency distribution in Table 3.2 shows 12 scores tallied in the interval 40–44, and we shall simply assume

TABLE 3.2 A Grouped Frequency Distribution Coded by Subtraction and Division

Scores	f	X	fX	$X'' = \frac{X - 37}{5}$	fX''
50–54	3	52	156	3	9
45–49	10	47	470	2	20
40–44	12	42	504	1	12
35–39	13	37	481	0	0
30–34	8	32	256	−1	−8
25–29	9	27	243	−2	−18
20–24	3	22	66	−3	−9
	$N = 58$		$\Sigma fX = 2176$		$\Sigma fX'' = 6$

$$M = iM'' + C \qquad M = \Sigma fX/N$$

$$M = 5\left(\frac{6}{58}\right) + 37 \qquad M = \frac{2176}{58}$$

$$M = 37.52 \qquad M = 37.52$$

for computational purposes that these 12 scores have their mean at 42, the midpoint of this grouping interval. If grouping interval midpoints are designated by X, and the frequency of scores in each interval by f, then fX is equivalent to ΣX for any grouping interval and the sum of all scores in the distribution will be given by ΣfX. This is exactly the logic by which we developed the formula for the mean of combined groups in the previous section. Here, for the sake of convenience, we have simply changed the symbols N_1M_1 to fX.

In Table 3.2, column f shows frequencies and column X shows interval midpoints. The sum of entries in column fX is 2176, and this yields a mean of 37.52 when divided by $N = 58$.

When data have been cast into a *grouped* frequency distribution we can reduce the labor of our calculations still further by subtracting a constant from each interval midpoint and then dividing the remainder by another constant. We shall use X'' to stand for the interval midpoint from which a constant (C) has been subtracted and the result divided by another constant (i).

$$\frac{X - C}{i} = X''$$ Defines X'', the coded midpoint.

$$X = X''i + C$$ Solve for X, the original midpoint (multiply both sides by i and then add C to both sides).

$$\Sigma X = \Sigma X''i + \Sigma C$$ Sum over all terms.

$$\Sigma X = i\Sigma X'' + NC$$ Simplify terms (Rule 2 and Rule 1).

$$\frac{\Sigma X}{N} = i\frac{\Sigma X''}{N} + C$$ Divide by N to obtain the mean.

$$M = iM'' + C$$

This small amount of algebra shows that if midpoints are coded first by subtraction and then by division, the mean of the original scores can be obtained by multiplying the mean of the coded midpoints by the dividend (i) and then adding the subtraction constant (C) to this product.

Again, returning to Table 3.2 and coding the interval midpoints by both subtraction and division, we can recalculate the mean of this distribution with considerably less arithmetic. From each interval midpoint (X), we shall subtract $C = 37$ and divide the remainder by $i = 5$. This yields the numbers in column X''. Multiplying these entries by f gives the entries in the fX'' column. The sum of this column, when divided by N, yields M'', or $(\Sigma fX'')/N$. We now multiply this value by $i = 5$

and then add $C = 37$ to yield 37.52, which is the mean of the original grouped distribution.

The rather elegant simplicity of all this hinges directly on the correct choice of the constants i and C. The constant i is *always* chosen to equal the size of the grouping interval. The constant C can be equal to *any* interval midpoint, but it will be quite helpful to choose an interval midpoint close to the mean of the distribution. Of course, we do not *know* the value of the mean, but its approximate location can be determined by inspection. The closer C is to the value of the mean, the smaller are the fX'' products, and the simpler it will be to obtain their sum. Remember that coding can be accomplished with C equal to *any* interval midpoint, but that i must be equal to the size of the grouping interval.

Table 3.3 contains another illustration of coding a grouped frequency distribution by both subtraction and division. Simply by looking at the distribution we select an interval close to the center. Subtracting the midpoint of that interval from the midpoints of all other intervals, and dividing the result by the size of the interval, we obtain the entries under the column headed X''. This procedure will always produce the simplified sequence of coded scores. Now we can determine the fX'' values and find $\Sigma fX''$ which is 26. Since $M = iM'' + C$, $M = 3 \times .313 + 106 = 106.94$. Again we have obtained the uncoded mean, M, with a minimum of arithmetic.

TABLE 3.3 A Grouped Frequency Distribution Illustrating Coding by Both Subtraction and Division

Scores	X	f	X''	fX''
114–116	115	6	3	18
111–113	112	10	2	20
108–110	109	20	1	20
105–107	106	26	0	0
102–104	103	12	−1	−12
99–101	100	7	−2	−14
96– 98	97	2	−3	−6
		$N = 83$	$\Sigma fX'' =$	26

$$M = i\,\frac{\Sigma fX''}{N} + C = 3 \times \frac{26}{83} + 106 = 106.94$$

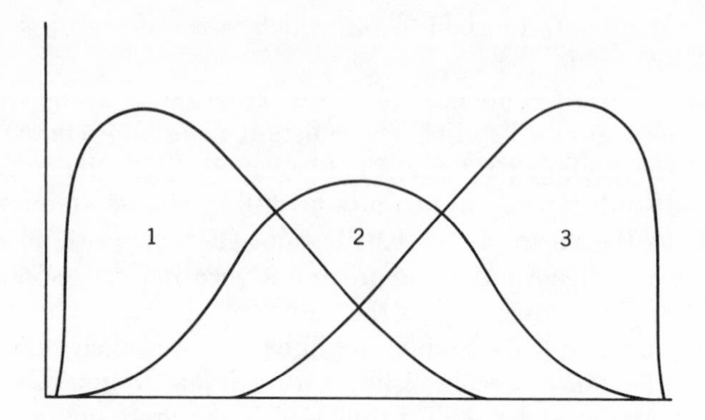

FIGURE 3.3 Distributions with positive, zero, and negative skew. *1*, Positive skewness. *2*, Zero skewness (symmetrical). *3*, Negative skewness.

SKEWNESS

When the scores in a distribution tend to pile up at the low end of the scale of measurement with just a few scores at the top, the distribution has a positive skew. If the scores pile up toward the high end with just a few low scores, the distribution is negatively skewed. If the distribution is symmetrical about the midpoint it is said to have zero skew (a symmetrical distribution). Figure 3.3 illustrates these three kinds of distributions. Although the mean is a very commonly used measure of central tendency, it can be misleading in skewed distributions. In a severely skewed distribution, the few very high or very low scores can exert a considerable impact on the mean and pull it off center so much that it is no longer a very good measure of central tendency.

MEDIAN

A better measure of central tendency for severely skewed distributions is called the median (Mdn). The median is the midpoint in a distribution of scores. It is the point above and below which fall exactly 50 percent of the measures. It is, in fact, the 50th centile to which we referred when discussing cumulative proportions. Suppose we have a positively skewed distribution consisting of the scores 5, 5, 6, 6, 6, 8, and 104. The mean of this distribution is 20; therefore, it is higher than 85 percent of the scores in the array. In such a distribution the mean would not be a very good measure of central tendency. The median, however, is 6.0 and better represents the central tendency for this distribution.

When data are presented as a frequency distribution with more than one score in each interval, the calculation of the median becomes somewhat more complicated. Suppose we must calculate the median for the distribution of data shown in Table 3.4. Here, we have grouped our measures into class intervals of size 1.0. The theoretical limits of these class intervals are recorded on the left side of the table under the heading "Theoretical limits." Under the column headed "Frequencies," all of the scores within each interval have been recorded as if they were spread

Table 3.4 Calculation of the Median from Grouped Data with $i = 1$

	Theoretical Limits	Frequencies	
MEASUREMENT SCALE	9.5		FREQUENCY SCALE
		Scores {9, 9, 9, 9} 4	
	8.5		
		Scores {8, 8, 8} 3	
	7.5		
		Scores {7, 7} 2	
	6.5		
		Score {6} 1	
	5.5		
		Score {5} 1	
	4.5		
		$N = 11$	

evenly throughout their interval. The median is the point on the *measurement scale* which divides the distribution of scores exactly in half. Since there are 11 cases, we must find the point on the measurement scale just opposite the middle of the sixth case counting from either end of the distribution. The point on the measurement scale which is exactly opposite the midpoint of the sixth case (5.5 cases from either end) will be the median since that point divides the distribution exactly in half.

If we count up from the lower end of the distribution, we find four cases below the theoretical lower limit (7.5) of the measurement interval 7.5–8.5. This interval itself contains three cases, but we need to proceed up through the frequencies column for only another $1\frac{1}{2}$ cases to obtain the point separating the two halves of the distribution. Since there are 3 cases spread evenly throughout the interval 7.5–8.5, and we need another $1\frac{1}{2}$ cases, we must proceed half (or .50) of the distance into the interval. This will take us to the midpoint of the frequency continuum. The median, located on the measurement scale, is exactly opposite this point. Since the scores are grouped in intervals of 1.0 on the measurement continuum we add .50 (times 1.0, the grouping interval) to the lower limit of the measurement interval, 7.50, to obtain the median. The median is 8.00. You may recognize the procedure as linear interpolation. It will be clearer after we have worked through several other examples.

Now we shall use an illustration involving grouped data when the measurement interval is *not* 1.0. We shall find the median for the data

TABLE 3.5 Calculation of the Median from Grouped Data, $i = 3$

Theoretical Limits	Score Limits	f	
	17–19	4	
	14–16	10	
13.5			
	11–13 } $i = 3$	12	
10.5			↑ $\frac{4}{12}$
	8–10	9	
	5– 7	7	18
	2– 4	2	
Mdn $= 10.5 + \frac{4}{12} \times 3 = 11.50$		$N = 44$	$N/2 = 22$

in Table 3.5. First determine $N/2$, the number of observations above which *and* below which the median must fall. $N/2$ for these data is 22. We therefore count up 22 observations and find that the median is in the measurement interval whose theoretical limits are 10.5–13.5. Since 18 observations fall *below* the *lower limit* of this interval, 10.5, we must proceed another 4 observations into the interval in order to encompass 22 observations. Since there are 12 observations in the interval and we must find a point which just surpasses 4 of them, we proceed $\frac{4}{12}$ (or .33) of the distance up into the interval. On the measurement continuum the grouping interval covers 3 units ($i = 3$), so we proceed $3 \times .33$ into the interval. This value (1.0) is added to the theoretical lower limit of the measurement interval (10.5) to give us the median, which is 11.50. We

TABLE 3.6 Illustrative Data for the Calculation of a Median

X	f
45–49	12
40–44	16
35–39	20
30–34	11
	$N = 59$

can check the procedure by working down from the top of the distribution. Counting down 22 cases, we find 14 above the upper limit of the 10.5–13.5 interval. This means that we must drop down 8 of the 12 observations in that interval. This $\frac{8}{12}$ is equivalent to .67 of the interval. Since the measurement continuum is divided into grouping intervals with $i = 3$, we need to drop $.67 \times 3$ below the upper limit of 13.5. This point will be the median. $13.5 - 3 \times .67 = 11.50$. A point on the score or measurement continuum which divides a distribution in half should be the same point whether we begin calculating from the top or the bottom.

We shall follow one more example using the data in Table 3.6. Again we determine $N/2$ which is 29.5 for the data in Table 3.6. Since 11 of the 29.5 observations are below 34.5, we must move 18.5 observations up into the next highest interval. We know there are 20 observations in this interval, and so we need to move 18.5/20 into the interval to find a point below which there are 29.5 scores. The decimal equivalent of 18.5/20 is .925. The score or measurement continuum is divided into grouping intervals of size 5, so we must move $5 \times .925$ of the distance into the interval whose limits are 34.5–39.5. Our median for these data is $34.5 + 5 \times .925 = 39.125$. Can you obtain the same answer by working down from the top of the distribution?

CENTILES

We have devoted considerable time to the problem of obtaining the median (or 50th centile) from grouped data, and the reason for this is that any given centile value can be obtained by using a similar procedure. A centile is a point on the scale of measurement below which a specified percentage of N is located, where N is the total number of measurements. The median is the 50th centile. It is the point which divides the distribution *exactly* in half. The 99th centile is the point on the measurement scale below which 99 percent of the distribution (N) may be found. If we wished to find the 75th centile we would approach the problem the same way as we did when finding a median, except that instead of finding $N/2$ as the first step, we would determine $.75N$. We would then proceed by counting up among the observations to the $.75N$ point and interpolate into the measurement continuum just as we did in obtaining the median. One caution is that we cannot check our result in quite the same way we checked a median, because counting *down* $.75N$ from the top will give us quite a different answer. It will, in fact, yield the 25th centile. We could, however, check the value obtained by counting down $.25N$ from the top of the distribution. In other words, the 75th centile can be obtained by moving up $.75N$ or down $.25N$.

As an illustration, we shall determine the 20th centile from the data in Table 3.6. First we determine $.20N$. This value is 11.8. Counting up the frequency scale we find 11.8 falls in the 35–39 interval. Actually we need only .8 of one score, and since there are 20 scores assumed to be evenly spaced in that interval, we move $.8/20$ of the distance into the interval. Since $.8/20$ is .04 we must move up .04 of the measurement interval. The measurement grouping interval is five units, so we must move $.04 \times 5$ or .20 units into the measurement interval. We add .20 to the lower limit of that interval to obtain the 20th centile, $34.50 + .20 = 34.70$. We know that 20 percent of scores in this distribution will fall below 34.70.

We can ask a related but somewhat different question. Up to this point we have described procedures to find the score point equivalent of a particular centile. Suppose we ask the question the other way around: at what centile is a score of X? If you understand the rationale of the procedure for obtaining the median, you should be able to solve such a problem. Using the data from Table 3.6, try to determine the centile value of a score of 44. The answer is the 77th centile.

First we find the proportionate distance into the interval of 39.5–44.5 occupied by a score of 44.0. Since 44.0 is 4.5 units above 39.5 and the total size of the grouping interval is 5.0 units, a score of 44.0 lies 4.5/5.0 (or .9) of the distance from 39.5 to 44.5. We now move over into the frequency continuum and find 16 cases in this interval. Nine-tenths of these cases would be 14.4. These are added to 31, which is the cumulative total below the interval. This sum is 45.4. There are a total of 59 cases, and a score of 44.0 is above precisely 45.4 of these; therefore it is at the 45.4/59 centile, or the 77th centile.

DIFFERENCE BETWEEN MEANS AND MEDIANS

Moving away from computation, let us be clear about what is represented by the median and by the mean. It is true that both are measures of central tendency; however, different kinds of distributions influence them quite differently. In the distribution of Table 3.6, suppose we have misread a score, erroneously entering 48 instead of 58. One might assume that such an error would make a substantial difference in the median of that distribution, but it will not make any difference at all! When we compute the median we are concerned only with having as many scores above as below some point in the score or measurement continuum. We are not concerned with how far a score is above, or below this point. Except for the score or scores that locate the median (the $N/2$ score), the actual size of all other scores play no part in its calculation. We could, in fact, calculate a median for a distribution of scores even if the top score were partially illegible. We would only need to be certain that it was above the interval containing the median.

The mean is quite different. ΣX is affected by *any* change in the value of a single X. Since $M = \Sigma X/N$, any error or change in the value of a score will change ΣX and consequently change the mean. Notice that 2, 4, 6, 8, and 10 as well as 2, 4, 6, 8, and 100 have the same median (6), but quite different means. The mean for the distribution 2, 4, 6, 8, and 10 is 6, the same as the median; but the mean of 2, 4, 6, 8, and 100 is 24.

You now know three ways to describe the central tendency of distributions: modes, means, or medians. The mean is usually preferable as a measure of central tendency, and it forms the basis for more advanced statistical treatments. On the other hand, if the data are badly skewed, the mean will be a very misleading description of central tendency and we might prefer to use the median.

REVIEW

Distributions differ in a variety of ways. One of these is central tendency. Three measures of central tendency are the mode (the most often occurring score), the mean (the sum of scores divided by their number), and the median (the point at which the distribution of scores is divided exactly in half). The calculation of the mean can be simplified by coding the measures themselves, or the midpoints of the intervals into which they fall. Either scores or midpoints of intervals may be coded by subtraction and, if need be, by division. Four rules of summation assist in the development of coding formulas. These will be used later in the text.

EXERCISES

1. Determine the mode for the data in Table 3.2.
2. Find the mean of the scores in Table 2.3, using a coding procedure.
3. Find the mean of the scores of the lecture group in Table 2.1, using a coding procedure.
4. Under what circumstances will $\Sigma fX''$ be negative?
5. Under what circumstances will M'' be zero?
6. Find the median, 25th centile and 75th centile for the distribution in Table 2.3.
7. (a) Under what circumstances will the mean, median and mode coincide?
   (b) As the negative skew of a distribution increases what happens to the distance between the mean and median? Which is higher?
8. Simplify $\Sigma(X^2 - 2MX + M^2)$. Remember, M is a constant for any given distribution.
9. A rat has had 8 trials in a runway, but has refused to leave the start box within four minutes on the first two trials. On the remaining 6 trials the running times were 160, 100, 40, 30, 30, and 10 seconds. What measure of central tendency is most appropriate *for the day's trials*, and what is its value in the distribution above?
10. An employer has a few highly skilled and well paid workers but the majority of his employees are unskilled women who earn no more than the minimum wage law permits. The plant union, seeking public sympathy, begins an "information program" and reports the median wage. The employer counters by reporting the mean wage. What difference does it make? Who is right?
11. Show that the sum of the deviations of a set of measures from their mean is zero.

12. For the distribution below, draw a cumulative proportion curve and graphically determine the 25th centile, median, and 75th centile. Determine the same points by a numerical solution. Use a coding procedure involving subtraction and division to determine the mean.

X	f	X	f
75–79	1	35–39	291
70–74	3	30–34	127
65–69	10	25–29	65
60–64	20	20–24	34
55–59	48	15–19	15
50–54	125	10–14	7
45–49	259	5– 9	6
40–44	370		

13. When data are cast into a grouped frequency distribution and we determine the mean by multiplying each interval midpoint by the frequency of scores in that interval, we assume that the scores in each interval are spread evenly throughout that interval. In point of fact this assumption is not tenable for most distributions, although the mean calculated from grouped data usually will be quite close to the mean calculated for ungrouped data. Explain why the assumption is not tenable and why this does not seem to matter very much.

CHAPTER 4

Measures of Variability

Distributions can differ in size, in central tendency, and they can also differ in variability. The scores in one distribution may be more disperse, more spread out, than the scores in another distribution. Figure 4.1 shows two distributions of the same size, with the same means, medians, and modes, but differing in variability.

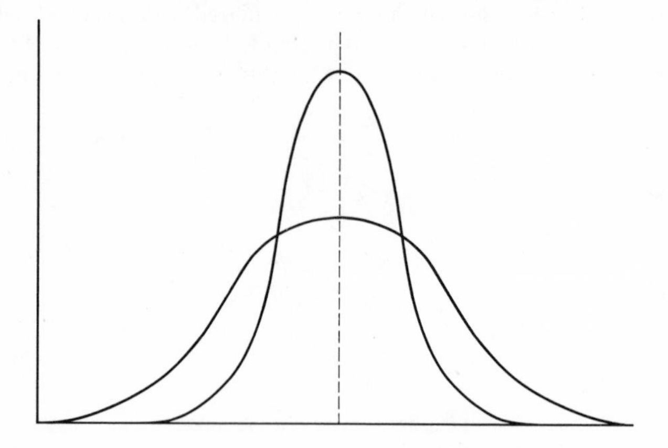

FIGURE 4.1 Two distributions with the same central tendency but differing in variability.

RANGE

One measure of variability is the range of scores (R), which is defined as the difference between the highest and lowest scores in the distribution.

$$R = H - L$$

FORMULA 4.1
Range

The range is useful because it provides a very quick measure of variability, but it may be misleading in distributions having a few extremely deviant scores.

SEMI-INTERQUARTILE RANGE

Although rarely used, another measure of variability is the semi-interquartile range, or Q, which is based on centiles. Q is defined as one-half of the distance between the 25th centile (Quartile 1) and the 75th centile (Quartile 3).

$$Q = \frac{Q_3 - Q_1}{2}$$

FORMULA 4.2
Semi-interquartile range

If scores are widely spread, Q will be large; in distributions with scores clustered closely about the median, Q will be small. Q is a fairly simple measure of variability, and it is sometimes used in manuals of psychological and educational tests to give the reader an idea of the relative dispersion of test scores for different norm groups.

VARIANCE

A much more useful measure of variability is called the variance (σ^2).

$$\sigma^2 = \frac{\Sigma(X - M)^2}{N}$$

FORMULA 4.3
Variance

To find the variance of the distribution 2, 4, 6, 8, and 10, we first determine the mean, which is 6.0. Then we subtract the mean from each score $(X - M)$, to obtain a distribution of deviation scores: -4, -2, 0, 2, 4. These are squared $(X - M)^2$, making the signs positive and yielding the squared deviation scores 16, 4, 0, 4, 16. The sum of these squared deviation scores $\Sigma(X - M)^2$ is 40. When this sum is divided by N we have $\Sigma(X - M)^2/N = 8.0$, which is the variance. Suppose the scores were less disperse and consisted of 4, 5, 6, 7 and 8. The mean would again be 6.0, but $\Sigma(X - M)^2$ would be 10, and the variance would be reduced to 2.0.

The formula for the variance is sometimes given in terms of deviation scores. If we define a deviation score $x = X - M$, then an algebraically equivalent formula for the variance will be $\sigma^2 = \Sigma x^2/N$. Unfortunately, both this formula and Formula 4.3 will be rather inconvenient if we wish to calculate the variance of a distribution of real data. The mean of such a distribution is rarely a whole number and consequently, to obtain the deviation scores, we shall be faced with a tedious series of subtractions and squares involving decimals. Using the defining Formula 4.3 and some algebra, we can develop a computing formula for the variance which will make its calculation from real data much easier.

$$\sigma^2 = \frac{\Sigma(X - M)^2}{N}$$

$$\sigma^2 = \frac{\Sigma(X^2 - 2MX + M^2)}{N}$$ Expanding $(X - M)^2$.

$$\sigma^2 = \frac{\Sigma X^2}{N} - \frac{2M\Sigma X}{N} + \frac{NM^2}{N}$$ Summing over all terms and dividing each term by N.

$$\sigma^2 = \frac{\Sigma X^2}{N} - 2M^2 + M^2$$ Substituting M for $\frac{\Sigma X}{N}$ and collecting terms.

$$\sigma^2 = \frac{\Sigma X^2}{N} - M^2$$ FORMULA 4.4 Computing formula for the variance

Formula 4.4 allows us to calculate the variance with only one subtraction. In order to use it, square each of the original scores, sum these, divide the result by N, and then subtract the squared mean. This process may not *look* much easier than that required by Formula 4.3, but if you have a table of squares (see Table S in the Appendix) and a pocket adding machine or a calculator, this computing formula will save a great deal of time.

If the data have been cast into a grouped frequency distribution, it is possible to reduce the labor of calculating the variance by using the same procedures we described for coding the mean. If the midpoint (X) of each interval has a constant (C) subtracted from it, and the remainder is divided by another constant (i), the variance of the resultant coded scores, X'', is developed below:

$$X'' = \frac{X - C}{i}$$ Defines the coded interval midpoint (X'') as a raw score from which a constant (C) has been subtracted with the result divided by a constant (i).

$$X = iX'' + C$$ Gives the coded equivalent of an uncoded interval midpoint.

$$X - M = (iX'' + C) - \left(i\frac{\Sigma X''}{N} + C\right)$$ Subtracting the mean from both sides and substituting the coded formula for the mean.

$$X - M \doteq iX'' + \not{C} - i\frac{\Sigma X''}{N} - \not{C}$$

Simplifying (the subtraction constant (C) now adds out, disappearing from the formula).

$$(X - M)^2 = i^2(X'' - M'')^2$$

First, i is factored; then both sides are squared.

$$\frac{\Sigma(X - M)^2}{N} = i^2\left(\frac{\Sigma X''^2}{N} - M''^2\right)$$

After summing and dividing by N, we expand and simplify terms in right parenthesis as in the development of Formula 4.4.

$$\sigma^2 = i^2\left(\frac{\Sigma fX''^2}{N} - M''^2\right)$$

FORMULA 4.5
Coded formula for the variance

We shall use Formula 4.5 to calculate the variance of the grouped frequency distribution shown in Table 4.1. Column X contains the midpoints of the grouping intervals. Column X'' lists the coded interval midpoints. These are the original interval midpoints from which a constant ($C = 37$) has been subtracted, and the result divided by $i = 5$. Column fX'' contains the frequencies multiplied by the coded midpoints, and column fX''^2 contains the frequencies multiplied by the square of the coded midpoints.

TABLE 4.1 Calculation of the Variance from a Grouped Frequency Distribution

Intervals	X	f	X''	fX''	fX''^2
50–54	52	2	3	6	18
45–49	47	6	2	12	24
40–44	42	10	1	10	10
35–39	37	16	0	0	0
30–34	32	13	−1	−13	13
25–29	27	9	−2	−18	36
20–24	22	5	−3	−15	45
15–19	17	3	−4	−12	48
		$N = 64$		*$(\Sigma fX'') = -30$ or $\Sigma X''$	$(\Sigma fX''^2) = 194$ or $\Sigma X''^2$

$$\sigma^2 = i^2\left[\frac{\Sigma fX''^2}{N} - M''^2\right] = 5^2\left[\frac{194}{64} - \left(\frac{-30}{64}\right)^2\right]$$

$$\sigma^2 = 25(3.03 - .22) = 70.25$$

* Please remember that $\Sigma fX''$ is equivalent to $\Sigma X''$. The f in the formula simply calls our attention to the fact that multiplication is no more than repeated addition.

The sums of these columns give all of the data we need to apply Formula 4.5. The formula and the appropriate substitutions are shown beneath Table 4.1. Notice the simplicity of the calculations required to determine the variance of this distribution consisting of 64 measures.

Formula 4.5 illustrates several interesting characteristics of measures of dispersion. When we obtain a mean for data coded by subtraction, we must add the subtracted constant to the mean of the coded scores to obtain the mean of the original measures. In the development of the formula for the *variance* of coded data, the constant of subtraction "adds out" and disappears from the formula. This seems quite reasonable; subtracting a constant is tantamount to sliding the entire distribution down the scale of measurement. Of course, this process will change the *mean* of the distribution, but it will not change its *shape*, and consequently it will not change the dispersion of the scores. Since subtracting a constant from each of the scores does not change the dispersion of the scores, we need not correct the variance if scores are coded by subtraction.

We have an altogether different matter once each score is divided by a constant. Dividing by a constant *reduces* the dispersion of scores, just as multiplying by a constant increases the dispersion. The variance of scores coded by division must therefore be corrected. The correction involves multiplying the variance of the coded scores by i^2, because variance is based on the *square* of the deviation of each score from the mean.

We shall calculate the variance for another set of measures, those shown in Table 4.2 below.

TABLE 4.2 Calculation of the Variance from a Grouped Frequency Distribution

Interval	X	f	X''	fX''	fX''^2
47–49	48	4	4	16	64
44–46	45	7	3	21	63
41–43	42	12	2	24	48
38–40	39	11	1	11	11
35–37	36	14	0	0	0
32–34	33	12	−1	−12	12
29–31	30	9	−2	−18	36
26–28	27	7	−3	−21	63
		$N = 76$		$\Sigma fX'' = 21$	$\Sigma fX''^2 = 297$

$$\sigma^2 = i^2\left[\frac{\Sigma fX''^2}{N} - M''^2\right] = 3^2\left[\frac{297}{76} - \left(\frac{21}{76}\right)^2\right] = 34.49$$

STANDARD DEVIATION

Although the variance is of great importance as a measure of variability, it is a measure in terms of *squared* scores or *squared* units of measurement. It is calculated by summing the *squared* deviation of each measure from the mean of its distribution. If we take the square root of the variance we have a measure of variability called the *standard deviation*, a measure expressed in terms of the *original* score units. The symbol for the standard deviation is σ, a lower case Greek sigma. The formula for the standard deviation is simply the square root of the formula for the variance.

$$\sigma = \sqrt{\frac{\Sigma(X - M)^2}{N}}$$

FORMULA 4.6
Standard deviation

The most convenient formula for computing σ is also found by taking the square root of the formula for computing the variance. This formula becomes:

$$\sigma = \sqrt{\frac{\Sigma X^2}{N} - M^2}$$

FORMULA 4.7
Computing formula for the standard deviation

The coding formula for σ is also the square root of the coding formula for the variance.

$$\sigma = i\sqrt{\frac{\Sigma f X''^2}{N} - M''^2}$$

FORMULA 4.8
Coding formula for the standard deviation

We shall calculate the standard deviation for the distributions in Tables 4.3 and 4.4. In the first table, the data are ungrouped, just as they might have come from an experiment; in the second table a different and larger set of data has already been grouped.

Throughout the text we shall find it convenient to explain a statistic with one formula but calculate it with another. We shall refer to these as "conceptual" and "computing" formulas. They will, of course, always be algebraically equivalent.

Formulas 4.6, 4.7 and 4.8 can all be used to calculate σ. Formula 4.6 is given because it most clearly illustrates the nature of σ as the square root of the mean of squared deviations from the mean. Formula 4.6 is useful as an aid to conceptualizing the standard deviation, but it is a clumsy formula to use for computation. Formula 4.7 is much more efficient as a computing formula, particularly when an automatic calculator is avail-

TABLE 4.3 Calculation of the Standard Deviation from Ungrouped Data

X	X^2	X	X^2
6	36	4	16
8	64	7	49
9	81	7	49
7	49	2	4
5	25	5	25
9	81	8	64
3	9	1	1
1	1	9	81
8	64	7	49
7	49	6	36
$N = 20$		$\Sigma X = 119$	$\Sigma X^2 = 833$

$$\sigma = \sqrt{\frac{\Sigma X^2}{N} - M^2} = \sqrt{\frac{833}{20} - \left(\frac{119}{20}\right)^2} = 2.50$$

TABLE 4.4 Calculation of the Standard Deviation from Grouped Data

Interval	X	f	X''	fX''	fX''^2
160–164	162	4	5	20	100
155–159	157	8	4	32	128
150–154	152	17	3	51	153
145–149	147	30	2	60	120
140–144	142	29	1	29	29
135–139	137	36	0	0	0
130–134	132	30	−1	−30	30
125–129	127	24	−2	−48	96
120–124	122	19	−3	−57	171
115–119	117	8	−4	−32	128
110–114	112	4	−5	−20	100
		$N = 209$		$\Sigma fX'' = 5$	$\Sigma fX''^2 = 1055$

$$\sigma = i\sqrt{\frac{\Sigma X''^2}{N} - M''^2} = 5\sqrt{\frac{1055}{209} - \left(\frac{5}{209}\right)^2} = 11.25$$

able. Most of these machines have an "automatic square" feature which allows the operator to determine ΣX and ΣX^2 quite easily. If no calculator is available and N is large, it is usually very efficient to cast the data into a grouped frequency distribution and use Formula 4.8 for the calculation of σ.

We have discussed four measures of variability: R, Q, σ^2 and σ, emphasizing the latter two. The reason for this emphasis is that σ^2 and σ have some very important mathematical properties, and these measures of dispersion are much more widely used than the simpler R or Q. The variance (and the standard deviation) is a function of the magnitude of *each* score in the distribution, while R and Q are determined by counting only. This means that R and Q cannot be used in algebraic operations—they cannot be combined with other functions. There are other, more specific, considerations regarding the variance that we shall discuss in detail later.

Now we examine one of the more interesting properties of σ, but to do so we must introduce a theoretical distribution.

THE NORMAL DISTRIBUTION

The distributions of measures used in all of our illustrative problems have been approximately "bell-shaped." Most of the measures tended to cluster about the center of their distribution and trail off with decreasing frequency toward the upper and lower ends, much like the curve in Figure 4.2.

When a distribution is both symmetrical and bell-shaped it is probably quite close to what is called a normal or Gaussian distribution. A great many distributions of psychological and biological measurements are "normally" distributed. Such diverse characteristics as sleeve lengths for adult males, intelligence test scores for school children, errors made by rats learning a maze, and achievement in statistics courses seem to follow the "normal" distribution.

Statisticians are not at all in agreement about *why* the normal curve seems to describe the distribution of so many psychological and biological variables. One line of reasoning makes use of an important theorem in statistics called the Central Limit Theorem. This theorem says that the distribution of sample *means* will be normal regardless of the way the parent population of measures is distributed provided that the samples are sufficiently large and that the measures composing the samples are randomly obtained. Let's see how this theorem might help account for the normal distribution of IQ's. Suppose we assume that an individual's IQ, like a sample mean, is the sum of a number of independent influences. It is reasonable to suppose that IQ depends upon many environmental and genetic components. If there are enough independent components, then IQ should be normally distributed, just as sample means will be if we have followed the conditions imposed by the Central Limit Theorem.

Actually, of course, it makes no difference whatever *why* the normal curve describes so much psychological data. The normal or Gaussian curve is a purely mathematical function. We could give the equation which defines this curve; many elementary books do, but little practical value results since most of the information we might wish to derive from it can be obtained more easily from Table N in the Appendix which we shall discuss shortly. The important point is this: to the extent that the distributions with which we work *approximate* that described by the mathematical equation of the normal curve, to that same extent the interesting mathematical properties of the curve can be applied to our real distributions.

This situation is not unique in mathematics. A plane is not a "real thing," but the principles of plane geometry work quite well for the surveyor. We even come to call any flat surface a plane although a mathematical plane has no thickness and could not actually exist. In a somewhat similar way we refer to any symmetrical, bell-shaped curve as a normal distribution even though it probably does not correspond exactly to the mathematical normal curve. Figure 4.2 illustrates such a distribution and some of its characteristics.

As Figure 4.2 shows, we have constructed a vertical line from the mean to intersect the curve at its highest point. Follow the curve in both directions and note where the slope changes from concave to convex. The points at which this change occurs are the points of inflection and they are always one standard deviation on either side of the mean. If we construct vertical lines at these points (-1σ and $+1\sigma$) we shall find approxi-

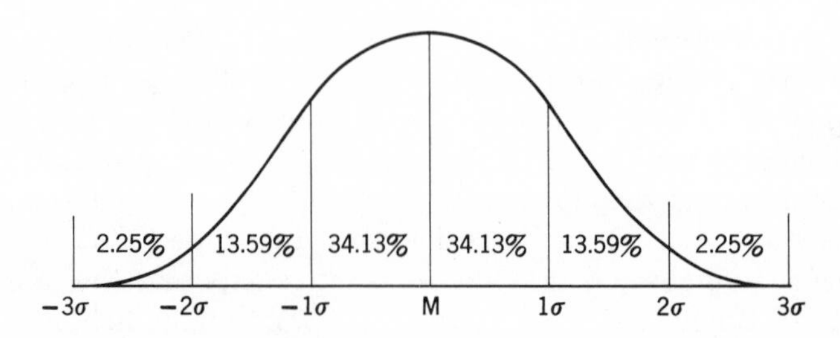

FIGURE 4.2 A normal distribution showing relative frequencies within given standard deviation distances from the mean.

mately 68 percent of the area under the curve falls between them. This will be true of any normal distribution regardless of the value of M, σ, or N. We can also construct vertical lines at -3σ, -2σ, $+2\sigma$ and $+3\sigma$. Figure 4.2 gives the percent of area which will fall between these lines in any normal distribution.

While the normal distribution in Figure 4.2 seems to extend from -3σ to $+3\sigma$, the theoretical normal distribution extends an infinite distance on either side of the mean. For all practical purposes however, we can permit the curve to intersect the abscissa once we have passed $\pm 3\sigma$, since only 26 measures in 10,000 will fall beyond these limits.

The characteristics described above will only be *exactly* true for distributions which are *exactly* normal, and we rarely find exactly normal distributions of real data. However, most distributions of psychological measures are approximately normal, close enough so that we can usually expect the characteristics of normal distributions to apply.

STANDARD SCORES

It is often helpful to transform raw scores into scores based on standard deviation units. These are called standard scores. One kind of commonly used standard score is called a z score. A z score is the deviation of a raw score from the mean of its distribution divided by the standard deviation of that distribution. Thus, a score which falls exactly one standard deviation above the mean of its distribution would be equivalent to a z score of 1.00; if it fell one standard deviation below the mean it would be equivalent to a z score of -1.00. The formula for z scores is

$$z = \frac{X - M}{\sigma}$$

FORMULA 4.9
z scores

If a normal distribution has $M = 46$ and $\sigma = 8$, the raw score 54 will yield $z = 1.00$. This z score will be equivalent to the 84th centile, a result we can deduce from Figure 4.2. Note that 50 percent of the curve falls below M, and an additional 34 percent falls between M and $+1\sigma$. Therefore, 84 percent must fall below $+1\sigma$. For any approximately normal distribution we can expect to find about 84 percent of the measures falling below $z = 1.00$.

Continuing with the same distribution, raw score = 38 will produce $z = -1.00$. From the percentages marked off by the various values of σ in Figure 4.2, it can be seen that $z = -1.00$ falls at the 16th centile. Thus $X = 38$ is at the 16th centile. Similarly, from Formula 4.9, $X = 62$

will yield $z = 2.00$, and Figure 4.2 shows that this score is equivalent to the 97.5 centile.

Verify the z equivalents of the raw scores for the following distribution: if $M = 60$ and $\sigma = 10$, $X = 55$ will be equivalent to $z = -.50$; $X = 49$, $z = -1.10$; $X = 74$, $z = 1.40$; $X = 60$, $z = .00$; and $X = 68$, $z = .80$.

Table N in the Appendix shows the percent of area lying between the mean and various values of x/σ for any normal distribution. The lefthand column of this table gives x/σ values to tenths and the top row of the table divides each of these tenths into ten parts. Thus, to find the percentage of the distribution between the mean and $.50\sigma$, we scan down the left column to the row of numbers opposite .5. The first entry in this row represents .50 and the tabled value is 19.15, which tells us that 19.15 percent of the distribution falls between the mean and $.50\sigma$ above the mean. Since the normal curve is symmetrical, 19.15 percent of the distribution will also fall between the mean and $.50\sigma$ below the mean.

To find the percentage of the normal curve falling between the mean and $.75\sigma$ we scan down the lefthand column to .7 then across to the column headed .05. The entry in the table is 27.34, telling us that 27.34 percent of a normal distribution falls between the mean and $.75\sigma$. From this we can determine that the centile value of $.75\sigma$ would be 77, since 50 percent of the normal distribution falls below the mean, and 27 percent falls between the mean and $.75\sigma$. The centile value of any z score can be found by similar procedures.

Both z scores and centiles provide a kind of common denominator, a way of locating a measure within a distribution independently of the distribution M, σ, and N. A score at the 75th centile, or $z = .68$, has exactly the same meaning in any approximately normal distribution regardless of the value of N, M, or σ for that distribution. This characteristic of z scores and centiles is very valuable when comparing scores from different distributions.

It is quite possible for a student to score 48 on one test and 26 on another. The raw scores alone would lead you to believe that his performance on test 1 was superior to his performance on test 2. A score of 48, however, might be at the 13th centile on test 1, but a score of 26 could be at the 81st centile on test 2.

While both z scores and centiles provide a similar kind of common denominator, allowing us to compare the relative standings of scores from different distributions, the z score procedure is usually preferred because centile scores have a serious shortcoming. Equal differences between centiles do not represent equal differences among the original units of measurement.

Figure 4.3 illustrates this problem with a distribution of IQ scores.

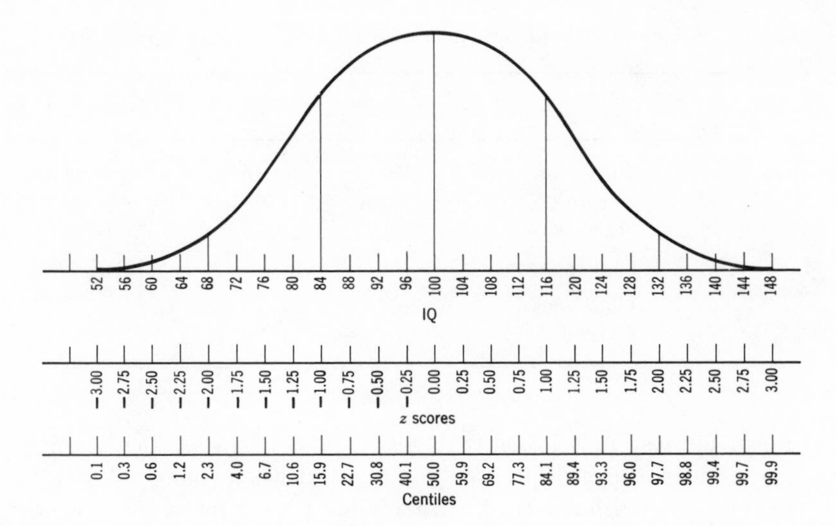

FIGURE 4.3 Comparison of raw scores, z scores, and centiles in a normal distribution.

The abscissa is also scaled in z score units and in centiles. Notice that a 4-point IQ spread (from 100 to 104) is equivalent to moving from the 50th to the 59.9th centile, while a 4-point spread from IQ 144 to IQ 148 is equivalent to moving from the 99.7th to the 99.9th centile. This is simply the result of the bulge in the middle of the normal curve. Since a majority of cases are located there, the centiles near the mean will be crowded tightly together; toward the tails, where there are fewer cases, the same raw score spread will produce a much smaller centile difference.

Since centiles are not linearly related to the original units of measurement, we shall not get the same result by averaging centiles that we get by averaging raw scores. For example, averaging IQ's of 56 and 88 yields an IQ of 72, which, from Figure 4.3, is equivalent to the 4th centile. However, if we convert IQ 56 to the 0.3th centile and IQ 88 to the 22.7th centile and then average the *centiles*, the result is about the 11th centile.

These difficulties *do not* arise when raw scores are converted to z scores, and the z scores are averaged. As you can see from the scale on the abscissa in Figure 4.3, z scores are linearly related to the original units of measurement. We shall give an additional example of the utility of z scores as a common denominator when comparing the results of several tests for several different people.

Look at the data in Table 4.5. The table illustrates the performance of two students on two different class tests. The data are exaggerated for illustrative purposes, but similar situations occur quite often. Notice that

TABLE 4.5 A Comparison of Raw Score Means and Standard Score Means

		Test A	Test B	Mean Score
Student 1	Raw score	50	75	62.5
	z score	2.00	.00	1.00
Student 2	Raw score	35	95	65.0
	z score	−1.00	1.00	.00
	Class mean	40	75	
	Class σ	5	20	

the mean for test A is lower than test B and has a smaller variance. Student 1 has performed quite well on this test when compared with the rest of the class, but has earned relatively few points due to the low mean and small variance of the distribution. His performance is average on the second test.

Student 2 did quite well on the second test (test B), but his relative performance is not as good as that of student 1 on test A. However, because of the differences between the distributions, he has a higher total point count on the two tests. The raw scores on these two tests are not directly comparable, but we can equate them by converting to z scores. The z scores place both performances on the same base line so that the comparison is made from the same point and with the same units of measurement. The points of origin are the means of the distributions, and the units of measurement are functions of the standard deviations of these distributions.

There are some disadvantages with z scores, however, because they are reported so that $M = 0.00$ and $\sigma = 1.00$. Consequently, all scores below the mean will be represented by negative numbers and most scores will involve decimals. It may also be awkward to explain why the class average is zero to a statistically unsophisticated parent who wishes to discuss his child's achievement. We can circumvent these problems by using other standard score systems which have more convenient values for M and σ. For example, we can use a standard score system based on $M = 500$ and $\sigma = 100$, or $M = 100$ and $\sigma = 20$. Either of these would avoid decimal notation and negative numbers.

We can give a formula for transforming any raw score to a standard score.

$$\text{Std. Score} = M_A + \sigma_A \left(\frac{X - M_X}{\sigma_X} \right)$$

FORMULA 4.10
Conversion to standard scores

M_A and σ_A represent the convenient values of M and σ in the standard score system we have chosen to use, and M_X and σ_X represent the true M and σ of the measures.

NORMALIZED STANDARD SCORES

Transforming raw scores into standard scores does not alter the shape of the parent distribution. If the original distribution is normal, the distribution of standard scores will be normal; if the original is skewed, the standard scores will be skewed.

There is, however, a method of normalizing a distribution of raw scores by converting them to what are called McCall T scores. The procedure is a relatively simple one although it can be time consuming when there are a great many different scores.

To normalize a distribution we find the centile equivalent of each score (remember we did this in the previous chapter). Then we find the z score equivalent of that centile from Table N. *These* z scores will be normally

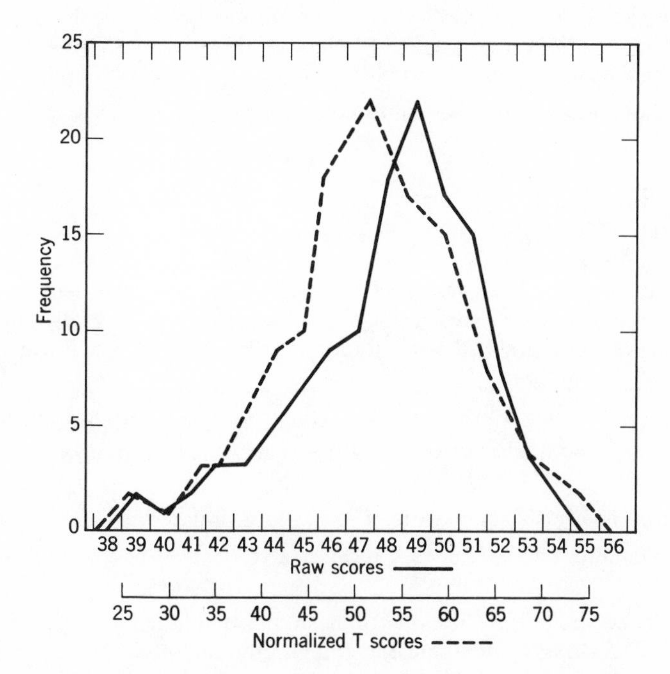

FIGURE 4.4 Frequency polygons of raw scores and normalized T scores for the lecture method data in Table 2.1.

distributed. The McCall T score procedure requires one further step; we convert the z scores to a distribution where $M = 50$ and $\sigma = 10$.

We can illustrate the effect of normalizing a distribution by applying this procedure to the data from which Figure 2.1 was drawn. This distribution of test scores had a decided negative skew, but as Figure 4.4 shows, the skew can be removed and the distribution normalized.

We simply find the centile equivalent of each score's midpoint, then determine the standard score equivalent of that centile from Table N. Frequency is then graphed against these standard scores rather than the original raw scores.

EXERCISES

1. Find Q, σ^2 and σ for the distributions in Table 3.2 and Table 2.3.
2. What is the z score equivalent of a raw score of 51 in Table 2.3? What is the centile equivalent of this z score from Table N of the Appendix? What is the centile value of $X = 51$ calculated directly from the distribution? Why are the two values slightly different?
3. Prove that the mean of a distribution of standard scores is zero.

$$\frac{\sum\left(\frac{X - M}{\sigma}\right)}{N} = 0$$

4. What differences would you expect between a cumulative frequency polygon of normally distributed data graphed with centiles spaced equally along the ordinate and one graphed with standard scores spaced equally along the ordinate?
5. Can two distributions with the same R have different standard deviations?
6. If $M = 64$, $\sigma = 8$, what z scores will be equivalent to raw scores of 51, 66, 70, and 80?
7. What are the centile equivalents of the z scores in Exercise 6?
8. What minimum z score is needed to place a student in the top 10 percent of his class?
9. T—F. Take each score in a distribution and divide it by 2. The distribution of $X/2$ will have the same σ as the distribution of X.
10. T—F. Distributions differing in variability must also differ in central tendency.

11.

	f
46–50	2
41–45	6
36–40	9
31–35	12
26–30	15
21–25	18
16–20	6
10–15	3

$M =$ 80th centile = z score equivalent of 12 =

Mdn = $\sigma =$ 14 =

32 =

41 =

12.

	f
12.26–12.50	3
12.01–12.25	9
11.76–12.00	18
11.51–11.75	34
11.26–11.50	21
11.01–11.25	15
10.76–11.00	10
10.51–10.75	6
10.26–10.50	3

$M =$ 80th centile = z score equivalent of:

Mdn = $\sigma =$ 12.27 =

10.46 =

11.51 =

CHAPTER 5

Measures of Association

In Chapters 3 and 4 we discussed procedures for measuring the central tendency and the variability of distributions. In this chapter we shall discuss ways of describing the degree of relationship, or association, between two variables. Consider, for instance, the data in Table 5.1 which lists the performances of eighteen children on an initial test of academic ability

TABLE 5.1 Ability and Achievement Test Scores for a Class of Children

Child	(X) Ability	(Y) Achievement
A	41	32
B	53	29
C	64	47
D	43	35
E	44	29
F	51	41
G	68	46
H	52	39
I	53	37
J	69	50
K	41	26
L	32	21
M	38	19
N	62	34
O	68	41
P	58	43
Q	45	44
R	59	49

and a subsequent test of achievement. How can we describe the degree to which "ability" and "achievement" are related in this example?

SCATTERGRAMS

It is possible to illustrate the relationship with a graph. Such graphs are called scattergrams and one constructed from these data appears in Figure 5.1.

Each child has an ability test score and an achievement test score. Each pair of scores is used to locate a single point on the scattergram. The point representing child *A*'s performance is located in the following way: we erect a line perpendicular to the abscissa from the child's ability score of 41 and another line perpendicular to the ordinate from his achievement score of 32. These two perpendicular lines meet to determine point *A* in the lower left quarter of Figure 5.1. Point *A* represents child *A*'s performance on the *two* measures in question. The two numbers, 41 (X) and 32 (Y), which locate point *A*, are called its coordinates.

The point representing child *P*'s scores has been plotted in the same way. This point is at the intersection of a line perpendicular to the ab-

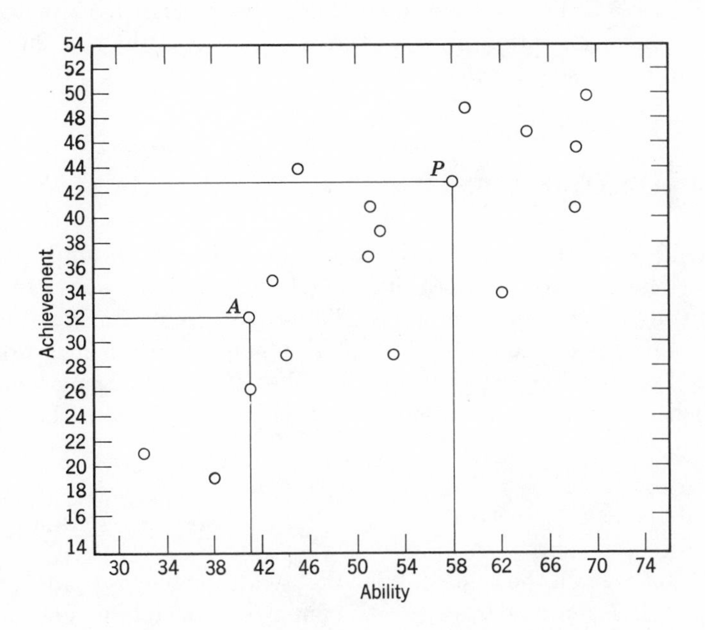

FIGURE 5.1 A scattergram for the data in Table 5.1.

scissa at ability score 58, and a line perpendicular to the ordinate at achievement score 43. The same procedure has been used to plot the remaining points in the scattergram. The two scores for each child supply the coordinates to plot one point on the scattergram.

This graph gives us a much better idea of the relationship between the two variables than we can get by just looking at the arrays of raw scores in Table 5.1. The scattergram illustrates several things about the relationship that otherwise would have been quite obscure. First, high academic ability is associated with high achievement and low academic ability is associated with low achievement. Second, this relationship, although reasonably good, is not perfect. Some students with high ability test scores also have high achievement test scores, but the achievement test performance of other high ability students is only average. Third, we can describe the general trend of this relationship by a straight line. A fixed increment of ability tends to be associated with the same fixed increment of achievement throughout the range of measurements.

Different kinds of relationships will yield different scattergrams. For example, we might find that two variables are very closely related, but the association is inverse instead of direct. An inverse relationship exists when high scores on one variable are associated with low scores on the other. A scattergram showing a strong inverse relationship appears at the top of Figure 5.2. The scattergram in the center of Figure 5.2 pictures a very weak inverse relationship, and the scattergram at the bottom of the figure shows a strong direct relationship.

THE CORRELATION COEFFICIENT

Although a scattergram provides a picture of the relationship between two variables, we can also *measure* the relationship just as we can measure central tendency or variability. The Pearson product moment correlation coefficient is the statistic most commonly used to measure the amount of relationship between two variables. This correlation coefficient is symbolized by r and the formula for its calculation is given below.

$$r = \frac{\Sigma xy}{N\sigma_X\sigma_Y}$$

FORMULA 5.1
The Pearson product moment correlation coefficient

Figure 5.3, based on the same data as Figure 5.1, will help us understand why Formula 5.1 provides an effective index of association. Let us first consider the Σxy term which makes up the numerator.

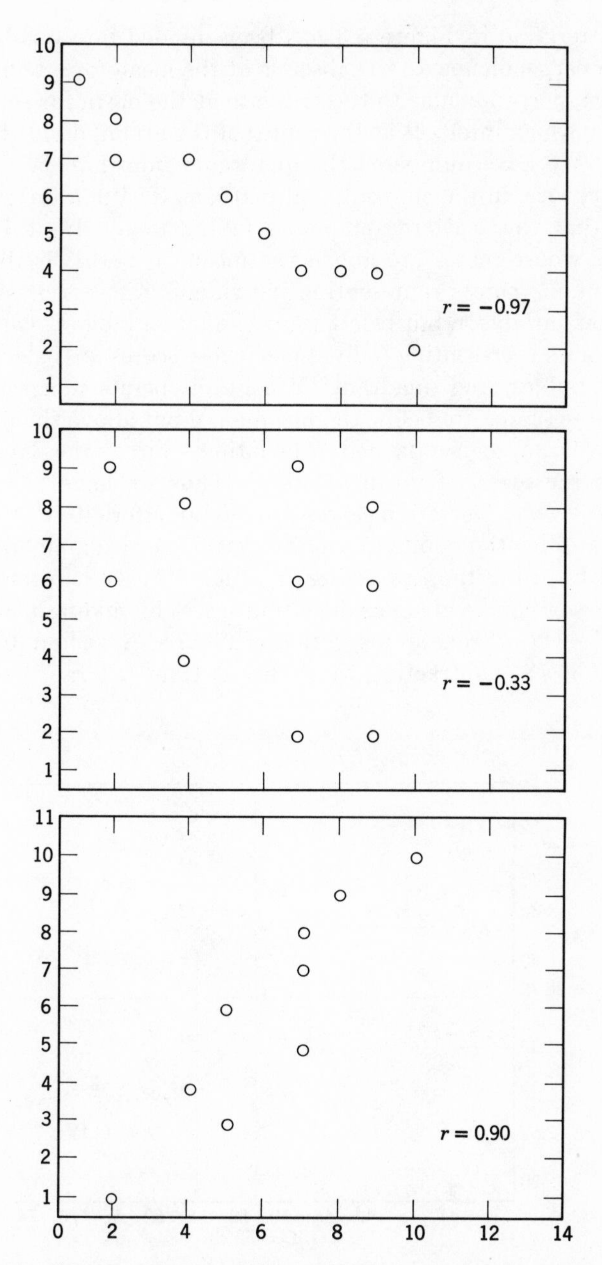

FIGURE 5.2 Scattergrams showing different degrees of correlation.

The scattergram in Figure 5.3 has been divided into quadrants by a line drawn perpendicular to the abscissa at the mean for intelligence, and a line drawn perpendicular to the ordinate at the mean for achievement. These lines, which intersect at the center of the graph, divide it into four quadrants. We have numbered the quadrants from I to IV, proceeding counterclockwise, in the conventional mathematical notation.

Notice that the scattergram points falling in quadrant I represent individuals whose scores are above the mean on both variables. Quadrant II contains points representing individuals whose scores are above the mean on variable Y but below the mean on variable X. Quadrant III contains points representing individuals whose scores are below the mean on both variables, and quadrant IV contains points representing individuals whose scores are below the mean on Y but above the mean on X.

In the last chapter we defined a deviation score as the deviation of a score from the mean of its distribution. Thus we have $x = X - M_X$, and $y = Y - M_Y$. Deviation scores provide an alternative to raw scores as coordinates for the points in a scattergram. For example, point P can be located by using the raw scores $X = 58$, $Y = 43$ as coordinates, or the coordinates can be given as deviation scores by finding $x = X - M_X$ and $y = Y - M_Y$. From an inspection of Figure 5.3 we find $M_X = 52.28$ and $M_Y = 36.78$. Subtracting these means from the raw score coordi-

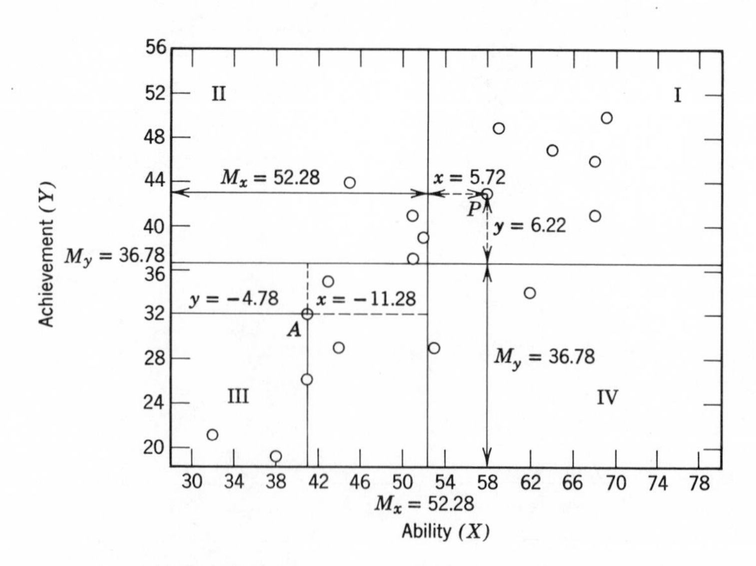

FIGURE 5.3 A scattergram of the data in Table 5.1 with the coordinates of two points shown as deviation scores.

nates, we have $x = 5.72$ and $y = 6.22$. These deviation scores can also be used to locate point P. Now find point A in Figure 5.3. Since this point falls below the means of both variables, the deviation score coordinates are both negative.

The numerator of Formula 5.1, Σxy, directs us to obtain the *product* of each pair of deviation score coordinates and then we must find the *sum* of these products. Notice that the sign of the products will be positive for all points falling in quadrants I and III. The sign will be negative for all points falling in quadrants II and IV. Consider what happens when we obtain the sum of these products for all of the points in the scattergram. If the relationship between the variables is direct and close, as in the lower scattergram of Figure 5.2, the majority of points will fall in quadrants I and III and will yield a relatively large *positive* value of Σxy. On the other hand, if the relationship is similar to that in the upper scattergram of Figure 5.2, most of the points will fall in quadrants II and IV producing a large *negative* value for Σxy. When little or no relationship exists, the points will be distributed haphazardly, falling in about equal numbers in each quadrant. In such a situation the sum of products with positive signs should be about equal to the sum of products with negative signs, and Σxy should be close to zero. Observe that the magnitude of Σxy indicates the magnitude of the relationship, and the sign of Σxy indicates the direction of the relationship. Thus, a large negative value for Σxy will indicate a strong inverse relationship, and a small positive value for Σxy will indicate a weak direct relationship.

As Formula 5.1 indicates, Σxy must be divided by $N\sigma_X\sigma_Y$ in order to obtain r, the correlation coefficient. The division by N permits us to compare correlation coefficients based upon different numbers of cases. As we have pointed out, the magnitude of Σxy depends upon the degree of association between the variables, but its magnitude also depends upon the number of pairs of measurements. A large N would supply more cross products, and with a very large number of cases a minor degree of correlation could produce a large positive or negative Σxy. When Σxy is divided by N the term becomes a *mean* of cross products and the measure of correlation becomes independent of sample size. This permits us to compare correlation coefficients based upon different numbers of cases.

The final terms in the denominator, $\sigma_X\sigma_Y$, provide for a common unit of measurement. Consider what will happen to the mean of the cross products of the deviation scores ($\Sigma xy/N$) if we change the scale of units by which ability and achievement are measured. Suppose we just change the way we *score* the ability and the achievement tests so that each question is "worth" ten points instead of one point. This will increase the value of each deviation score by a factor of ten. Then the *product* of each pair of deviation scores will increase by a factor of 100, and so will the mean of these cross products, $\Sigma xy/N$. Keep in mind, however, that the relative positions of the points in the scattergram have *not* changed: the correlation between the two variables remains exactly as it was, but by changing the units of measurement we have effected a considerable change in $\Sigma xy/N$. The problem is to find a method of expressing the deviation scores that will be independent of the units by which the variables are measured. In short, we need a common denominator for units of measurement.

Remember that in the last chapter we ascribed this property to z scores. We said that z scores were comparable from one distribution to another because they were *independent* of the mean and standard deviation of the original units of measurement. If the measurement of both variables is reduced to z scores, we should have the common denominator we seek.

We have defined a z score on some variable, X, as $z_X = (X - M_X)/\sigma_X$. Similarly, we shall define a z score on some variable Y as $z_Y = (Y - M_Y)/\sigma_Y$. Notice that z scores are just deviation scores divided by the standard deviation of the distributions from which they came. Consequently, we can convert the "raw" deviation scores to z scores if we divide $\Sigma xy/N$ by $\sigma_X\sigma_Y$. The result (Formula 5.1) is *still* the mean of the products of the coordinates of all points, but the coordinates are now expressed as z scores. Perhaps this fact will be clearer if we point out that an algebraically equivalent expression for Formula 5.1 is $r = \Sigma z_X z_Y/N$.

Note that $r = \Sigma xy/N\sigma_X\sigma_Y$ is independent of the units used to measure the correlated variables. If we use Formula 5.1 and expand the units of measurement by a factor of 10, the hundredfold increase in the numerator will be paralleled by an equivalent hundredfold increase in the denominator since σ_X and σ_Y will each be increased by a factor of ten. Using either Formula 5.1, or its equivalent, will give us a measure of association which is not influenced by N or by the units in which the individual variables are measured.

The Pearson r can vary between $+1.00$ and -1.00. High positive coefficients will occur when the association is direct and close. When the correlation coefficient assumes its maximum positive value, when $r = 1.00$, all of the points in the scattergram will lie exactly on a straight line extending from the lower left quadrant through the upper right quadrant. High negative coefficients will occur when the association is inverse and close. When the correlation coefficient assumes its maximum negative value, when $r = -1.00$, all of the points will again lie on a straight line, but this time the line will extend from the upper left quadrant through the lower right quadrant. When r is near zero, the correlation between the variables is negligible and we can expect the points to be dispersed about equally among the four quadrants. Figure 5.2 shows scattergrams of several distributions for which r has been calculated. Once again, it is worth mentioning that the sign of the correlation coefficient has nothing to do with the closeness of the association between the variables. Sign indicates only the direction of the relationship; the absolute value of the correlation coefficient indicates the magnitude of the relationship.

Although it is frequently used, the Pearson product moment correlation coefficient is by no means universally applicable. It is only appropriately used to describe the correlation between linearly related variables. This is the situation in which the general trend of points in a scattergram can best be described by a straight line. *Only under these circumstances* is the calculation of r appropriate. When the relationship between X and Y is other than linear, the Pearson product moment correlation coefficient underestimates the amount of relationship between the variables. For this reason it is usually desirable to construct a scattergram before calculating r so that the assumption of linearity can at least be subjected to a visual check.

A nonlinear relationship is shown in the scattergram of Figure 5.4. Notice that in these hypothetical data the degree of adjustment increases as intelligence increases only to about IQ 130–150; beyond that range

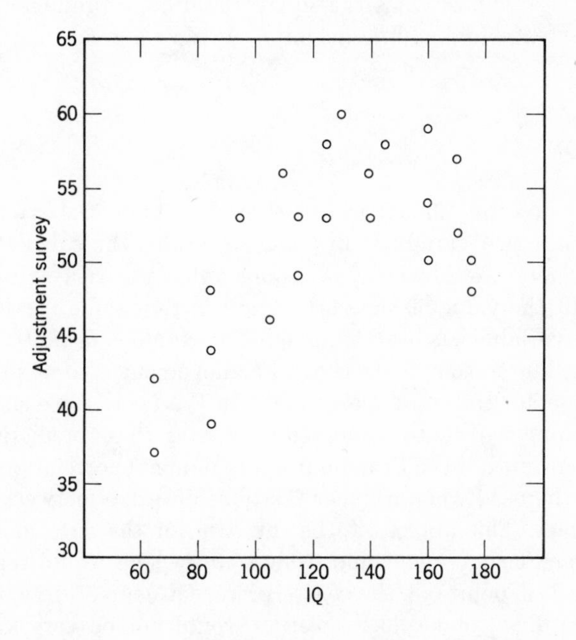

FIGURE 5.4 An illustration of a nonlinear relationship.

there is a slight decrease in adjustment. Since these data are best described by a curved line, the Pearson r would not be appropriate to describe this relationship. A treatment of nonlinear correlation can be found in most intermediate or advanced texts.

It is possible to use either $r = \dfrac{\Sigma xy}{N\sigma_X\sigma_Y}$ or $r = \dfrac{\Sigma z_X z_Y}{N}$ to calculate r, but there is a more helpful computing formula.

$$r = \frac{\Sigma XY - \dfrac{\Sigma X \Sigma Y}{N}}{\sqrt{\left[\Sigma X^2 - \dfrac{(\Sigma X)^2}{N}\right]\left[\Sigma Y^2 - \dfrac{(\Sigma Y)^2}{N}\right]}}$$

FORMULA 5.2
Computing formula for r

Formula 5.2 is another algebraic equivalent of Formula 5.1. It looks more complicated, but it is actually easier to use, particularly if a calculator is available. Formula 5.2 permits us to work directly with raw scores while Formula 5.1 requires us to convert each raw score to a deviation score. The quantities ΣX, ΣX^2, ΣXY, ΣY, and ΣY^2 for Formula 5.2 are easily obtained with a calculator. When a calculator is not available, a table of squares will be a great convenience, though calculating the ΣXY term will still prove tedious. Once the squares and cross products (ΣXY) are obtained, a small hand adding machine can save a great deal of time.

PREDICTION

For many investigations it will be enough to know the extent of correlation between two variables; in these situations the calculation of r is sufficient. There are, however, occasions when we wish to make a *prediction* about the value of one variable given that some particular value of the other variable has been obtained. For example, using the data from our previous illustration, what level of achievement is to be predicted for a child whose aptitude test score is 60? In this section we shall see how the correlation coefficient is involved in making these predictions.

We said earlier that the Pearson product moment correlation coefficient should only be used when a linear relationship exists between the correlated variables. This appears to be the case for the data in Figure 5.1. Using a straight edge, one could draw a line which would represent the general trend of points in the scattergram. Of course, drawing the line solely on the basis of visual inspection would not be very satisfactory. The straight line that might best represent the data for one person might not be the line preferred by another. *The line of best fit* needs an objective definition.

We shall define the line of best fit, also called *the regression line of* Y *on* X, as that straight line from which the sum of squared deviations of the Y scores is minimal. In Figure 5.5 we have replotted the scattergram for the data recorded in Table 5.1. Line A on this scattergram is the regression line of Y on X. (We shall discuss line B later.) We shall label the deviation of any value of Y from this regression line as y'. The regression line, line A, is that straight line for which $\Sigma y'^2$ is less than it would be for any other straight line. This property *defines* the regression line and leads us to refer to it as fulfilling a least squares criterion of best fit.

Once the regression line has been drawn on the scattergram, we can use it to predict a Y value for any given value of X. As an example, using a graphical solution, we shall find the value of Y to be predicted given $X = 60$. In Figure 5.5 we constructed a vertical line at $X = 60$ and extended it upward to the intersection of the regression line of Y on X, line A. Then, from that point, we constructed a horizontal line to the ordinate. The predicted value of Y is given by the intersection of this horizontal line with the ordinate. From Figure 5.5 we see that $Y = 42$ will be our predicted value. The same procedure is used to predict a value

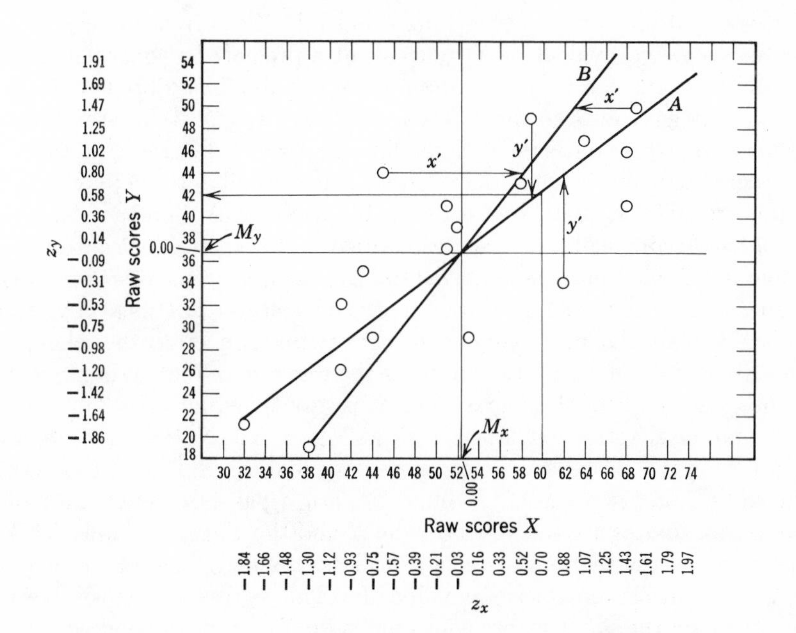

FIGURE 5.5 Regression of Y on X (line A) and X on Y (line B).

of Y given any other value of X. Remember that these are *predicted* values of Y. They won't be perfectly accurate unless $r = \pm 1.00$. Any other value of r will result in some "errors" of prediction. The advantages of basing our prediction on the regression line are that it permits us to estimate the magnitude of these errors and we are assured, by definition, that the sum of squared errors of prediction will be less than they would have been had we used any other procedure.

Now, this method of making predictions is all very fine *if* the regression line has already been drawn on the scattergram, but we cannot draw it until we know its equation. The equation for any straight line takes the form $Y = a + bX$ where a is the Y intercept, or the value of Y when $X = 0$, and b is the slope of the line, or the unit increase in Y for each unit increase in X.

The equation for the regression line of Y on X can be simplified if the raw scores on both variables are reduced to z scores. Under these circumstances the equation for the regression line of Y on X (now z_Y on z_X) becomes:

$$\hat{z}_Y = rz_X$$

FORMULA 5.3
Equation for the regression line of z_Y on z_X

As this equation shows, we need only multiply the correlation coefficient by the z score equivalent of X to arrive at a predicted z equivalent of Y. We shall use the caret (^) to represent the variable being predicted.

You may have noticed the difference between $Y = a + bX$, the general form of our equation for a straight line, and $\hat{z}_Y = rz_X$, the equation for the regression line when z scores are used. Actually, the equations are equivalent. The substitution of $\hat{z}_Y$ for Y and z_X for X simply reflects the use of z scores instead of raw scores. Notice, however, that the constant a, the value of the Y intercept, seems to have disappeared. This is because, when z scores are used, the Y intercept (now z_Y intercept) is always zero. We will explain this more fully in the next paragraph. The other change is the substitution of r for b (the slope of the regression line). When z scores are used the slope of the regression line is exactly equal to r.

We can clarify this if you will look at Figure 5.5. Notice that we have used both a raw score scale and a z score scale for each axis of the graph. The regression line of Y on X (line A) *always* passes through the point whose coordinates are the means of the X and the Y distributions. This is a property of the regression line and will always be the case regardless of the scale of measurement or the value of r. The regression line will always pass through the point whose coordinates are M_Y and M_X just as it does in Figure 5.5. However, when z scores are used, M_Y and M_X are also the origins, or zero points of the z score scales. Thus, since the regression line passes through this point, the z_Y intercept must be zero. From the equation $\hat{z}_Y = rz_X$, you can see that when z_X is zero, $\hat{z}_Y$ will be zero regardless of the value of r.

The practical result of this fact is that if an individual has a score at the mean of one distribution ($z_X = .00$) and we are required to predict his score on a second distribution, the predicted score will also be at the mean of that second distribution ($\hat{z}_Y = .00$). This will be true regardless of the correlation existing between the two sets of scores. This follows directly from the equation for the regression line, $\hat{z}_Y = rz_X$.

Now let us confirm our earlier prediction of $Y = 42$ given $X = 60$ by using the regression equation instead of the graphical solution we used earlier. Since we shall be converting raw scores to z scores we shall need the following information in addition to the slope of the regression line which is $r = .79$: we have $M_X = 52.28$, $\sigma_X = 11.0$, $M_Y = 36.78$, $\sigma_Y = 9.1$. Converting $X = 60$ to its z score equivalent we shall have $z_X = \dfrac{X - M_X}{\sigma_X} = \dfrac{60.00 - 52.28}{11.0} = .70$. Since $\hat{z}_Y = rz_X$ we have $\hat{z}_Y = .79 \times .70 = .55$. Converting this z score on Y to its raw score equivalent we solve $\hat{z}_Y = \dfrac{\hat{Y} - M_Y}{\sigma_Y}$ for $\hat{Y}$. This yields $\hat{Y} = \hat{z}_Y\sigma_Y + M_Y = .55 \times 9.1 + 36.78 = 41.78$, which is quite close to 42, the predicted value of Y we obtained by a graphical solution.

There is a raw score formula for predicting Y directly from X and it is fairly easy to derive from Formula 5.3 by algebra. The formula is:

$$\hat{Y} = r\frac{\sigma_Y}{\sigma_X}(X - M_X) + M_Y$$

FORMULA 5.4
Raw score formula for the prediction of Y from X

So far we have considered only the prediction of Y from X. However, it is possible to predict X from Y by essentially the same procedures; when we predict X from Y we shall make use of line B (Figure 5.5). This is the regression line of X on Y. The regression line of X on Y is defined as that straight line from which the sum of squared deviations of the X scores are minimal. If we let x' represent the deviation of any X value from the regression line of X on Y, then the regression line is chosen to minimize $\Sigma x'^2$. The equation for this regression line is given by:

$$\hat{z}_X = rz_Y$$

FORMULA 5.5
Equation for the regression line of z_X on z_Y

As with the regression line of Y on X we can give an alternative formula for use directly with raw scores.

$$\hat{X} = r\frac{\sigma_X}{\sigma_Y}(Y - M_Y) + M_X$$

FORMULA 5.6
Raw score formula for the prediction of X from Y

It is evident from inspection of Figure 5.5 that line B passes through the origin of the z scores. This means that the z_X intercept will be zero for line B, just as the z_Y intercept was zero for line A. In fact, both regression lines will always have one common set of coordinates, M_Y and M_X. The slope of the regression line of X on Y is also equal to r, as was the slope of the regression line of Y on X. However, the slope of the regression line of X on Y (line B) is calculated with respect to the Y axis. You should recall that for the regression line of Y on X (line A) the slope was calculated with respect to the X axis.

We shall see what happens to the relationship between these regression lines and to our predictions when $r = .00$ and when $r = \pm 1.00$. When $r = .00$ the slopes of the regression lines are zero and the regression lines are perpendicular to each other. Line A becomes a horizontal line parallel to the abscissa at M_Y (or $z_Y = .00$). Line B becomes a vertical line parallel to the ordinate at M_X (or $z_X = .00$). When $r = .00$, the equation $\hat{z}_Y = rz_X$ yields $\hat{z}_Y = .00$ regardless of the z_X value from which the prediction is made. Similarly, $\hat{z}_X = rz_Y$ will yield $\hat{z}_X = .00$ regardless of the z_Y value from which that prediction is made. In summary, when no relationship exists between X and Y, that is, when $r = .00$, our best prediction for Y will be the mean of the Y distribution. Similarly, when $r = .00$ our best prediction for X will be the mean of the X distribution. If no relationship exists between two variables ($r = .00$), the prediction of values from one variable will be independent of values on the other from which the prediction was made.

Suppose that $r = \pm 1.00$, what is predictable then? In this situation the regression lines will coincide. A line through the origin with a slope of 1.00 will be the same line whether the slope is calculated with respect to the X or the Y axis. When $r = 1.00$, the regression equations become $\hat{z}_Y = z_X$ and $\hat{z}_X = z_Y$. In this case *all* of the plotted points will fall precisely on a diagonal line and the prediction will be perfect. When $r = -1.00$ the same situation results, with both regression lines falling in quadrants II and IV.

In your work with regression lines it is particularly important to remain aware that predictions of Y from X involve a different regression line, and hence a different equation, from that involved in predicting X from Y. For example, when we obtained the most probable value of Y for $X = 60$, we erected a line perpendicular to the regression line of Y on X in Figure 5.4, and then constructed a horizontal line to the Y axis to obtain a predicted value of $Y = 42$. What happens if we now assume that we have been *given* $Y = 42$, and we are asked to find the best prediction for X? We would begin by reversing the process and constructing a line perpendicular to the Y axis until it intersects the regression line. But the appropriate regression line is *now* X on Y, not Y on X; and the value of X below the point of intersection with line B, the regression line of X on Y, is $X = 57$.

The same situation can be illustrated algebraically. If we wish to find the predicted $\hat{z}_Y$ when $z_X = 1.00$ and $r = .80$, we have $\hat{z}_Y = rz_X$, or $\hat{z}_Y = .80 \times 1.00 = .80$. However, if we are given $z_Y = .80$ and wish to *predict* the most probable $\hat{z}_X$, we must use the equation for a different line, $\hat{z}_X = rz_Y$. This equation yields $\hat{z}_X = .80 \times .80 = .64$. In summary: given a value of X, the best prediction for Y involves the use of line A. Given a value of Y, the best prediction of X involves the use of line B.

THE CONCEPT OF REGRESSION

When the scale of measurement for both variables has been transformed to z scores, the slope of the regression line of Y on X is given by the correlation coefficient. This coefficient cannot exceed 1.00, and in practice it is usually a good deal less than 1.00. Whenever we wish to predict z_Y from z_X, Formula 5.3 shows that we simply multiply the z score from which the prediction is to be made by the correlation coefficient. Since the correlation coefficient will usually be less than 1.00, the predicted z score will have a smaller absolute value than the z score from which the prediction was made. If the predicted z score has a smaller absolute value than the predictor z score, it follows that the predicted z score will always be closer to the mean. This is called regression toward the mean; this simply reflects the fact that a predicted value is always closer to the mean of its distribution than is the value from which the prediction was made.

The amount of regression depends upon the magnitude of the correlation coefficient. When r is high we will find relatively little regression toward the mean. For example, if $r = .90$ and the predictor is $z_X = 2.00$, we shall still predict a fairly extreme value for $\hat{z}_Y$, for example, $\hat{z}_Y = 1.80$. However, when $r = .10$ and the predictor is $z_X = 2.00$, we have $z_Y = .20$, a value substantially closer to the mean of Y. In general then, when r is low or at best modest, as it usually is in behavioral research, we shall expect to find a good deal of regression in our predictions.

The phenomenon of regression is very important because if it is not taken into account it can easily lead to serious errors in the interpretation of data. Suppose you have decided to investigate the effects of brain-penetrating war wounds on intelligence. Specifically, you wish to find out if such wounds changed the intellectual level of individuals receiving them, and you hypothesize that any effect will be a function of the subject's *original* intelligence. It is reasonable to suppose that very bright soldiers would suffer a greater deficit after a head wound than would soldiers of average intelligence. Suppose you select, from all of the soldiers with head wounds, a group whose average IQ was 100 prior to their injury, and then retest them with the same intelligence test. Let us assume that their mean score is again 100. You also select a group whose mean score was 125 prior to injury, and in retesting them you find their mean score has declined to 115. Assuming equivalence in the severity of injury, why would it be incorrect for you to conclude from these data that the effect of the wounds on the intellectual performance of these men was a function of their preinjury intelligence?

The answer to this question involves the concept of regression. Quite apart from any real effect produced by wounds, the obtained difference could be accounted for by differences in the absolute amount of regression expected for the two groups. Assume that the correlation between the two administrations of the intelligence test is .80. Assume further that there is *no effect* on intelligence test performance as a result of the wounds. How much change shall we predict for the soldier whose intelligence test score is at the mean of the distribution on the initial test? A score at the mean yields a $z_X = .00$, so the predicted retest score will be $\hat{z}_Y = .00$. We shall not predict any change in relative standing for a soldier whose performance was at the mean on the initial test. On the other hand, consider the soldier who has an initial test score 1σ above the mean. His $z_X = 1.00$ and, since $r = .80$, we have $\hat{z}_Y = .80$. Our prediction is that this soldier's retest score will drop by $.2\sigma$ from its initial level. Note that the phenomenon of regression alone leads us to expect at least some retest difference for the brighter soldier, and this effect is completely independent of *any* hypothetical changes in the soldiers between the first and second test administrations.

The experimental question posed here is an interesting and legitimate one. What additional observations would be needed in order to conclude that the effect of brain-penetrating war wounds was related to initial level of intelligence?

ERRORS OF PREDICTION

When the correlation is less than 1.00 there will be errors of prediction, and the amount of error will increase as the correlation decreases. Figure

5.6 can be used to illustrate the problem. In this figure we have used an ellipse to enclose an imaginary set of scattergram points, and we have drawn the regression lines of Y on X (line A), and X on Y (line B). Let us assume that the correlation between X and Y is .80.

If an individual whose score is included in these data obtains $z_X = 1.50$, our best prediction for his z_Y score will be $\hat{z}_Y = .80z_X$, or $\hat{z}_Y = 1.20$. This prediction follows from Formula 5.3, or it can be deduced from inspection of the regression line in Figure 5.6.

Unfortunately, as the width of the scattergram ellipse (directly above $z_X = 1.50$) indicates, subjects with $z_X = 1.50$ will actually *obtain* a rather wide array of Y scores. In fact, a careful inspection of Figure 5.6 shows that their z_Y scores will vary from about $z_Y = .00$ to $z_Y = 2.40$. The elliptical figure should make it clear that a certain variability of *obtained* values of z_Y will be found regardless of the values of z_X from which predictions are made. For example, if we are given $z_X = .50$, we shall predict $\hat{z}_Y = .40$, but we see from the ellipse that obtained z_Y values will vary from $z_Y = -1.30$ to $z_Y = 2.00$. The extent of this variability, when it is measured from the regression line for all values of z_X, is used as a mea-

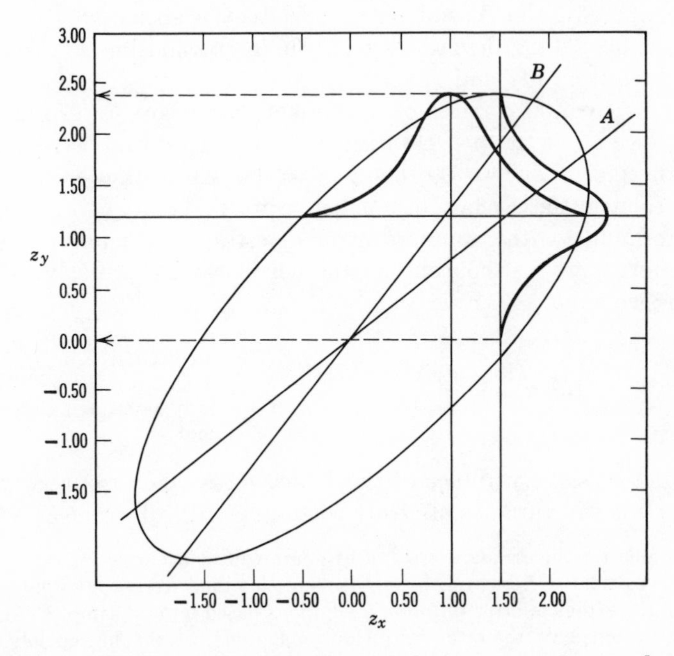

FIGURE 5.6 An illustration of regression toward the mean when predictions are made with $-1.00 < r < +1.00$.

sure of error. When prediction is quite accurate this variability should be minimal, reaching zero when $r = \pm 1.00$ and all Y values fall directly on the regression line. When prediction is quite inaccurate, as when r is low, this variability will be very great, reaching its maximum when $r = .00$.

The variability of a distribution can be measured by its standard deviation. Normally of course, the standard deviation is calculated by taking the deviations of scores from their mean. In this instance, we shall take the deviations of Y scores from the regression line. We shall find $\sqrt{\Sigma y'^2/N}$ where y' represents the deviation of a Y score from the *regression line.* This kind of "standard deviation" is called the standard error of estimate and symbolized by $\sigma_{Y \cdot X}$. A convenient formula for calculating the standard error of estimate is:

$$\sigma_{Y \cdot X} = \sigma_Y \sqrt{1 - r^2}$$

FORMULA 5.7[1]
Standard error of estimate: Y from X

Note from Formula 5.7 that when $r = \pm 1.00$, $\sigma_{Y \cdot X} = .00$ and when $r = .00$, $\sigma_{Y \cdot X} = \sigma_Y$. Let us see why this should be so.

We have already pointed out that when $r = .00$, our best estimate of Y is M_Y. This follows from the equation $\hat{z}_Y = rz_X$. If $r = .00$, the regression line of Y on X will be parallel to the abscissa at M_Y. Thus, when $r = .00$, taking deviations from the regression line is the same as taking them from M_Y and $\sigma_{Y \cdot X} = \sigma_Y$.

When $r = \pm 1.00$, prediction is perfect; there are no errors of prediction since all of the points in the scattergram will fall exactly on the regression line. If $\Sigma y'^2 = .00$ then $\sigma_{Y \cdot X} = .00$. An inspection of Formula 5.7 will confirm these relationships between r and $\sigma_{Y \cdot X}$.

The formula for the standard error of estimate can provide a somewhat different way of conceptualizing correlation. If we solve Formula 5.7 for r^2 we have:

$$r^2 = 1 - \frac{{\sigma_{Y \cdot X}}^2}{{\sigma_Y}^2}$$

FORMULA 5.8
r^2 as a percentage of variance in X associated with variance in Y

The term σ_Y^2 is the variance of the Y measures taken from their mean, while $\sigma_{Y \cdot X}^2$ is the variance of errors when predicting Y from X. We have

[1] This formula for the standard error of estimate will be accurate for the data from which it is calculated. However, in problems of this kind we are more interested in *estimating* the extent of error in *future* predictions based on new samples of subjects. If we wish to estimate the error for future predictions, we should multiply $\sigma_{Y \cdot X}$ by $\sqrt{N/N - 2}$. This will correct for sampling bias, a topic we shall discuss more fully in a subsequent chapter.

said that σ_Y^2 provides us with a measure of the initial error, or variability, in the Y distribution. This is the error which would exist if we had to predict values of Y *without* any knowledge of X (or if $r = .00$). If $\sigma_{Y \cdot X}^2$ is the error remaining in the distribution of predicted Y values *given* knowledge of X, then $\sigma_{Y \cdot X}^2/\sigma_Y^2$ becomes a ratio of the error remaining in the system relative to the initial error. If this ratio is subtracted from 1.00 it will give us a ratio of error *reduction* relative to the initial error in the system. This reduction is, of course, occasioned by our knowledge of X. When r^2 is multiplied by 100 this ratio is changed to a percentage. It is often thought of as the percentage of variance in Y which is accounted for by our knowledge of X.

If the correlation coefficient between variables A and B is .70, it may be said that 49 percent of the variance in A is associated with, or explained by, the variance in B. The reverse can also be said. This is a rather sobering interpretation, but an important one to keep in mind. We quite often find correlations as low as .30 between the variables with which we work; it is well to consider that such a correlation indicates that only 9 percent of the variance in one variable has been accounted for by variance in the other.

MULTIPLE CORRELATION

There are situations in which we have more than one predictor. We may for example, wish to know the correlation between a criterion, such as achievement (Z) and the most efficiently weighted linear combination of two predictors, such as intelligence (X), and perhaps motivation (Y). The multiple correlation coefficient ($R_{Z \cdot XY}$) for such a problem is given by:

$$R_{Z \cdot XY} = \sqrt{\frac{r_{ZX}^2 + r_{ZY}^2 - 2r_{ZX}r_{ZY}r_{XY}}{1 - r_{XY}^2}}$$

FORMULA 5.9
Multiple correlation

It is possible to illustrate some of the logic of multiple correlation by the three groups of circles shown in Figure 5.7. The areas of the circles represent the variance of the variables with which we are concerned. The area of overlap between two circles represents r^2, the proportion of variance common to both variables. (Refer to Formula 5.7.)

In each of the situations illustrated by Figure 5.7 we let r_{XY} and r_{ZY} equal .50, and we shall observe what happens to $R_{Z \cdot XY}$ as r_{XY} varies from almost 1.00 to zero. In other words, we shall assume that intelligence and motivation each correlate .50 with achievement, and we shall see what happens to R as the correlation between the two *predictor* variables varies.

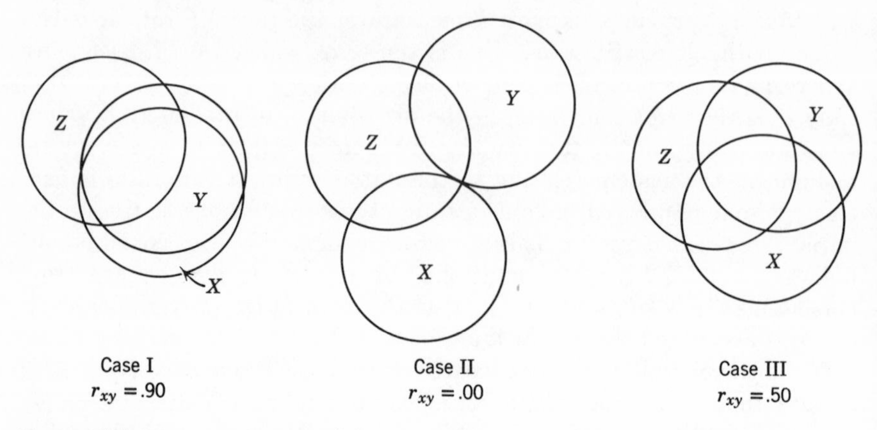

FIGURE 5.7 Variance accounted for in Z when $r_{ZY} = .50$, $r_{ZX} = .50$ and r_{XY} varies from .90 to .00.

In case I we shall assume that $r_{XY} = .90$. This means that variables X and Y are really measures of just about the same thing. We have illustrated this by drawing the X and Y circles so that they almost completely overlap each other. Since X and Y are highly intercorrelated neither variable can account for any appreciable proportion of the variance in Z which is not already accounted for by the other. Prediction of Z from both X and Y should not be much improved over predictions based on either X or Y alone. This is confirmed when we find that Formula 5.9 yields $R_{Z \cdot XY} = .51$ for this situation.

In case II we shall assume that $r_{XY} = .00$. This means that the predictor variables have no variance in common, that they are really measures of two distinctly different characteristics. This is illustrated by the failure of circles X and Y to show any overlap. However, each predictor variable continues to have variance in common with the criterion measure, variable Z, but since X and Y have no variance in common they must each be accounting for distinctly separate components of the criterion variance. Prediction of Z from both X and Y should be considerably better than if either is used alone. This is confirmed by finding that Formula 5.9 yields $R_{Z \cdot XY} = .70$.

Case III illustrates what happens when $r_{XY} = .50$. As you would no doubt expect from the logic developed in the first two examples, when $r_{XY} = .50$ the prediction of Z is improved by the use of both X and Y, but it is not a very dramatic improvement. In this case $R_{Z \cdot XY} = .57$.

This nice orderly state of affairs holds only when r_{ZX} and r_{ZY} are equal and moderate in size. Some very complex and curious effects can be generated when r_{ZX} and r_{ZY} are unequal. For example, if $r_{ZX} = .50$, $r_{ZY} = .00$, and $r_{XY} = .80$, we have $R_{Z\cdot XY} = .84$ which is a higher multiple correlation than exists when both r_{ZX} *and* r_{ZY} are .50. These complexities involve the operation of suppressor variables and their treatment is beyond the scope of this text. The student is referred to Guilford,[2] who provides a particularly lucid discussion of these and other more complex correlational topics.

RANK ORDER CORRELATION

The Pearson product moment correlation coefficient presupposes at least interval scales of measurement for both variables. When data are the result of ordinal scales and when N is not large a correlation coefficient may still be calculated. A statistician named Spearman has shown that the Pearson r may be calculated from ranks. The resulting coefficient is called rho (ρ), the rank order correlation coefficient.

$$\rho = 1 - \frac{6\Sigma D^2}{N(N^2 - 1)}$$

FORMULA 5.10
Spearman's rank order correlation coefficient, rho

D is the absolute difference between ranks and N is the number of *pairs* ranked. Rho will have the same characteristics as r; it will vary from -1.00 through zero to 1.00.

As an example of the use of rho, consider a hypothetical ranking of seven coeds on beauty and pleasantness—characteristics not easily measured by more sophisticated scales.

[2] Guilford, J. P. *Fundamental Statistics in Psychology and Education*, 4th ed. New York: McGraw-Hill, 1965.

TABLE 5.2 Hypothetical Ranking of Coeds on Beauty and Pleasantness

Coed	Beauty	Pleasantness	D^2
A	1	3	4
B	2	4	4
C	3	6	9
D	4	1	9
E	5	2	9
F	6	7	1
G	7	5	4
			$\Sigma D^2 = 40$

$$\rho = 1 - \frac{6\Sigma D^2}{N(N^2 - 1)}$$

$$\rho = 1 - \frac{240}{336}$$

$$\rho = .29$$

For this group of coeds the rank order correlation between beauty and pleasantness is quite low. We have obtained $\rho = .29$.

The use of rho is not limited to ordinal scales of measurement. Since ρ is easier to calculate than r, particularly when N is small, it is often used to estimate r, even when the data are based on interval or ratio scales. In this situation, simply convert the raw scores to ranks and calculate ρ. In the case of tied scores (or tied ranks), give each subject involved in the tie the average rank of the tied set. For example:

Subject	*Score*	*Rank*
A	31	1
B	25	3
C	25	3
D	25	3
E	22	5
F	17	6.5
G	17	6.5
H	10	8

Note that the last subject in a group of N should have the nth rank unless that subject is involved in a tie.

THE ϕ COEFFICIENT

Instead of scales of measurement we may find that our observations permit only a dichotomous classification. A dichotomous variable consists of two mutually exclusive and exhaustive classes; for example, one can be either a college graduate or a noncollege graduate, male or female, brown-eyed or not brown-eyed. If we wish to measure the extent of association between two such dichotomous variables as male-female and Republican–non-Republican, we might classify each of 150 registered voters as male or female and as Republican or non-Republican. The resulting table is called a contingency table, in this case a 2×2 contingency table. An

	Republican	Non-Republican	
Male	a 25	b 50	75
Female	c 50	d 25	75
	75	75	150

inspection of the table suggests a relationship between these variables: female is associated with Republican and male with non-Republican. The extent of this correlation is measured by the phi coefficient (ϕ) where

$$\phi = \frac{bc - ad}{\sqrt{(a + c)(b + d)(a + b)(c + d)}}$$

FORMULA 5.11
The correlation between two dichotomous distributions

For the data above, $\phi = .33$. The phi coefficient is a measure of association. It is an approximation of r when dichotomous data must be used. Its interpretation is basically the same as we gave for r and ρ except that the sign of ϕ is sometimes arbitrary. In our example it is equally logical to arrange the contingency table with the non-Republicans on the left; when this arrangement is used, the two largest cells will occupy the opposite diagonals and $\phi = +.33$ becomes $\phi = -.33$. When the dichotomies are not orderable, we must know the arrangement of the cells in the contingency table before we can interpret the sign of ϕ.

REVIEW

The research worker often needs to describe the relationship between two sets of measurements on the same subjects. The correlation coefficient (r) provides such a measure. The absolute size of this coefficient, which can vary from -1.00 to $+1.00$, indicates the degree of relationship and the sign indicates the direction of the relationship. The formula for the correlation coefficient is $r = \Sigma xy/N\sigma_X\sigma_Y$, or the mean of the products of the z scores for the subjects in the distribution. The correlation coefficient can also be used to predict a score on one variable from knowledge of an individual's score on another variable. When the correlation is linear, the line of best fit by a least squares criterion is given by the formula $\hat{z}_Y = rz_X$. The standard deviation of the errors of prediction, or standard error of estimate, is given by the formula $\sigma_{Y \cdot X} = \sigma_Y \sqrt{1 - r^2}$. It is also possible to calculate the multiple correlation (R) which is the correlation between a criterion and the best combination of two predictors. Other correlation coefficients, ρ for rank ordered data, and ϕ for dichotomous data are also discussed.

EXERCISES

1. Find r for the following data:

Person	A	B	C	D	E	F	G	H	I	J	K	L	M	N
Score X	3	7	8	10	12	6	4	3	3	8	9	7	10	16
Score Y	6	10	12	16	14	6	8	2	3	12	16	9	14	20

2. Transform the data above into ranks and calculate ρ.
3. At a certain college the correlation between college entrance examination board scores and grade point average at the end of the freshman year is .50. John Jones has a college board score (CEEB) of 600. What is our best prediction for his grade point average (GPA) at the end of his freshman year? $M_{CEEB} = 500$, $M_{GPA} = 2.20$, $\sigma_{GPA} = .60$, $\sigma_{CEEB} = 100$.
4. We wish to give scholarships only to those students for whom we can predict a grade point average of 3.00. What CEEB score will we require? What percentage of all entering students would receive scholarships?
5. A teacher points out that Sam Smith with a college board score of 500 is achieving a 3.00 grade point average. Is this an unusual case or will it happen fairly often?
6. For a certain test, we want to know if there is a relationship between the ability to answer Exercises 1 and 2 correctly. The contingency table is given below. Calculate ϕ. What is your conclusion?

		Question 1		
		Correct	Incorrect	
Question 2	Incorrect	5	15	20
	Correct	15	5	20
		20	20	40

7. Will a correlation of .60 result in twice as much reduction in the error of prediction as a correlation of .30? Explain.
8. Assuming that the correlation between the first and second tests in this course is .80, what z score would be needed on test 1 for the regression equation *to predict* a $\hat{z} = 1.00$ on test 2? If $z = 1.00$ is obtained on test 2, what does the regression equation predict for that individual's $\hat{z}$ score on test 1? Why are the answers different?
9. Which correlation coefficient allows prediction with the least error?

   1. $-.87$ 2. $+.65$ 3. $-.21$ 4. $+1.44$

10. If all of the frequencies in the contingency table of question 6 are doubled, what is the effect on ϕ?

CHAPTER 6

The Nature of Research

In Chapter 1 we showed, with an illustration from the animal laboratory, how random variables can disguise the presence of a systematic relationship between the independent and dependent variables of a research study. We now return to this problem as a basis for a continued discussion of the nature of behavioral research and the role of inferential statistics in the analysis of research data.

INDEPENDENT AND DEPENDENT VARIABLES

The typical experiment, whether in chemistry or psychology, is designed to answer the question, "What is the effect of X on Y?" The nature of X and Y will vary with the discipline, but the form of the question is much the same in every science. In the behavioral sciences these questions are usually answered by assigning a group of subjects to each of several levels of the independent variable, X, and then recording the performance of the subjects on some measure of the dependent variable, Y. The illustrative experiment in Chapter 1 considered the effect of different amounts of reward on running time in a maze; amount of reward was the independent variable, and running time was the dependent variable.

If we were to investigate the effect of special training in frustration tolerance on the aggressive behavior of six-year old children, we might proceed with a similar design. Some children would be selected to serve in an experimental group and receive training in frustration tolerance. Others would be assigned to a control group and receive no special attention. After the training period each child would be subjected to a mildly frustrating experience and his aggressive responses observed. If we found that the experimental group responded less aggressively than the control group, we might conclude that the difference resulted from the special training of the experimental group. For this conclusion to be valid, the groups of children would have to be initially similar and then treated in the same way *except* for the differences imposed by the independent variable.

Observe that we *assigned* subjects to conditions in both the amount of reward and the frustration tolerance studies. We imposed the various conditions of the experiment on our subjects. This is the mark of a true experiment, and this must be the case before we can conclude that the independent variable actually exerted an effect on the dependent variable.

In other types of research we may wish to investigate relationships without imputing a direct effect to the independent variable. Such studies might investigate the relationship between physique and temperament, or the relationship between membership in a fraternal organization and the probability of voting a straight party ticket. Research of this type does not have a true independent variable because the variables in question are characteristics of the subjects themselves and cannot be manipulated apart from other subject characteristics. Such research can establish the presence of a relationship between variables, but not the effect of one variable upon another. For example, we may wish to compare the personality patterns of underweight and overweight individuals. To conduct this study we probably would have to select people who are already overweight or underweight. However, any personality differences between underweight and overweight subjects could easily be the result of other variables: perhaps hormonal imbalance produces *both* weight differences and personality differences.

Consider another example. Suppose we wish to determine if membership in some fraternal organization tends to influence one's choice of political party. We can rarely impose membership in a fraternal organization on one group of subjects and deny it to another, so we conduct the study by selecting two groups of otherwise equivalent subjects, one group drawn from members in the fraternal organization and the other drawn from nonmembers. If these two groups differ in their political party affiliation, it might seem reasonable to attribute that difference to the effect of membership in the fraternal organization. Unfortunately, such a conclusion cannot be justified. We can only conclude that membership in the fraternal organization is *related* to party affiliation; we cannot conclude that it affects party affiliation.

The explanation of this distinction is straightforward enough. When we conduct an *experiment* from which we wish to infer a cause-effect relationship, subjects must be equivalent on all variables except the independent variable. The subjects in this illustration, in addition to their differences in affiliation with some fraternal organization, must also *lack* equivalence on whatever variables predisposed the membership group to join the fraternal organization and the nonmembership group to stay out. In short, it is not really possible to select two groups of subjects which are equivalent except for membership in a fraternal organization if the potential subjects have already made the decision to join or not to join the organization before they are selected for the "experiment." Experimental designs of this type are called *ex post facto* and can establish relationships but not causes.

In order to be reasonably sure that differences between groups are produced by differences in the independent variable, we must control or inhibit *all* other systematic differences which might affect the dependent variable. If we are investigating the effect of reward magnitude on the running time of rats in a runway, we must be certain that the experimental and control animals are equivalent in weight, age, periods of food deprivation, handling, and housing. No systematic differences are permitted except for the differences imposed by the independent variable. When systematic differences between groups do exist, we will be unable to determine if they or the independent variable have produced the effect on the dependent variable. In such experiments the independent variable is said to be "confounded" with the other variables which are not controlled.

However, even if an experiment seems to be free of systematic errors, the investigator still cannot be sure that a difference between his experimental and control groups was produced by the independent variable. The reason for this uncertainty is a phenomenon called sampling error.

POPULATIONS AND SAMPLES

In order to understand the concept of sampling error we must first define the terms "population" and "sample." A *population* is defined as an exhaustive group of objects having one or more common characteristics. All fifth-grade children in the United States constitute a population, as do all male albino rats weighing 140–160 grams, or all students with at least a B average and at least junior standing in this statistics class. Populations can be of any size, from the millions of fifth-grade children to the dozen or so B average juniors in this statistics class. Populations can even be imaginary in the sense that they can be defined as all male adults, age 25–34, given .3 cc of a certain drug. This is a legitimate way of defining a population, even if the drug has not yet been given to anyone. A population can be composed of any number of any kind of object as long as all of the objects have a common characteristic and all such objects are included.

On the other hand, when we come to define some *particular* population, the definition must become very restrictive. A particular population must be defined so that any individual's membership in it is unequivocal. If we specify that attractive young female psychology majors are the population of interest, then we must be very careful to define exactly what we mean by each of these terms so that there can be no ambiguity about which individuals are in the population and which are not.

The behavioral scientist is not normally able to conduct research on populations; his investigations usually must be confined to samples. A *sample* is defined as any portion of a population less than the total. When all the members of a population are not available for the investigation, the research worker studies the phenomena in a sample drawn from the population. Then, if the circumstances warrant, he estimates the characteristics of the population from his observations of the sample. For instance, he estimates the proportion preferring brand X in a population of consumers from the proportion actually preferring brand X in a sample of consumers. The preference for political candidates among a population of registered voters is inferred from the results obtained by questioning samples of voters. Conclusions about the effect of spaced trials on learning in the rat are based on samples of rats.

Statisticians distinguish between means, variances, correlation coefficients, and other distribution characteristics calculated from samples and those calculated from populations. A descriptive measure based on a sample is called a statistic; when it is based on the entire population it is called a parameter. If the population of observations consists of the weights of all women students enrolled at a university, then the mean and variance of these weights are parameters of that population. However, if we obtain a sample of 200 weights, then the mean and variance of this sample are sample statistics.

If the research worker wishes to know population parameters, then usually he will estimate them from statistics derived from samples. Thus, to know the mean difference (if any) in maze running times of rats rewarded with one or two food pellets, we would base our estimate of this parameter on the difference in mean running times for two samples of rats. If we wished to know the mean income of all families with undergraduates who plan to enter medical school (a parameter) we would estimate this mean from the mean income of a sample (a statistic) of undergraduates. Sample means are used as estimates of population means, sample proportions of population proportions, and sample variances (with some alterations) as estimates of population variances.

A single sample mean, although it is an estimate of the population mean, will probably not coincide with the population mean. However, if samples are unbiased, the mean of a sufficiently large number of sample means will coincide with the population mean. This situation occurs *only* if the samples are unbiased. In biased samples the mean of the statistic in question will not coincide with the value of the parameter.

Bias in a sample usually occurs when every individual in the population does not have an equal chance of appearing in the sample. For instance, if we wish to estimate the mean IQ for a population consisting of all high school students, we will have a biased sample if we select the students entirely from among those enrolled in high school physics. To the extent that high school students enrolled in physics constitute a brighter than average group, our sample will underrepresent the less intelligent student. The mean of a large number of such samples will therefore tend to *overestimate* the mean IQ of the population. The same kind of bias can mislead us if we want to sample student opinion and question only those students who are in the student union, or if we wish to predict the vote in a general election and question only the registered voters who are at home in the evening. In the first instance students who have heavy schedules and cannot afford to spend time in the student union will not be adequately represented, and in the second instance, people who work evenings will not be represented in proportion to their numbers. Bias in a sample cannot be compensated for by increasing the size of the sample.

We can never be certain that a sample is unbiased, but there are methods which give at least reasonable assurance that the sample will represent the population from which it was drawn. Most of these procedures involve a process called random sampling. Random sampling is a procedure for selecting a sample which insures that all possible samples of a specified size are equally likely to be chosen from a population. If each individual, or element, of the population has an equal chance of appearing in the sample, and if the selection of any element is in no way dependent on the selection of any other element, then we have a random sample.

Suppose we wish to select a random sample of 15 people from a population of 150 people. We might write everyone's name on separate slips of paper, mix them well in some container, and then withdraw 15 of them. These names will probably constitute a random sample from the population of 150. But this is a rather tedious procedure, and it is subject to error. We must be certain that the slips of paper don't stick together, that they are well mixed, and that we don't peek at the contents as we are selecting the slips.

Random sampling is made much more convenient (and more likely) by the use of a table of random numbers similar to the one in the Appendix. A random numbers table is a list of the digits 0 through 9 arranged in random order. This means that any digit is equally likely to appear in any position. We can enter the table at any point and progress in any direction using any set of digits, and the sequence of numbers will be random.

Suppose we have a population consisting of 5000 individuals and we wish to select a random sample of 100 from the population. We might number the 5000 names and then combine four adjacent columns of digits from the table of random numbers to provide a series of four digit random numbers. We could then select the number 0017, 6952, 3176, 2651, 8912, etc., from the table, but use only 0017, 3176, 2651, etc. which are within the range of our population of numbers. The first 100 individuals whose numbers are selected by this procedure will constitute a random sample from the population of 5000.

Several other methods are commonly used to select unbiased samples for survey research. One of these methods is called area sampling. This is a technique in which a sample of blocks, or other subdivisions of a geographical area, are selected randomly and then every adult living there is interviewed. Another procedure is called stratified sampling. Here the percentages of the population in various categories are known beforehand, and the sample is obtained in such a way that those percentages appear in the sample. If we know that a population of 100,000 men is composed of 20% professional workers, 40% small business and managerial, 30% blue collar, 8% retired, and 2% unemployed, then we shall insist that these percentages also appear in the sample. Of course the subjects *within* these categories must still be obtained by random selection. One should not forget that the random selection of subjects is a means to an end and not an end in itself. However, in the long run, a random sampling procedure is most likely to produce representative samples.

Bias in the selection of subjects is not only a potential problem in survey research; it can also be an important source of systematic error in experiments. For example, if we are interested in the effects of special training on achievement we cannot place the first volunteers in one group and the last volunteers in another because there are probably systematic differences between the motivation of "early" and "late" volunteers which could influence the dependent variable. A systematic error can even result when rats are removed from a shipping crate for assignment to treatment groups. Suppose the first animals to be removed are assigned to one group and the remainder assigned to another. The first animals to be removed from the crate may be the easiest to catch and consequently the least active or the least aggressive. This will provide a source of systematic error if our dependent variable is influenced by either of these subject differences.

Systematic errors of this type can usually be eliminated if each subject has an equal chance of appearing in any of the several groups which constitute an experiment. When subjects are to be assigned to experimental and control groups we can use the table of random numbers to assign them. Subject designations can be written on slips of papers, placed in a container and mixed well. We then decide arbitrarily to let either odd or even numbers in our random number table represent the control group and the remaining numbers the experimental group. Since the table of random numbers can be entered at any point, closing our eyes and stabbing gently with a pencil may be an acceptable (if inelegant) way of gaining an entry place. The number closest to the pencil may be a seven, in which case, Jones, the first name drawn from the container, is assigned to the control group. We can now move in any direction previously agreed upon, up, down, or across and the numbers will remain randomly distributed. The second digit in the column (or row) may be an eight, indicating that Smith, the second name drawn, is assigned to the experimental group. If we have imposed the usual limitation that half of our subjects are to be assigned to one group and half to the other, we will continue this procedure until one group is filled and then arbitrarily assign the remaining names to the other group. If we are to investigate the effect of four levels of drug concentration we may wish to assign subjects randomly to four treatment groups. We can let the digits 1 and 2 represent assignment to group I, digits 3 and 4 represent group II, digits 5 and 6 group III, and digits 7 and 8 group IV. When digits 0 and 9 occur we simply ignore them. This procedure will result in the random assignment of subjects to four treatment groups.

Regardless of the care we take to assure random assignment of subjects, keep in mind that random assignment does not *eliminate* differences between groups. Two groups of 10 numbers selected from the table of random numbers will very probably produce different sample means. This difference, or the difference between any two sets of measurements derived from subjects who have been randomly assigned to conditions, is an example of sampling error. It is to this problem that we now turn our attention.

SAMPLING ERROR AND SAMPLING DISTRIBUTIONS

If we draw several random samples of behavioral measurements from almost any population, one feature will be immediately evident: samples differ. These differences are the result of variation in the population from which the samples have come.

Differences among samples give rise to differences among the values of any statistic calculated from these samples. There will be variation among sample means, sample variances, and sample correlation coefficients. In spite of the fact that population parameters are estimated from sample statistics, the sample statistics tend to vary about the population parameters rather than coincide with them. It is this tendency for sample statistics to vary about their respective population parameters which is called sampling error.

We have already pointed out that systematic error can provide a trap for the unwary investigator. He may assume that an effect was produced by an independent variable when in fact it was the result of systematic error contributed by some uncontrolled variable. Sampling error can also mislead the investigator. He may observe a difference due entirely to sampling error and mistake it for the influence of his independent variable. On the other hand, if the effect of the independent variable is subtle, sampling error may conceal it. The advantage of sampling error over systematic error is that there are statistical techniques which can help us estimate the magnitude of sampling error and thus help us to separate its effects from those of the independent variable.

Before proceeding further we shall distinguish between $\overline{X}$, a symbol used to represent a sample mean, and M, a symbol used to represent a population mean. We shall also use n to stand for the number of cases in a sample (or subgroup) and reserve N for the number of cases in a population (or total number of cases). Thus $\overline{X} = \Sigma X/n$ while $M = \Sigma X/N$.

We shall illustrate sampling error by constructing a sampling distribution. This is a frequency distribution of the sample means (or any statistic) calculated for a series of samples randomly obtained from some population. The population we shall use for this illustration is provided by the digits in the table of random numbers. We draw a sample of ten digits, compute the sample mean, tally it, and repeat the process for each of 100 samples. The sampling distribution which results appears in Table 6.1. Notice that in this sampling distribution, where $n = 10$, there is considerable sample to sample variation. We would probably not wish to put much faith in the accuracy with which the mean of a single, randomly obtained sample could be used to represent the population mean. However, the sampling distribution does provide information about the amount of variability to be expected among sample means. For example, we can see that $\overline{X} = 3.5$ is not at all unlikely, but that $\overline{X} = 8.4$ is highly improbable. In fact, no sample means above 6.9 were obtained in 100 samples.

When random samples of a given size are drawn from any population,

TABLE 6.1 A Sampling Distribution of 100 Means ($n = 10$) Drawn from a Table of Random Numbers

6.0–6.9	卌 //
5.0–5.9	卌 卌 卌 卌 卌
4.0–4.9	卌 卌 卌 卌 卌 卌
3.0–3.9	卌 卌 卌 卌 ////
2.0–2.9	卌 ////
Mean of means = 4.44	

some sample means are much more likely to occur than others. Sample means close to the population mean will occur rather frequently, but sample means which depart substantially from the population mean will be quite rare. In fact, as you can see from Table 6.1, this sampling distribution looks much like a normal distribution. This will generally be the case for sample means if n is sufficiently large.

Let us see how these concepts can be applied to a research situation. Suppose that the sampling distribution in Table 6.1 is actually derived from a population of aggressiveness scores for a strain of adult rats instead of being derived from a table of random numbers. We can see from the sampling distribution that the typical mean aggressiveness scores for samples with $n = 10$ will range from about $\overline{X} = 2.5$ to $\overline{X} = 6.5$. Now imagine that we have a hypothesis which predicts that a particular kind of early training will influence the aggressiveness of adult mammals. We can test this hypothesis by selecting a random sample of 10 weanling rats, subjecting them to the training procedure, and then determining their mean aggressiveness scores as adults. If the independent variable is effective this sample of 10 specially trained rats will have come from a hypothetical population with a *different* mean aggressiveness index than the untrained population. If the experimental group's mean is 5.6 what can we conclude? On the basis of these data there seems to be little evidence that the training procedure is actually effective. The sampling distribution of mean aggressiveness scores for *untrained* rats shows that sample means of 5.6 can easily occur from sampling error alone. Since sampling error alone can easily account for our observations, we have no convincing evidence that the independent variable (training) exerted any effect on the dependent variable (aggressiveness).

On the other hand, if the experimental group's mean is higher than 6.9 or lower than 2.9, we have a very improbable event *if* we assume that the deviant sample mean is simply the result of sampling error. If the experimental sample mean is as high as 8.9, it seems most reasonable to discount sampling error as the cause and conclude that something other than sampling error is responsible. If the experiment is well controlled, that something should be the independent variable, namely, the type of early training.

The effect of an independent variable must always be evaluated against a background of sampling error. If the independent variable's effect is subtle, sampling error may obscure it entirely, or sampling error may itself be mistaken for an experimental effect when none exists. The research worker is constantly faced with deciding between two alternatives —whether a given difference is the result of his independent variable or the result of sampling error.

THE NULL HYPOTHESIS

In most research the investigator's interest lies in providing evidence to support some experimental hypothesis. He may wish to show that a drug aids memory, or that some process reduces aggressiveness. Such experimental hypotheses are usually tested by comparing an experimental and a control group. If the difference between group means is so large that sampling error cannot reasonably account for it, the investigator concludes that the means of the two populations differ, and that this difference results from the effect of the independent variable.

The experimenter really has two different hypotheses. One, an experimental hypothesis which he wishes to confirm, that his independent variable exerts an effect; and a second, or statistical hypothesis that he hopes to reject, that the experimental and control groups come from populations with the same mean. The statistical hypothesis, called a null hypothesis, specifies that the population means are the same and that any observed difference between sample means is merely the product of sampling error. If he can *reject* the statistical hypothesis the investigator will accept the experimental hypothesis that his independent variable has exerted an effect.

The investigator cannot prove or disprove the null hypothesis. All he can do is report the probability of an observed outcome assuming that the null hypothesis is true. If the probability is sufficiently small, the null hypothesis (random variation) is rejected, and the experimenter concludes that the independent variable exerted an effect.

The null hypothesis is literally a hypothesis of "no difference" among the parameters of two or more populations. For example, a null hypothesis may state that the proportions of six-year old children learning to read will be the same for populations taught by methods A, B, C, or D. Even if this hypothesis is true, we would still expect to find some variation among samples of children taught by the various methods simply as a result of sampling error. When variation among the sample proportions becomes larger than can reasonably be attributed to sampling error, we reject the null hypothesis and conclude that the teaching methods are not equally effective. This, of course, is what we have really expected to find when we set out to do the experiment.

Since the investigator does not have access to entire populations he cannot "prove" or "disprove" any hypothesis about population parameters. The investigator can only decide if hypothesized population parameters are *credible* given the statistics he has calculated from his experimental and control group samples.

Suppose we have a hypothesis that an acquaintance is cheating in a coin-tossing game in which he wins a dollar from us every time a coin comes up heads. This is the same as saying that we suspect he is influencing the fall of the coin, that something other than "chance" is affecting the outcome. Our experimental hypothesis is that something other than chance is affecting the outcome of the coin-tossing series; that our opponent is using a biased coin! This experimental hypothesis can be accepted on the basis of statistical evidence only if we can reject the null hypothesis. The null hypothesis, in this situation, is that there is no difference between the number of heads our opponent throws and the number that might reasonably be expected on the basis of chance.

Suppose our opponent has just tossed his coin ten times. This series can be viewed as a sample of ten tosses drawn from an infinitely large population of a ten-toss series. If the coin is unbiased, heads and tails are equally likely. If only chance is operating, the theoretically expected number of heads on a series of ten tosses will be 5. However, everyday experience with sampling error would not lead us to expect 5 heads on every series of ten tosses. Some series might produce 6 heads, occasionally 7 heads, less often 8 heads, rarely 9 heads, and very rarely 10 heads. In fact, *any* number of heads from 0 to 10 *could* occur entirely by chance when an "honest" coin is tossed ten times.

The experimenter is left to deal with uncertainty by using probabilities. If his opponent has made 1000 tosses, and the experimenter has won about 51% of them himself, it is perhaps unreasonable to assume that anything other than chance is operating. Certainly he has no evidence to reject the null hypothesis, but that does not mean it is true. Similarly, if the experimenter has lost 1000 tosses out of 1000, he is either phenomenally unlucky or the victim of fraud. Phenomenal bad luck does not *dis-*

prove the null hypothesis, it only reduces its credibility. Since random sampling remains a possible, though extremely unlikely explanation, the null hypothesis would be rejected, but not disproved.

TWO KINDS OF ERRORS

The experimenter, obliged to evaluate the effect of his independent variable against a background of sampling variation, will make one of two types of correct decisions, or one of two types of errors. Using the example of coin tossing, we shall review these alternatives.

First, the experimenter may be playing a dishonest opponent and decide that the improbable nature of the events warrants withdrawing. He has in this instance, at least implicitly, rejected the null hypothesis of random sampling, and he was correct in doing so. This outcome falls in cell D of Table 6.2.

Second, the proportion of heads thrown by his opponent may be well within reasonable expectation on the basis of chance, and he may correctly assume that the game is honest. He has not rejected the null hypothesis, and the results are due only to sampling error. Since his opponent is honest, this is also a correct decision. This outcome falls in cell A of Table 6.2.

Third, he may observe a very unlikely run of heads and reject the null hypothesis when, in fact, his opponent is honest. Since the null hypothesis has been rejected when the highly deviant observations were actually just the result of random sampling, an error has been made. This is called

TABLE 6.2 Varieties of Errors and Correct Decisions

		In the true situation the difference is due to	
		Sampling error	Independent variable
Experimenter's conclusions about the difference	Sampling error	A Correct decision	B Type II error
	Independent variable	C Type I error	D Correct decision

a Type I error. Whenever the null hypothesis is true but a very deviant sample has led to its rejection, a Type I error has been made. This outcome falls in cell C of Table 6.2.

Fourth, the experimenter may observe a proportion of heads sufficiently common so that he does not become suspicious. If he fails to reject the null hypothesis and is, in fact, being victimized by a biased coin, he has committed a Type II error. A Type II error occurs when the null hypothesis is falsely accepted. When the experimenter accepts sampling error as an explanation for a difference when, in fact, random sampling is not entirely responsible, a Type II error has occurred. This outcome falls in cell B of Table 6.2.

LEVELS OF SIGNIFICANCE

At this point in our analysis we have avoided giving a technical definition of probability, since a technical definition has not been required to follow the discussion. Now, however, we shall give a provisional definition of probability. The probability of an event A is given by the number of equally likely outcomes favorable to A, divided by the total number of equally likely outcomes. It is usually expressed as a decimal fraction. If 100 beans are in a jar and one is red while the rest are green, then the probability of selecting the red bean by chance in one draw is .01 ($p = .01$). If an unbiased coin is tossed, the probability that it will fall heads is .50 ($p = .50$), since there is a head on only one side of the coin and there are two sides altogether. The probability of selecting a spade at random from a deck of cards is .25 ($p = .25$), since there are 13 spades among the 52 cards in a normal deck. Probability can be seen as varying between impossibility ($p = .00$) and certainty ($p = 1.00$).

The experimenter's decision to accept or reject the null hypothesis depends upon his assessment of the probability that his observations have arisen as a result of random sampling from some specified population. If the observations are very improbable from the standpoint of random sampling from that population, then the null hypothesis is rejected and the experimenter concludes that the sample arose from some other population. We begin by *assuming* that the null hypothesis is true and that any apparent discrepancy is simply the result of sampling error; then, if we find that our observations would occur only 1% of the time or less, we reject the null hypothesis.

The level of probability required for the rejection of the null hypothesis is called alpha (α); it is normally equal to or less than .01 ($\alpha \leqq .01$).

When observations are this improbable, given that the null hypothesis is true, we reject the null hypothesis. The results of the experiment are then said to be significant at the 1% level. If the observations are so very rare that their probability of occurrence under the null hypothesis is less than .001, they are said to be significant at the .1% level.

Let us return to the coin-tossing example. If a coin is unbiased and we are not being cheated, then the probability of obtaining a head should be .50. The null hypothesis is that no difference exists (other than that produced by sampling error) between the probability of heads when the stranger tosses his coin, and the probability of heads for any unbiased coin. In essence, the null hypothesis claims that our opponent's tosses are a randomly obtained series from a population in which the probability of heads is .50. If, over a series of tosses, we find that our opponent has thrown an improbably large number of heads ($\alpha \leqq .01$), then we shall reject the null hypothesis that the series is simply a random selection from a population in which the proportion of heads is .50.

Suppose we ask our opponent to toss his coin ten times and he tosses ten heads. The event of ten heads in ten tosses of an unbiased coin will occur by chance only about once in one thousand sets of ten tosses if the coin is unbiased and chance alone is operating. Since sampling error alone can be expected to produce this result on far less than 1% of all ten-toss series, we shall reject the null hypothesis and conclude that our opponent is using a biased coin.

The level of α, the probability of rejecting the null hypothesis if it is true, is not immutably set at the 1% level for all research although the 1% level is a common convention. If the experimental findings seem to follow logically those of other investigators, the experimenter may require a lower level of significance, perhaps the 2% level or the 5% level. Results less significant than the 5% level are rarely acceptable in the scientific community, but higher levels of significance may be required. Before we accept some finding which is completely at odds with earlier research, we shall probably insist upon significance well beyond the 1% level as well as a careful look at the research design!

An important point of terminology must now be made; the level of statistical significance is not necessarily related to practical significance. We might find that a new method of teaching arithmetic is superior at the .0001 level of significance, but this would not necessarily require an adoption of the new method by all school districts. A statistically significant difference can be small and perhaps trivial. "Significant," when used in the statistical sense, means a difference unlikely to have been the result of random sampling from some specified population.

When the null hypothesis is true and we reject it on the basis of an improbable sample, we commit a Type I error. The more improbable we require the sample to be before we reject the null hypothesis, the fewer Type I errors we make. Consequently, the level of α which the experimenter sets for the rejection of the null hypothesis determines the relative frequency or probability of Type I errors. If α is set at the 1% level, then on 1% of all occasions when the null hypothesis is true we shall falsely reject it. Remember that sampling error operates whether the null hypothesis is true or false. When the null hypothesis is true and α has been set at .01, we shall expect to find samples sufficiently deviant to lead to the rejection of the null hypothesis on about 1% of all tests. When α has been set at .02 or .05, the probability of Type I errors rises to .02 and .05. It would seem reasonable then to keep α as low as possible (to reject the null hypothesis only when the results are highly significant) in order to reduce the frequency of Type I errors. Unfortunately, this cannot be done without increasing the probability of making Type II errors. This probability is given by beta (β).

A Type II error occurs if we fail to reject the null hypothesis when it is actually false. This situation is usually found when an independent variable's effect is subtle, perhaps creating such a small difference between an experimental and a control group that sampling error seems a most reasonable explanation. If sampling error is accepted and we fail to reject the null hypothesis of no difference between the groups, we have committed a Type II error. The probability of making a Type II error is partially determined by α, the probability of making a Type I error, but it is also related to several other aspects of the situation. We shall discuss some of these later, but for the moment let us discuss the relationship between α and β more fully.

If α is set at .01, it may be that the effect of a weak independent variable, such as a very slightly biased coin, cannot, over a few tosses, produce a sufficiently startling departure from chance expectation to reach the required level of significance and occasion the rejection of the null hypothesis. If the coin's bias is slight, the results of a series of tosses may be well within the limits of sampling error and, failing to reject the null hypothesis, we shall commit a Type II error. If, however, α is lowered to .05, a slightly biased coin may produce results which exceed these narrower limits of tolerance; the null hypothesis will be rejected and a Type II error avoided. Of course, when we change α from .01 to .05 in order to decrease Type II errors, we increase the probability of Type I errors.

Setting an appropriate level for α requires careful consideration of the relative importance of Type I and Type II errors. To carry our coin-tossing example further we may want to minimize the probability of a Type I error, perhaps by setting $\alpha < .001$, if we are tossing a coin with an acquaintance. We may be more concerned about reducing β if we are short of funds and gambling with a stranger. In the latter situation we might materially reduce β by lowering the level of α to .05 or .10.

Before a cure for cancer is announced, α should be extremely small, but before a potential cure is discarded as useless, β should be minimized. Selection of appropriate values for α and β is ultimately a matter of the experimenter's judgment, and is obviously of considerable importance.

EXAMPLES

The ideas we have been discussing are some of the most difficult ones in the text. It is, therefore, wise to review them before proceeding to the example research situations described later in this chapter. Below is a brief glossary of terms.

Independent Variable. The variable in an experiment whose effect on another variable is the focus of interest; as in the effect of time since learning on amount retained, or the effect of noise level during learning on trials to learn. Time and noise level would be independent variables in these experiments.

Dependent Variable. The variable in an experiment upon which the independent variable is presumed to exert its effect. In the examples above the dependent variables are amount retained and trials to learn.

Systematic Error. A difference between the experimental and control groups of a study other than that which may arise from random assignment of subjects or from the effect of the independent variable. If we are investigating the effect of time since learning on amount retained, and if we assign more difficult material to one group, or choose more highly motivated subjects for one group, a difference between the experimental and control groups' retention could result. Such a difference would not be the result of the independent variable, or of the random assignment of subjects. Systematic error is contrasted with random or sampling error.

Sampling Error. The random variation among samples drawn from the same population. Two sets of numbers drawn from a table of random numbers will usually have different means. Such a difference is attributable to sampling error. Sampling error may mask the operation of an independent variable, or it may be mistaken for the operation of an independent variable.

Sampling Distribution. A frequency distribution of a statistic based on samples drawn randomly from some population. If we draw a large number of samples each consisting of 24 cases, where each case consists of a two-digit number selected from a table of random numbers, we could calculate medians, standard deviations, means, etc. for each such sample. A frequency distribution of the particular statistic calculated from these samples forms a sampling distribution.

Null Hypothesis. Literally, a hypothesis of "no difference" with the implication that any observed departure from "no difference" is the result of sampling error. The meaning of the null hypothesis is sometimes expanded to include any hypothesis specifically set up for attack. The null hypothesis is rejected if it can be demonstrated to be sufficiently *improbable.* It cannot be proved or disproved.

Type I Error. The rejection of the null hypothesis when it is actually true. When the discrepancy between the experimenter's data and the observations expected on the basis of the null hypothesis are found to be improbable on the basis of sampling error alone, the experimenter rejects the null hypothesis. He may then attribute the discrepancy to the action of the independent variable. If the observed effects *are* simply the result of sampling error and not due to any effect of the independent variable, then the experimenter will have committed a Type I error by rejecting the null hypothesis.

α *(The Probability That a Type I Error Will Occur).* The probability of rejecting the null hypothesis when it is true. The experimenter must decide the "point of improbability" at which he is willing to reject sampling error as an explanation for his observations. He may decide to reject sampling error as an explanation if he can show that his observations would have occurred as a result of sampling error less than one time in a hundred. In such an experiment, α would equal .01. However, using this level of α, the experimenter will find that in 1% (.01) of all experiments *in which the independent variable really exerts no effect,* and in which sampling alone operates to produce the observations, sampling error is rejected as an explanation. This is a Type I error, and α, the level set by the experimenter for the rejection of the null hypothesis, will determine the proportion of such errors.

Significant. A term used to describe experimental results which have led the experimenter to reject the null hypothesis. If the null hypothesis is rejected with α set at .01, the results are said to be "significant at the one percent level." The level of significance is determined by the level of α at which the null hypothesis can be rejected. Thus, results which lead to the rejection of the null hypothesis with $\alpha = .001$ are said to be "highly significant." Significant is used here in a technical sense and should not be considered synonymous with important.

Type II Error. The failure to reject the null hypothesis when it is false. When the results of an experiment fail to reach the required level of significance, the experimenter is not justified in rejecting chance as an explanation. However, perhaps chance was not entirely responsible; perhaps the independent variable did exert an effect. If so, by failing to reject sampling error as an explanation of the event, the experimenter will have committed a Type II error.

β *(The Probability of Committing a Type II Error).* The probability of failing to reject the null hypothesis when it is false. The probability that the independent variable has really exerted an effect, but that we have mistaken it for sampling error is given by β. The level of β is determined partially by the level of α set by the experimenter and partially by other variables which will be described later. The effect of α on β can be clarified by an example.

If we are investigating the effect of drug X on visual acuity, and we wish to minimize Type I errors, we shall require a large difference between the visual acuity of drug groups and no-drug groups before rejecting the null hypothesis. However, if the drug effects are very slight we are likely to commit a Type II error by attributing the small true difference in visual acuity to the operation of sampling error. On the other hand, if we change α from .01 to .05 and thus accept a higher risk of a Type I error, we shall be more likely to reject the null hypothesis if the drug's effects are subtle. Increasing α thus decreases β.

We shall now describe two quite different research studies to further illustrate some of the complex and difficult ideas we have just discussed.

Suppose a drug firm believes it has a preventative for the common cold. To test this hypothesis it randomly selects two samples of 1000 individuals, gives the control group isotonic saline and gives the experimental group the new vaccine. The subjects are examined weekly, and at the end of 16 weeks it is found that 46 experimental subjects and 164 control subjects have developed colds. The experimental hypothesis is that the independent variable, vaccine X, has an effect on the dependent variable, the tendency to develop colds. If the vaccine is effective, then we have drawn samples from populations with different proportions of people tending to develop colds over a 16 week period. If the vaccine is ineffective, then the samples have come from the same population with respect to the proportion of people developing colds, and the difference between these sample proportions is just the result of sampling error. The null hypothesis for this study would specify that there was no difference in the proportion of cold sufferers for the vaccinated population and the unvaccinated population. The difference between the sample proportions is then assumed to be the result of random sampling. We shall return to this problem in Chapter 9. There we shall discuss the statistical procedure the drug company might actually use to determine the efficacy of the vaccine.

In the next example we shall conduct a mock experiment using a table of random numbers to supply our data. Suppose we wish to investigate the effect of two different conditions of problem presentation on the number of problems solved. We randomly assign 10 subjects to each of two groups. Then we give 9 problems to one group under condition A, and the same 9 problems to the other group under condition B. Our experimental hypothesis states that the conditions will differentially affect the number of problems solved by the two groups. This experimental hypothesis will be accepted *only* if we can reject the null hypothesis that the difference between the mean number of problems solved by the two groups (2.8) is the result of random sampling from a common population.

In this mock experiment the scores (problems correct) for all subjects have actually been selected from a table of random numbers so, of course, the null hypothesis is true; the two means have come from a common population. The data appear below.

Group A	*Group B*
7	9
9	5
2	1
9	4
8	1
2	2
7	1
5	5
3	2
7	1
$M = 5.9$	$M = 3.1$

Our experimental hypothesis has specified that the independent variable (method of presentation) had an effect on the dependent variable (number of problems solved). We can accept this experimental hypothesis only if we cannot reasonably expect this large a difference between means as a result of sampling error. If this null hypothesis is rejected, and if our experiment is well controlled, we will accept the experimental hypothesis and conclude that the method of presentation affects the number of problem solutions obtained.

To proceed with this analysis, we must answer the following question: "When random samples with $n = 10$ are drawn from the same population, what is the probability that we shall observe a difference as great or greater than 2.8 between successive pairs of means?" We can get some idea of the relative frequency of mean differences of 2.8 by constructing a sampling distribution of mean differences based on samples of ten cases, each obtained from the table of random numbers. The procedure is similar to the one we described earlier for developing a sampling distribution of the mean. A sampling distribution of mean differences can be obtained by drawing pairs of samples where each sample consists of ten cases selected from the table of random numbers. We shall arbitrarily subtract the second sample mean from the first, and plot these differences in a frequency distribution.

If the first and second drawn samples have come from populations with a common mean (as they did in this illustration, since both were drawn from a table of random numbers), then we will expect about half positive and half negative differences with the mean of these differences falling close to zero. The distribution of differences will be approximately normal, or bell-shaped. Small positive or negative differences will occur rather often, but large positive or negative differences will occur very rarely. The difference of 2.8, which we obtained between the mean of our two groups, can be evaluated against this background of sampling error.

A sampling distribution of mean differences based on 100 randomly selected pairs of samples appears in Table 6.3. The mean of this sampling distribution is −.02, which is quite close to the theoretically expected mean difference of zero. The distribution is also approximately normal.

On the basis of this distribution we can estimate the probability of a mean difference such as we have obtained between the means of Group A and Group B. With a Group A mean of 5.9, and a Group B mean of 3.1, we have observed a mean difference of 2.8. In our illustrative experiment we hypothesized a difference, but we did not specify the direction of the difference. A difference can have either a positive or negative sign, so we must consider the frequency of extreme values in *both* tails of the sampling distribution when determining the probability of obtaining differences as great or greater than this one. The observed difference of 2.8 is exceeded by only 4 sample mean differences in the sampling distribution of 100[1] mean differences; thus a mean difference of 2.8 is safely significant at the 5% level.

Since this mean difference is demonstrably rare on the basis of random sampling, we would probably reject the null hypothesis that random sam-

[1] For purposes of illustration we have limited our sampling distribution to 100 mean differences. The mean difference of 2.8 is "significant" at the 5% level for this sampling distribution of 100, but it might not be for the next sampling distribution of 100. When the investigator makes his decision to reject or not to reject the null hypothesis, he does it on the basis of a theoretical sampling distribution consisting of an infinite number of samples. The forms of these sampling distributions are known and will be described later.

TABLE 6.3 A Sampling Distribution of 100 Mean Differences where Samples ($n = 10$) are Drawn from a Table of Random Numbers

Interval	Tally
2.5 to 3.4	///
1.5 to 2.4	卌 卌 /
.5 to 1.4	卌 卌 卌 卌 //
−.5 to .4	卌 卌 卌 卌 卌 /
−1.5 to −.6	卌 卌 卌 卌 ////
−2.5 to −1.6	卌 卌 /
−3.5 to −2.6	/
−4.5 to −3.6	/

pling accounted for it. We would probably conclude that such a large mean difference was in part the result of method of instruction, the independent variable. Such a conclusion would be *wrong;* it would be a Type I error because the mean difference of 2.8 is in fact due to nothing more than random variation in mean differences. We know this to be true in this instance because we selected both samples from the table of random numbers.

Now imagine that our independent variable had operated to *reduce* the problem solving efficiency of Group A. We can illustrate this by subtracting three points from each "score" in the experimental group. It will then have a sample mean of 2.9. The mean difference between Group A and Group B is now 0.2. This mean difference could hardly lead to the rejection of the null hypothesis, and yet the null hypothesis is now false! If we fail to reject it, and we should not reject it on the basis of these data, we have committed a Type II error.

The effect of any independent variable must be evaluated by comparing the variation it produces with the variation to be expected from chance alone. This is much like throwing a pebble into a lake; if the lake is perfectly smooth the smallest disturbance is noticeable against the calm. As the overall turbulence increases, so must the size of the pebble if its effect is to be noticeable. Similarly, the magnitude of the independent variable's effect must be substantial if it is to produce a significant result against a background of substantial sampling error.

In the final analysis, the experimenter must be able to make accurate statements about the probability that particular events will occur as a result of sampling error. The study of probability is therefore necessary before we can develop the study of inferential statistics in more detail; we shall continue the discussion of probability in the next chapter.

CHAPTER 7

Probability

There are several definitions of probability and its thorough study, like the thorough study of statistics, involves some very complex mathematics. The probability problems presented in this chapter are relatively simple ones and do not require principles beyond those necessary for understanding the statistical tests of significance presented later in the text.

A DEFINITION OF PROBABILITY

The probability of an event A, $p(A)$, may be defined as the number of mutually exclusive and equally likely events favorable to its occurrence, divided by the total number of mutually exclusive, equally likely events possible. The probability of an event is therefore expressed as a proper fraction, or a decimal, which can vary between 0 and 1—between impossibility and certainty. The probability of heads occurring when an unbiased coin is tossed is 1/2; there is a head on one side of the coin and a total of 2 sides. The probability of drawing an ace of hearts from a well shuffled deck of cards is 1/52; there is 1 ace of hearts and a total of 52 cards. The probability of drawing any heart from a well shuffled deck of cards is 13/52, or 1/4; there are 13 hearts among the 52 cards. While the probability of an event may be expressed either as a decimal or as a proper fraction there is some preference for stating probabilities as decimals. Thus, we state the probability of obtaining a heart from a well shuffled deck as .25 rather than 1/4, although either form is technically correct.

In the preceding paragraph we gave an a priori or theoretical definition of probability. It was based upon the ratio of events favorable to some outcome divided by the total number of events. One can also use an empirical definition of probability. This too is a ratio of events, but it is a ratio based on experience over a very large number of trials rather than on a rational consideration of equally likely alternatives. For example, if we find that a coin comes up heads 70,000 times in 100,000 tosses, then the probability of obtaining a head on a given trial with this particular coin must be approximately .70. Since this probability differs from .50, we would have to assume that this particular coin is biased, that when tossing this coin heads and tails are not equally likely outcomes.

While a priori and empirical definitions of probability seem fundamentally different, they are usually just different ways of looking at the same thing. If the alternatives are mutually exclusive and equally likely, and if enough trials are given, we shall find the a priori and empirical definitions yielding exactly the same probability for any particular event.

THE MULTIPLICATIVE RULE

Sometimes we are required to determine the probability of a compound event composed of several constituent events. For example, we may wish to determine the probability of throwing two consecutive heads with an unbiased coin. The probability of the compound event is given by the *product* of the probabilities for the constituent events. When the constituent events are independent this multiplicative rule may be stated as follows: the probability that *all* of a group of independent events will occur is the product of the probabilities that the events will occur separately; that is, $p(A \text{ and } B) = p(A) \cdot p(B)$.

Independent events are, of course, unrelated to each other. When events are independent it means that the occurrence of one event does not in any way influence the occurrence of the other event. For example, tossing a head on the first throw of a coin has no effect on the outcome of the second toss. The coin does not remember what it has done. If the coin is unbiased and we have thrown 10 consecutive heads, the probability of obtaining a head again on the eleventh toss is still .50. The coin feels no obligation to compensate for a string of heads by beginning a string of tails.

Let us consider some examples in which the product rule is applied to compound events composed of independent constituent events. The probability of throwing a head on the first toss of a coin is 1/2. The probability of throwing a head again with that same coin is also 1/2. Therefore the probability of the compound event, obtaining a head on the first toss of a coin *and* a head on the second toss, is $1/2 \cdot 1/2$ or .25. By similar reasoning we can find the probability of obtaining aces on two consecutive draws from a deck of cards. The probability that the top card of a well shuffled deck will be an ace is 4/52, or 1/13. If the top card is replaced and the deck reshuffled, the probability of obtaining an ace again is also 1/13. The probability of the compound event consisting of drawing two consecutive aces, assuming replacement of the first drawn card, will be $1/13 \cdot 1/13$ or 1/169.

Suppose we wish to obtain the probability that a coin, when it is tossed three times, will fall in the order: heads, tails, heads. These are each independent events, and the probability of each is 1/2. The probability of the compound event will, therefore, be the product of these separate probabilities $p(A \text{ and } B \text{ and } C) = p(A) \cdot p(B) \cdot p(C) = 1/2 \cdot 1/2 \cdot 1/2 = .0125$.

Not all compound events are composed of constituent events which are independent. In the previous illustration we determined the probability of dealing two aces from a deck, and we specified that the first card dealt was to be reshuffled into the deck before the second card was dealt. If we had *not* specified the replacement of this card we would have had a problem in dependent or conditional probabilities. If the cards are simply dealt one, two, from the top of the deck, then the probability that the first card will be an ace is 4/52 (or 1/13) just as before, but the probability that the next card will be an ace is conditional upon the outcome of the first card dealt. If the first card is an ace, then there are only 3 remaining aces and 51 remaining cards. Therefore, the probability of drawing an ace on the second card, given that the first card was an ace, is 3/51. This is called a conditional probability because its value depends upon the occurrence or nonoccurrence of another event; in this case an ace on the first card. The probability of the compound event, drawing two aces without replacement, is the product of the unconditional probability of obtaining the first ace multiplied by the conditional probability of obtaining the second ace: $4/52 \cdot 3/51 = 1/221$.

We shall distinguish conditional probabilities by the use of a slightly different symbol. The probability of obtaining B given that A has occurred is written $p(B/A)$ and read, "the probability of B given A." Now we can give a general formula for the probability of compound events regardless of the independence of the constituent events. The general multiplicative rule is given below.

$$p(A \text{ and } B) = p(A) \cdot p(B/A)$$

FORMULA 7.1
Probability of a compound event illustrating the multiplicative rule

Let the compound event (A and B) be drawing two red cards from a normal deck without replacement. The probability that the first card will be red, $p(A)$, is 26/52 or 1/2. If the first card is red there are only

25 red cards left so $p(B/A)$ will be 25/51. Thus,

$$p(A \text{ and } B) = 26/52 \cdot 25/51 = 325/1326.$$

We shall also apply the multiplicative rule to the determination of the probability of being dealt *three* successive aces from a standard deck of cards. Since the combined event is composed of three elements, we shall extend Formula 7.1 to: $p(A \text{ and } B \text{ and } C) = p(A) \cdot p(B/A) \cdot p(C/A \text{ and } B)$. We let event A represent an ace on the first draw, event B represent an ace on the second draw, and event C represent an ace on the third draw. We have $p(A) = 4/52$, $p(B/A) = 3/51$ and $p(C/A \text{ and } B) = 2/50$. Thus, $p(A \text{ and } B \text{ and } C) = 1/5525$.

It may be apparent that the product rule for independent events, which we discussed at the beginning of this section, is really just a special case of Formula 7.1, the general product rule. When A and B are independent the probability of B is not altered by the occurrence of A, nor is the probability of A altered by the occurrence of B. This means that when the components of a combined event are independent $p(B/A) = p(B)$ and $p(A/B) = p(A)$. If, in the problem involving three aces, we had reshuffled each card dealt before dealing again, the events would have been independent. That is, the occurrence of an ace on the first draw would not have altered the probability that an ace might be obtained again on the second draw, and, obtaining an ace on the first and second draws would not have changed the probability that an ace might be obtained on the third and final draw. In that situation, the formula for conditional probability simplifies to $p(A \text{ and } B \text{ and } C) = p(A) \cdot p(B) \cdot p(C)$. In formula form, this is the multiplicative rule for independent events which we gave at the beginning of this section.

THE ADDITIVE RULE FOR MUTUALLY EXCLUSIVE EVENTS

The probability that either of a pair (or any number) of mutually exclusive events will occur is the sum of their separate probabilities. This is called the additive rule. If the probability of an ace of hearts is 1/52, and the probability of an ace of spades is 1/52, then the probability of an ace of hearts *or* an ace of spades being drawn from an ordinary deck in 1 draw is $1/52 + 1/52$ or 1/26.

If we wish to determine the probability of drawing *any one* of four aces, the probability will be $1/52 + 1/52 + 1/52 + 1/52$, or 1/13. The situations requiring the additive rule are those which ask for the probability of events A, *or* B, *or* C, *or* D, etc. In such cases we add the probabilities of the separate events to determine the probability of the disjunctive event.

Any discrete event, such as drawing a spade, can either occur or not occur. We can either draw a spade or a "no spade"; there are no other possibilities. We are *certain* to draw *either* a spade or a "no spade," so we would expect the additive rule to show that the sum of these probabilities equals 1.00; one or the other of the two events is certain to occur. We have specified that the probability of occurrence for any event (A) will be symbolized by $p(A)$. We shall now specify that the probability of *non*-occurrence for event A will be symbolized by $q(A)$. Since any discrete event can either occur or not occur, $p(A) + q(A) = 1.00$, and $q(A) = 1 - p(A)$. If n mutually exclusive events are possible, the sum of their probabilities must also equal 1.00. Thus, the sum of the probabilities of all possible draws from a deck of cards is

$$1/52 + 1/52 + \cdots + 1/52 = 52/52 = 1.00.$$

It is also possible to have a problem requiring the probability of either A or B when these events are not mutually exclusive. For example, suppose we find the probability that the top card of a deck will be an ace (A), or a heart (B). These are *not* mutually exclusive events, so we cannot obtain the probability that either will occur by summing their separate probabilities. Notice that we can obtain an ace which is not a heart, a heart which is not an ace, or an ace which is also a heart. If we were to follow the rule for mutually exclusive events and sum the separate probabilities for aces and for hearts, we would have included the last event twice. That is, the probability of an ace of hearts would have been included under aces and included again under hearts. This being the case, we must subtract a correction term which is the probability that the card can be *both* a heart and an ace. The formula becomes:

$$p(A \text{ or } B) = p(A) + p(B) - p(A \text{ and } B)$$

FORMULA 7.2
Probability of a disjunctive event A or B illustrating the additive rule

If $p(A)$ is the probability of an ace and $p(B)$ is the probability of a heart we have $p(A) = 4/52$ and $p(B) = 13/52$. The probability of an ace *and* a heart $p(A \text{ and } B)$, that is, the ace of hearts, is calculated from the multiplicative rule for combined events which we discussed in the last section. If $p(A) = 4/52$ and $p(B) = 13/52$, then $p(A \text{ and } B) = 1/52$. We now have all of the individual probabilities needed to obtain the answer; $p(A \text{ or } B) = 4/52 + 13/52 - 1/52 = 16/52$. We can check this answer easily enough by counting the cards in a deck which meet the criteria "an ace or a heart." There are of course 13 hearts and 3 additional aces making a total of 16. The probability is 16/52, just as we have calculated it from Formula 7.2.

We began this section with the additive rule for mutually exclusive events, and then developed the more general case of an additive rule to apply when events were not mutually exclusive. If we have mutually exclusive events we can still use Formula 7.2, but the final term will be zero. If events A and B are mutually exclusive, then by definition the occurrence of both A and B on the same trial has a probability of zero.

EXAMPLES USING BOTH THE ADDITIVE AND MULTIPLICATIVE RULES

We find the probability of throwing an ace and a deuce when rolling dice by using both the multiplicative and the additive rules. (First, we should explain that a die is a cube, the six faces of which bear the numbers from 1 to 6 in the form of dots. The plural of die is dice.) Each cube has an ace on one face and a deuce on one face. The probability of obtaining any face is 1/6, since a cube has six sides and we are assuming the dice are unbiased or "true." The probability of obtaining both an ace on the first die *and* a deuce on the second die is $1/6 \times 1/6$ or 1/36 by the multiplicative rule. Note that the events are independent; one die's behavior has no effect on the behavior of the other die. We can also satisfy the conditions by obtaining a deuce on the first die and an ace on the second. That probability is found in the same way; it is also 1/36. Either of these mutually exclusive events will satisfy the condition "an ace and a deuce." Obtaining the probability that either of these events will occur requires the additive rule and yields $1/36 + 1/36$, or 1/18.

What is the probability that we can throw a head *within* three tosses of one coin? We can break this problem into three parts: a head occurring on the first toss, a head failing to occur on the first toss but occurring on the second, and finally, a head failing to occur on either the first or the second toss but occurring on the third. We must find the sum of the probabilities for these three events since any one of them satisfies the requirement of "a head within three tosses." We will assume that the coin is unbiased so p, the probability of heads, is equal to q, the probability of tails. Now let us find the probabilities of the individual events. The probability of a head occurring on the first toss is 1/2, ($p = 1/2$). The probability of "no head" on the first toss (q) and a head on the second toss (p) is given by the product qp. Since $q = p = 1/2$, the value of this term is 1/4. The probability of failing to throw a head on the first two tosses and then throwing a head on the third toss is also obtained by using the multiplicative rule; it will be the product of $q \times q \times p$, or q^2p. Since $q = p = 1/2$, the value of this term is 1/8. The problem is to find the probability of a head occurring within three tosses, so we add the probabilities of throwing a head on the first toss, failing on the first and hitting on the second, and failing on the first two and hitting on the third. Thus, we have $p + qp + q^2p$, and its value is $1/2 + 1/4 + 1/8 = 7/8$.

Suppose we extend the problem one more step to find the probability of obtaining a head within four tosses. To do this we must add another term to the series; this term must represent the probability of three successive tails followed by a head. Adding this term we have the series $p + qp + q^2p + q^3p$, which has a value of 15/16. We shall list below the sequence of tosses and the probability of occurrence for each event represented by these terms.

1. *H* 1/2 The probability of a head on the first toss.
2. *TH* 1/4 The probability of no head on the first toss, head on the second toss.
3. *TTH* 1/8 The probability of no head on the first or second toss, head on the third toss.
4. *TTTH* 1/16 The probability of no head on first, second or third toss, head on the fourth toss.

Since *any* one of these sequences can satisfy the conditions we have imposed, the *sum* of these probabilities will give the answer to the problem. Notice that no matter how far we expand the series, its sum never reaches 1. There is no fixed number of tosses which can guarantee a head.

PERMUTATIONS AND COMBINATIONS

Suppose we have written the letters *A*, *B*, *C*, *D*, and *E* on folded slips of paper, mixed them well, and then asked someone to draw them from a container in the proper alphabetical order. To calculate the probability that he will be successful, we must know the number of possible orders or arrangements, called permutations, which five letters can take. Only one of these permutations, *ABCDE*, is correct. To find the probability of obtaining 5 letters in alphabetical order, we must determine the number of possible orders or permutations of n objects (letters) when $n = 5$.

The formula is $n!$ and is read as "n factorial." The factorial sign (!) means that n is multiplied by all positive integers less than itself. In our problem $n! = 5 \times 4 \times 3 \times 2 \times 1 = 120$. Consequently, there is one chance in one hundred-twenty that an individual can draw the five letters in the correct sequence.

Suppose we change the problem and ask the subject to draw the letters ABC, in that order, from the letters A, B, C, D, E. Now the problem requires us to determine the number of possible permutations of three objects drawn from five. In this problem, only one of these permutations, ABC, is correct. The formula for finding the number of permutations when r objects are obtained from n objects is:

$$\frac{n!}{(n-r)!}$$

FORMULA 7.3
Permutations of r objects drawn from n objects

(When $r = n$, as it did in the preceding problem, Formula 7.3 simplifies to $n!$ because $(n - r)! = 0!$, and $0!$ is defined as 1.) Applying Formula 7.3 to the preceding problem we have $r = 3$ and $n = 5$, and making the appropriate substitutions in Formula 7.3, we have

$$\frac{5!}{(5-3)!} = \frac{5 \times 4 \times 3 \times \cancel{2} \times \cancel{1}}{\cancel{2} \times \cancel{1}} = 60.$$

There are 60 possible permutations of three objects selected from five objects. The probability that the subject will select the one correct permutation; that is, the correct arrangement of the correct group of three letters (ABC) is 1/60.

The rationale for the formula for permutations is relatively easy to understand. If we wish to know the number of ways 3 different objects can be drawn from 5, we might begin by determining the number of ways the *first* object can be chosen. If there are 5 objects, then 5 different first choices are possible. For each different first choice there are 4 possible second choices. For each different second choice there are 3 possible third choices. Consequently there are $5 \times 4 \times 3$, or 60 different permutations of 3 objects drawn from 5.

If we wish to determine the number of permutations which results when two objects are drawn from three, Formula 7.3 yields

$$\frac{3!}{(3-2)!} = \frac{3 \times 2 \times \cancel{1}}{\cancel{1}} = 6.$$

There are three different first choices, A, B, or C, and for each of these there are two different second choices. This yields a total of $3 \times 2 = 6$ permutations which are listed below.

$$\begin{matrix} AB & BC & AC \\ BA & CB & CA \end{matrix}$$

If the *arrangement* of objects within a set is not relevant, and we wish to know only the number of different sets of objects, we are speaking of the number of *combinations* rather than permutations. For example, we may want to know the probability of selecting the letters A, B, and C, without regarding order, from the group $ABCDE$. Note that in this problem we are *not* insisting that the letters be drawn in alphabetical order; we will accept either CBA, ABC, BAC, ACB, CAB, or BCA as having satisfied the conditions of the problem. The formula for the number of combinations of r things drawn from n is:

$$\frac{n!}{(n-r)!\,r!}$$

FORMULA 7.4
Combinations of r objects drawn from n objects

For the preceding problem we have $n = 5$ objects, and we wish to know the number of different sets of $r = 3$ objects which can be drawn from 5. The formula yields:

$$\frac{5!}{(5-3)!\,3!} = \frac{5 \times 4 \times \not{3} \times \not{2} \times \not{1}}{(2 \times 1) \times (\not{3} \times \not{2} \times \not{1})} = \frac{5 \times 4}{2 \times 1} = 10.$$

We can construct 10 different three-letter sets from five different letters, so the probability of selecting the letters ABC, regardless of order from the letters $ABCDE$, is 1/10.

Note that Formula 7.4 is Formula 7.3 divided by $r!$. Why should permutations divided by $r!$ give combinations? The permutations of r objects drawn from n includes all possible arrangements (permutations) of every different set of r objects (combinations) that can be obtained from the n objects. If we divide this by the permutations *per set* of r objects, we will have the number of different sets of r objects which can be obtained from n objects. Since the number of permutations of r objects is $r!$ the formula for combinations is simply the formula for permutations divided by $r!$. In a previous problem we found that three letters drawn from five can produce 60 permutations. However, a particular three letter combination will produce 6 ($r!$) permutations. Therefore, the number of combinations is given by dividing the total number of permutations by the number of permutations for each combination. You will recall that 10 combinations result when three letters are drawn from five.

As another example, when two objects are drawn from three we find six permutations possible: AB, BA, BC, CB, CA, and AC. If we wish to obtain just the number of combinations, AB and BA are considered one unit; so are BC and CB, as well as CA and AC. There are three possible combinations when drawing two objects from three. These are AB, BC, and CA. To find the number of *combinations* we divided the 6 permutations by $r!$ (2×1). This is because $r!$ gives the permutations of each set of two objects, and each set of two permuted objects is one combination.

Let us consider some illustrative problems. If we are tossing four coins, we might want to know the number of ways we can obtain two heads and two tails. Since the arrangement (the order in which heads must appear) is *not* specified, the problem requires the formula for combinations. Formula 7.4 yields

$$\frac{4!}{(4-2)!\,2!} = \frac{4 \times 3 \times \not{2} \times \not{1}}{(2 \times 1)(\not{2} \times \not{1})} = 6.$$

If the four coins are labeled A, B, C, and D, then the following six different sets of heads are possible: AB, AC, AD, BC, BD, and CD. In this example the order of selection *within pairs* is unimportant. Since we are only concerned with the number of *different* pairs, we are dealing with combinations.

Suppose we throw six dice. How many combinations are possible in which three of the six dice show an ace? Again, we have a combination problem because the order in which the aces appear within the sets of three is not relevant. From Formula 7.4 we have

$$\frac{6!}{(6-3)!\,3!} = \frac{6 \times 5 \times 4 \times \not{3} \times \not{2} \times \not{1}}{(3 \times 2 \times 1)(\not{3} \times \not{2} \times \not{1})} = 20.$$

There are 20 different combinations when three dice are selected from six and, consequently, twenty ways in which an ace can occur on three dice when six dice are thrown.

An experiment has four conditions which can be presented to subjects in any order. We want to have three subjects assigned to every possible order. How many subjects will we need? This problem requires the formula for permutations. In how many different ways can the four treatments be ordered? The answer is simply 4!, or 24. If we assign three subjects to each arrangement of conditions we shall need 72 subjects.

When we speak of permutations we are considering a unit as consisting of different sets of objects *and* different arrangements or orders of objects within sets. When we speak of combinations, we are considering only groups composed of different sets of objects; the order of objects within sets is ignored.

THE BINOMIAL EXPANSION AND PROBABILITY

You may have learned in an algebra course that the expression $(p + q)$ is called a binomial, because it is the sum of two terms, p and q. The binomial $(p + q)$ may be squared, and the result is found just as it would be if any number were squared. For example:

$$\begin{array}{r}12\\ \underline{12}\\ 24\\ \underline{12(0)}\\ 144\end{array} \qquad \begin{array}{r}10 + 2\\ \underline{10 + 2}\\ 20 + 4\\ \underline{100 + 20\phantom{{}+4}}\\ 100 + 40 + 4\end{array} \qquad \begin{array}{r}(p + q)\\ \underline{(p + q)}\\ pq + q^2\\ \underline{p^2 + pq\phantom{{}+q^2}}\\ p^2 + 2pq + q^2\end{array}$$

When the binomial has been squared or raised to any power, it is said to have been "expanded" to that power. The expanded binomial has very important applications to probability problems.

Suppose we wish to find the probability of every possible outcome when two coins are tossed. The possible events consist of two heads, one head and one tail (disregarding order), and two tails. We shall let p stand for the probability of heads, and q stand for the probability of "no head," or tails.

Our first concern will be to determine the probability of obtaining two heads. If the probability of obtaining a head when one coin is tossed is represented by p, then the probability of obtaining two heads when two coins are tossed will be $p \times p$ or p^2, by the multiplicative rule.

We can also obtain one head and one tail (disregarding order). There are two ways this can happen: the first coin can be a head and the second a tail, or the first can be a tail and the second a head. The probability of the first of these sequences is given by pq, since the multiplicative rule is again involved. The probability of the second sequence is given by qp, again the result of applying the multiplicative rule. The event, one head and one tail, can be satisfied by obtaining *either* sequence HT or TH. The probability that *either* of these will occur is the sum of their separate probabilities, or $pq + pq$, which is $2pq$. (Remember that in a series of multiplications the order of operations is immaterial.)

There is a final possibility; the two coins can both show tails. If the probability of obtaining one tail is q, then the probability of obtaining tails on both coins will be $q \times q$ or q^2, again by using the multiplicative rule.

We can consolidate this information into the table below.

Event	*Symbolized by*	*Probability*
HH	p^2	1/4
HT or TH	$2pq$	1/2
TT	q^2	1/4
	$\Sigma = p^2 + 2pq + q^2 =$	1.00

Since we have listed all the possible events which can occur when two coins are tossed, the sum of the separate probabilities must equal 1.00, which it does!

Notice that in the process of obtaining the terms symbolizing the probabilities of these events we went through *exactly* the operations required to expand the binomial $(p + q)^2$. We multiplied $p \times p$ to get p^2 for the probability of HH; we multiplied $p \times q$ to get the probability of HT; and $q \times q$ (or q^2) was the probability of TT.

Not only do the terms of the expanded binomial yield the probabilities of the particular events each term represents, but the number of terms yields the number of different events that are possible. When $(p + q)^2$ is expanded we find that three terms result, and we have shown that (disregarding order) there are three mutually exclusive outcomes when two coins are tossed.

In addition, the coefficient of each term corresponds to the number of combinations that constitute the event. The coefficient of the p^2 term is understood to be 1 (that is, $1p^2$ is simply written as p^2); thus, two heads can be obtained from two coins in just one way. The coefficient of the pq term is 2, and this indicates that a head and a tail can occur in two ways. The coefficient of the q^2 term is 1; two tails can occur in one way. We must emphasize that this correspondence is not accidental. It is the direct result of using the same mathematical operations to expand the binomial as we perform to obtain the probabilities of the outcomes when tossing two coins.

The correspondence of the mathematical operations required to expand a binomial, and those required to determine probabilities, is not limited to squared binomials and the probabilities of events which occur when just two coins are tossed. If p and q represent the probabilities of occurrence and nonoccurrence of any event, whether or not p is equal to q, and if they are used to compose a binomial where the number of events (n) is used as an exponent or power to which the binomial is raised, then the operations used to expand the binomial will *always* correspond to those required to determine the probabilities of the different events. We will examine this correspondence more closely in the next example.

Suppose we solve the same kind of problem, but with 4 coins. To obtain the required terms we must expand the binomial $(p + q)^4$, where p is the probability of heads and q is the probability of tails. The expansion of $(p + q)^4$ is somewhat tedious and short cut methods will be demonstrated later, but we shall undertake it here to illustrate the correspondence of operations.

$$
\begin{array}{rrrrrl}
 & & & p & +\,q & = (p+q)^1 \\
 & & & p & +\,q & \\
\hline
 & & & pq & +\,q^2 & \\
 & & p^2 & +\,pq & & \\
\hline
 & & p^2 & +\,2pq & +\,q^2 & = (p+q)^2 \\
 & & & p & +\,q & \\
\hline
 & & p^2q & +\,2pq^2 & +\,q^3 & \\
 & p^3 & +\,2p^2q & +\,pq^2 & & \\
\hline
 & p^3 & +\,3p^2q & +\,3pq^2 & +\,q^3 & = (p+q)^3 \\
 & & & p & +\,q & \\
\hline
 & p^3q & +\,3p^2q^2 & +\,3pq^3 & +\,q^4 & \\
p^4 & +\,3p^3q & +\,3p^2q^2 & +\,pq^3 & & \\
\hline
p^4 & +\,4p^3q & +\,6p^2q^2 & +\,4pq^3 & +\,q^4 & = (p+q)^4
\end{array}
$$

The completed expansion is:

$$p^4 + 4p^3q + 6p^2q^2 + 4pq^3 + q^4.$$

The different events which can occur when four coins are tossed are shown in Table 7.1. They consist of 4 heads, 3 heads and 1 tail, 2 heads and 2 tails, 1 head and 3 tails, and 4 tails. The probability of obtaining 4 heads is $p \times p \times p \times p$, or p^4, by the multiplicative rule for independent events. These are also the operations by which we arrived at the first term of the expanded binomial $(p + q)^4$. The term p^4 from this expansion can be written as $1p^4q^0$, but coefficients of 1 are dropped in practice, and any number to the zero power is one, so we ordinarily write $1p^4q^0$ as p^4. However, thinking of it as $1p^4q^0$ illustrates some important relationships. First, the exponents of p and q indicate that these symbols stand for the probability of 4 heads and 0 tails; second, the coefficient 1 indicates that 4 heads, or p^4, can occur in just one way. If $p = 1/2$, then $p^4 = 1/16$, and this is the probability of obtaining 4 heads when four coins are tossed.

Another possible result when four coins are tossed is three heads and one tail. Three heads and one tail can be obtained in four ways. There are four possible combinations, since any one of the four coins could be a tail and the rest heads. The probability of each combination is obtained from the multiplicative rule. These are: $q \times p \times p \times p$, $p \times q \times p \times p$, $p \times p \times q \times p$, and $p \times p \times p \times q$. Each of these terms may be rewritten p^3q, and since any one of these combinations satisfies the condition "three heads and one tail," the additive rule applies. We sum the four terms to obtain $4p^3q$. If $p = q$, then the probability of three heads and a tail is $4(1/2)^3(1/2) = .25$.

TABLE 7.1 Correspondence of Outcomes when Four Coins Are Tossed to the Terms of the Binomial Expansion $(p + q)^4$

Outcomes	Term
HHHH	$p^4 = (\frac{1}{2})^4 = \frac{1}{16}$
THHH *HTHH* *HHTH* *HHHT*	$4p^3q = 4(\frac{1}{2})^3(\frac{1}{2}) = \frac{1}{4}$
HHTT *HTHT* *THHT* *TTHH* *HTTH* *THTH*	$6p^2q^2 = 6(\frac{1}{2})^2(\frac{1}{2})^2 = \frac{6}{16}$
TTTH *TTHT* *THTT* *HTTT*	$4pq^3 = 4(\frac{1}{2})^3(\frac{1}{2}) = \frac{1}{4}$
TTTT	$q^4 = (\frac{1}{2})^4 = \frac{1}{16}$

$$\Sigma(p^4 + 4p^3q + 6p^2q^2 + 4pq^3 + q^4) = \Sigma(\tfrac{1}{16} + \tfrac{4}{16} + \tfrac{6}{16} + \tfrac{4}{16} + \tfrac{1}{16}) = 1.00$$

When the binomial $(p + q)^4$ is expanded, the second term, $4p^3q$, also results from summing $q \times p \times p \times p$, $p \times q \times p \times p$, $p \times p \times q \times p$, and $p \times p \times p \times q$—just as it did when we calculated the probability of three heads and a tail.

A third possible result when tossing four coins is to find two heads and two tails. This event can occur in six different ways. The combinations are listed below.

HHTT	*HTTH*	*THTH*
HTHT	*THHT*	*TTHH*

The probability that any one combination will occur is given by the multiplicative rule; it is p^2q^2. However, since each of the six combinations will satisfy the conditions of two heads and two tails, the additive rule applies and yields $6p^2q^2$. This is also the third term of the binomial expansion, and when $p = q = 1/2$, it yields $6(1/2)^2(1/2)^2 = 6/16$.

Since $p = q$, the probability of obtaining one head and three tails is exactly the same as the probability of obtaining three heads and one tail. This event is produced by any one of four possible combinations, depending on which coin comes up tails. The probability of each combination is pq^3, and the probability that any one will occur is the sum of the four, or $4pq^3$. The value of $4(1/2)(1/2)^3 = 1/4$.

The remaining result, four tails, can occur in only one way. The probability of this event is simply the product of $q \times q \times q \times q$, or q^4. The value of $(1/2)^4 = 1/16$.

We have now seen that the expansion of $(p + q)^4$, or $p^4 + 4p^3q + 6p^2q^2 + 4pq^3 + q^4$, can symbolize the probabilities of each event which can occur when four coins are tossed. Each of the five different terms in the expansion represents a separate event, and the sum of the probabilities generated by these terms will equal 1.00. These results are summarized in Table 7.1.

Expanding binomials beyond the fourth power is a tedious proposition, so we shall describe a short cut method for determining the coefficient and exponent of any term for any expansion through $(p + q)^{15}$. We shall consider how to determine the exponents of the terms first, and then the coefficients.

The exponents of the terms in any binomial expansion are quite regular; always beginning p^n + (coefficient) $p^{n-1}q$ + (coefficient) $p^{n-2}q^2$ + (coefficient) $p^{n-3}q^3 + \cdots +$ (coefficient) $pq^{n-1} + q^n$, where n is the power to which $(p + q)$ must be raised. The exponents of the first term for any expansion are always p^nq^0, regardless of the value of n. (Since $q^0 = 1$, p^nq^0 is written p^n.) After the first term, the value of the exponent for p *decreases* by a unit a term; at the same time the exponent of q *increases* by a unit a term through the $n + 1$ terms of the expansion. The exponent of p for any term is always $n - t + 1$, where n is the power of the expansion, and t is the number of the term in the expansion. Similarly, the exponent of q for any term is always $t - 1$. Notice the way this rule applies to the exponents for the expansion of $(p + q)^6$ given below.

$$p^6 + 6p^5q + 15p^4q^2 + 20p^3q^3 + 15p^2q^4 + 6pq^5 + q^6$$

The easy progression of the exponents is not found in the coefficients, consequently these are given directly in Table B of the Appendix through $(p + q)^{15}$. Suppose we wish to expand $(p + q)^5$. We read down the first column of Table B to the 5th row where the coefficients of all the appropriate terms are located. The coefficients for the terms of the expansion are 1–5–10–10–5–1 and the expanded binomial is

$$p^5 + 5p^4q + 10p^3q^2 + 10p^2q^3 + 5pq^4 + q^5.$$

The coefficients of the terms in the binomial expansion are the number of ways (combinations) in which the events represented by the terms can occur. Although the values of the coefficients are given in Table B, we can calculate them directly from the formula for combinations. For example, how many ways can six heads and three tails occur when nine coins are tossed? The answer is given by the coefficient of the p^6q^3 term in the expansion of $(p + q)^9$, or we can use Formula 7.4 to determine the number of combinations when six objects are drawn from nine (or three objects drawn from nine). We have

$$\frac{9!}{(9-6)!\,6!} = \frac{9 \times 8 \times 7 \times \cancel{6} \times \cancel{5} \times \cancel{4} \times \cancel{3} \times \cancel{2} \times \cancel{1}}{(3 \times 2 \times 1)\cancel{6} \times \cancel{5} \times \cancel{4} \times \cancel{3} \times \cancel{2} \times \cancel{1}} = 84.$$

There are a variety of probability problems which can be solved by evaluating the individual terms from a binomial expansion. For example, what is the probability of finding exactly half heads and half tails when 14 coins are tossed? The binomial to be expanded will be $(p + q)^{14}$, but we need to evaluate *only* the middle term. When a binomial $(p + q)^n$ is expanded there are $n + 1$ terms, so the middle term of the expansion $(p + q)^{14}$ will be the eighth term. We determine from Table B that when $n = 14$ the coefficient of the eighth term is 3432 and, using the formula for exponents, we have $3432p^7q^7$.

Since $p = 1/2$, we must find the value of $3432(1/2)^7(1/2)^7$. When multiplying terms consisting of the same base raised to different exponents, we retain the base and *add* the exponents; therefore, $3432(1/2)^7(1/2)^7 = 3432(1/2)^{14}$. We must now obtain the value of $(1/2)^{14}$, which may also be written $(1)^{14}/(2)^{14}$. The 14th power of 1 is 1; the 14th power of 2 may be found in Table B and is 16384. The expression $3432(1/2)^7(1/2)^7$ therefore simplifies to 3432/16384, or .209. This is the probability of tossing 14 coins and finding exactly half of them heads.

Another example may help to clarify the use of Table B when evaluating specific terms of the binomial. Suppose we wish to know if the probability of half heads when 10 coins are tossed is greater than the probability of half heads when 14 coins are tossed. We have just determined that the latter probability is .209. In order to find the probability of obtaining 5 heads when ten coins are tossed, we must evaluate the middle term in $(p + q)^{10}$. Since there are $n + 1$, or eleven terms in the expansion, we need the sixth. When $n = 10$, the coefficient of the sixth term from Table B is 252; the sixth term will then be $252p^5q^5$ which we can rewrite $252(1/2)^{10}$. The 10th power of 2 is given by Table B as 1024, so the value of $252p^5q^5$ is 252/1024, or .246. This is somewhat greater than the probability of obtaining half heads when 14 coins are tossed.

BINOMIAL PROBABILITIES WHEN $p \neq q$

The binomial expansion also provides us with a method of calculating probabilities when p is not equal to q ($p \neq q$). We may wish to determine the probability of throwing three sixes with four dice. Since a six appears on only one of the six faces of a die, the probability of obtaining a six with an honest die is 1/6. The probability of "no six" is therefore $1 - 1/6$, or 5/6. With four dice we will need to expand and evaluate $(p + q)^4$, where $p = 1/6$, and $q = 5/6$. The term $4p^3q$ gives us the required probability, and $4(1/6)^3(5/6) = 20/1296 = .015$. Three sixes out of 4 throws is a very unlikely event.

What is the probability that 4 dice can be thrown *without* throwing a six? This is the same as throwing 4 "no sixes," and for the answer we must evaluate q^4, the final term of the expansion of the same binomial. Using Table B we find that $(5/6)^4 = 625/1296$, or .482—somewhat less than half.

We must often answer questions about the probability of obtaining more than, or less than, some number of events by chance. Suppose we wish to conduct an experiment to determine if a subject (S) has precognition. We might instruct S to guess the number which will appear on the face of a die before the die is thrown. If S throws the die five times, how many "hits" will he need before we can reject chance as an explanation for his performance? We shall determine the number of hits required for rejecting the null hypothesis at the 5% and at the 1% levels of significance.

The probability that S can make 5 hits out of five attempts by chance is given by p^5 where $p = 1/6$, which is 1/7776, or .00013. We will surely reject the null hypothesis if S produces this improbable event, but we do not need to be this demanding to achieve the 1% level of significance. The probability that S can achieve 4 hits is given by $5p^4q = 5(1/6)^4(5/6)$, which is 25/7776, or .0032. It seems that we should reject the null hypothesis if either 4 or 5 hits occur, for the probability that either event will occur by chance is the sum of the two probabilities or 26/7776, or .0033, and still well under .01.

Suppose S gets only 3 hits, will this still be sufficiently improbable so that chance can be rejected? The probability of obtaining exactly 3 hits is $10p^3q^2 = 10(1/6)^3(5/6)^2$ which is 250/7776, or .0321. Achieving either 3, 4, or 5 hits gives a probability of 276/7776, or .0355. Since this probability is no longer less than .01, we can accept only 4 or 5 hits as evidence for the rejection of the null hypothesis at the 1% level of significance. If α had been set at .05, a less rigorous demand, we could reject the null hypothesis if either 3, 4, or 5 hits occur. In the next chapter we shall continue our discussion of the binomial expansion and its direct application to research situations.

Exercises

1. What is the probability that *HTHT* will occur in that order when a coin is thrown four times?
2. Why is this statement incorrect? "The probability of a head is 1/2; therefore, the probability of getting a head on either the first *or* the second toss of a coin, by the additive rule, is 1."
3. T—F. If only two events are possible they are equally likely.
4. What is the probability of selecting a red card, or a face card, in one draw from a well-shuffled deck?
5. What is the probability of obtaining the numbers 1, 2, 3, 4 in exactly that order from the table of random numbers?
6. What is the probability of getting 8 items right and 2 wrong, by chance, on a 10-item T—F test?
7. What is the probability that a randomly selected student in this class will be above the median on the next statistics test?
8. Can you calculate the probability that a student who is above the median on the first test will also be above the median on the second test?
9. Using Table B, and the rules for exponents, determine which of the following, if any, are incorrect.
   (a) The fourth term of the expansion $(p + q)^{10}$ is $120p^7q^3$.
   (b) The eighth term of the expansion $(p + q)^{14}$ is $3432p^7q^7$.
   (c) The third term of the expansion $(p + q)^8$ is $28p^4q^3$.
   (d) The twelfth term of the expansion $(p + q)^{15}$ is $1365p^4q^{11}$.
   (e) The second term of the expansion $(p + q)^9$ is $9p^8q$.
10. What is the probability of obtaining more than 9 questions right, by chance, on a 12-item T—F test?
11. At year's end we survey an exclusive suburb and find that 20% of all families have had incomes in excess of $18,000 for the year, and that 50% of all families have purchased new cars. We conclude therefore, that the probability of a family having an income in excess of $18,000 for the year *and* purchasing a new car will be .10. Why is this conclusion almost certainly erroneous? Under what circumstances would it be correct?
12. The proportion of migrant workers who have telephones is .30. The proportion of migrant workers who have electric lights is .70. Therefore, all migrant

workers have either telephones *or* electric lights. Why is this conclusion almost certainly erroneous? Under what circumstances would it be correct?

13. We want each student in a social psychology class to observe a different group of four children. How many different groups of four can be selected from among 12 children who have volunteered to act as subjects in the study?
14. What is the probability of being dealt three cards and finding that two of them are aces?

CHAPTER 8

The Binomial Distribution and Its Normal Approximation

In this chapter we shall apply the concepts of sampling and probability to a variety of problems involving binomial populations. These ideas are fundamental to statistical inference, and they provide a continuously recurring theme for the remainder of the text.

THEORETICAL SAMPLING DISTRIBUTIONS BASED ON BINOMIAL POPULATIONS

A binomial population is composed of two mutually exclusive classes of discrete events where all events within each class have the same probability of occurrence. If a coin is tossed 500 times, one class of events (or outcomes) is heads, the other tails. These classes exhaust the possibilities and they are mutually exclusive. If the probability of heads is the same on every toss, and if the probability of tails is the same on every toss, then we have a binomial population. All of the Republicans and Democrats in the United States constitute a binomial population; so does the infinite number of possible die tosses where the outcome of each toss is classified as either a "6" or a "no 6." Notice that a binomial population can be composed of a finite *or* an infinite number of events, and that the probability of obtaining an event in one of the classes is not necessarily the same as the probability of obtaining an event in the other class.

Imagine a binomial population consisting of an infinite number of tosses of an unbiased coin. Suppose we select samples of four tosses ($n = 4$) from such a population, and we continue to select such samples until we have drawn an indefinitely large number. Each sample of four coins will contain either 0 tails, 1 tail, 2 tails, 3 tails, or 4 tails; there are, then, five different possibilities. In the last chapter we showed how the relative frequency, or probability, of each of these outcomes can be determined from the expansion of $(p + q)^4$. Once these relative frequencies have been calculated they can be used to construct a histogram representing the expected sampling distribution of heads or tails.

The histograms in Figure 8.1 represent the theoretical sampling distribution of successes (in this case tails) when samples of $n = 4$, $n = 8$ and $n = 12$ are randomly obtained from a binomial population in which $p = q$. For example, the expansion of $(p + q)^4$ yields $p^4 + 4p^3q + 6p^2q^2 + 4pq^3 + q^4$. If $p = q$, and we allow q to represent the probability of tails when four coins are tossed, we find that the probability or relative frequency of samples with no tails is $(1/2)^4 = .0625$; the probability of samples with one tail is $4p^3q = .2500$; two tails is $6p^2q^2 = .3750$; three tails is $4pq^3 = .2500$; and four tails is $q^4 = .0625$. Each of the histograms in Figure 8.1 has been constructed in a similar way from expanded binomials of the form $(p + q)^n$ where $p = q$ and n, representing sample size, takes values of 4, 8, and 12.

The relative frequency distributions shown in Figure 8.1 are properly called *theoretical* sampling distributions because they are based on the relative frequencies to be expected when we obtain an infinitely large number of samples from a binomial population. If we draw only 16 samples of $n = 4$ coins, it is rather unlikely that we will have *exactly* one sample with 0 tails, four samples with 1 tail, six samples with 2 tails, and so on. The histograms derived from the binomial expansion really tell us the relative frequencies to be expected when we obtain an *infinite number of samples.* An empirical sampling distribution, especially one based on a small number of samples, will probably not have relative frequencies which are very close to the theoretical sampling distribution given by the expanded binomial. Sampling distributions based on a small number of samples are themselves samples, and, like other samples, are subject to sampling error.

However, as the number of samples increases, the empirically obtained relative frequencies will approach the theoretical relative frequencies given by the appropriate binomial expansion. If we obtain 16,000 samples of 4 coin tosses, we can expect to find a much closer correspondence with the relative frequencies of the theoretical sampling distribution than if we had obtained only 16 samples.

The histograms of theoretical sampling distributions shown in Figure 8.1 are all symmetrical because they are all derived from binomial populations in which $p = q$. In binomial populations where p is *not* equal to q $(p \neq q)$ the theoretical sampling distributions will normally be skewed. Figure 8.2 shows a histogram, based on a theoretical sampling distri-

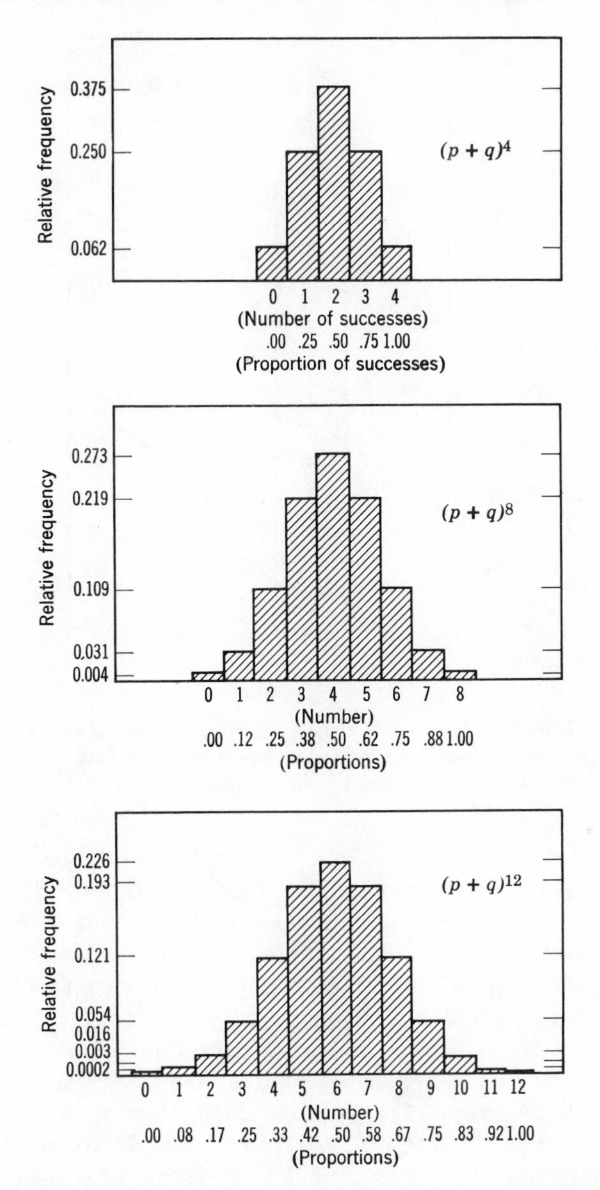

FIGURE 8.1 Theoretical relative frequency distribution of successes where the probability of success $(q) = .50$.

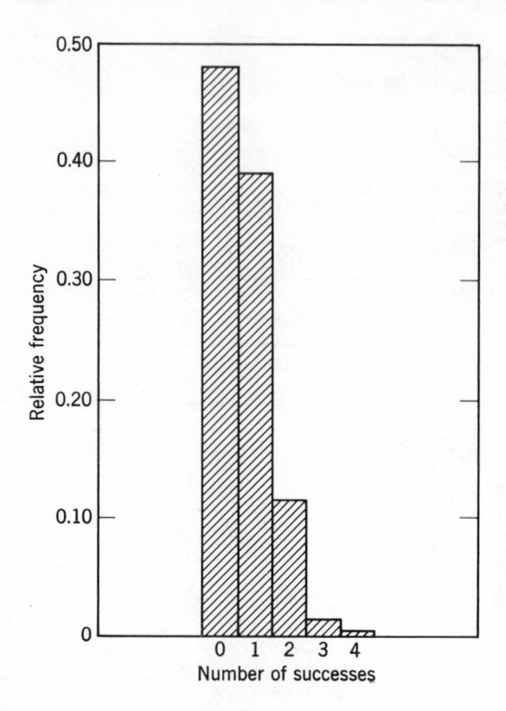

FIGURE 8.2 Theoretical relative frequency distribution of successes in blocks of four trials, where p (success) = 1/6, and q (no success) = 5/6. These relative frequencies (probabilities) were obtained from expanding $(q + p)^4$.

bution, with $n = 4$ drawn from a binomial population in which $p = 1/6$ and $q = 5/6$. This theoretical sampling distribution might represent the relative frequencies of obtaining different numbers of aces when 4 dice are tossed. Since a cube has six faces, the probability that a die will show an ace on any toss is 1/6, and the probability that it will not show an ace is 5/6. When these values are substituted into the binomial expansion of $(q + p)^4$, it[1] yields the following relative frequency for aces: 0 aces, $(5/6)^4 = .4822$; 1 ace, $4(5/6)^3(1/6) = .3858$; 2 aces, $6(5/6)^2(1/6)^2 = .1157$; 3 aces, $4(5/6)(1/6)^3 = .0154$; and 4 aces, $(1/6)^4 = .0008$. Note that the histogram in Figure 8.2, which represents this theoretical sampling distribution, is rather badly skewed. The skew will become even

[1] Note that we have *reversed* the order of p and q so that the expanded binomial is $(q + p)^4$. This was done so that the *number* of successes represented by the terms of the expanded binomial would increase as we read from left to right. This means that the first term gives the probability of zero successes and the last term the probability of n successes; otherwise, the procedures are just as they were before.

more pronounced if the difference between p and q is increased and n, the sample size, remains constant.

There is a factor which can reduce the skew in the theoretical sampling distribution when $p \neq q$. As the sample *size* increases, theoretical sampling distributions become more and more symmetrical regardless of the difference between p and q. We can illustrate this by a comparison of the histogram in Figure 8.2 with the one in Figure 8.3. Both represent theoretical sampling distributions drawn from binomial populations where $p = 1/6$ and $q = 5/6$, but the highly skewed histogram in Figure 8.2 is based on sampling with $n = 4$, while the minimally skewed histogram in Figure 8.3 is based on sampling with $n = 15$. If we increase sample size further, perhaps to $n = 60$, the skew would not be noticeable.

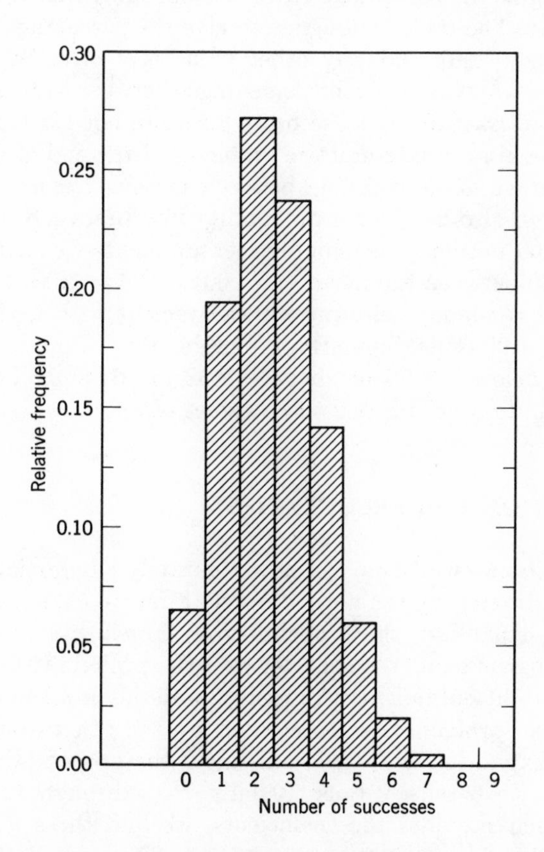

FIGURE 8.3 Theoretical relative frequency distribution of successes in blocks of 15 trials, where p (success) $= 1/6$, and q (no success) $= 5/6$. These relative frequencies (probabilities) were obtained from expanding $(q + p)^{15}$.

As n increases, the theoretical sampling distribution derived from any binomial not only becomes more symmetrical, but its form approaches that of a normal curve. This is true even when p and q are quite different. As a general rule of thumb if np (or nq, whichever is smaller) is at least 5 (if np and $nq \geqq 5$), the theoretical sampling distribution will be sufficiently close to the normal curve so that the properties of the normal curve may be used to solve problems involving the binomial. This is a very great convenience because we have already learned to use Table N to find the percentages of area encompassed by various sections of the normal curve. When the binomial is approximately normal, Table N can also be used to estimate areas of binomial sampling distributions.

You should recall from our discussion in Chapter 4 that Table N can be used to determine the percentage of the normal curve falling between any two values of z. The table is designed to give the percentage of the curve between M ($z = .00$) and any other value of z directly. Percentages above or below any value of z, or percentages between values of z, can be obtained by subtraction. If these procedures are not clear, review them and review the appropriate practice problems at the end of Chapter 4.

The percentage of cases falling between various z scores of a normal distribution can also be given a probability interpretation. For example, if we have 1000 normally distributed measurements we shall find 340 of them, or 34%, between the mean ($z = .00$) and $z = 1.00$; therefore, the probability of randomly selecting a measurement with a value of z between .00 and 1.00 is .34. Similarly, the probability of randomly selecting a value of z below -2.58 or above $+2.58$ is .01, since only 1% of a normal distribution lies beyond these values of z.

APPROXIMATION FOR FREQUENCIES

In this section we will show just how accurately a binomial distribution can be approximated by the normal curve. First we shall calculate, from the binomial expansion, the probability of throwing 8 *or more* tails in 12 tosses of a true coin. We shall let p = the probability of heads and q = the probability of tails. For the solution to the problem we must find the sum of the probabilities represented by the 9th through the 13th terms of the expansion $(p + q)^{12}$. This gives the probabilities of obtaining samples with 8 tails, 9 tails, 10 tails, . . . through 12 tails. Using Table B, which provides the coefficients, we find these five terms are $495p^4q^8 + 220p^3q^9 + 66p^2q^{10} + 12pq^{11} + q^{12}$. The *sum* of the probabilities for the series is .194.

Now we shall solve the same problem by using Table N and the normal curve approximation. The procedure is fairly straightforward: the point in the binomial sampling distribution representing 8 tails is converted into a z score. Then we find the proportion of area to the right of such z scores in a normal distribution. If the binomial is approximately normal this proportion should be very close to the exact probability of obtaining 8, or more tails, which we earlier calculated directly.

It is sometimes helpful in problems of this sort to sketch the binomial distribution and superimpose a normal curve upon it. Figure 8.4 shows such a sketch. Now, for this sampling distribution, we must find the z score equivalent to 8 tails. Since the formula for z is $(X - M)/\sigma$ we must find the mean and standard deviation of the number of tails in this binomial distribution.

In any binomial distribution the mean number of events for which p gives the probability is np; the mean number of events for which q gives the probability is nq. The formula is given below.

$$M_f = np \quad \text{or} \quad nq$$

FORMULA 8.1
Mean number of events in a binomial distribution

In this sampling distribution, where a sample consists of tossing 12 unbiased coins, the mean frequency of tails is $M_f = nq = 12 \times .5 = 6.0$. The accuracy of this answer should be apparent from an inspection of Figure 8.4. Keep in mind, however, that we have calculated the mean of a *theoretical* sampling distribution. This is the mean frequency of tails

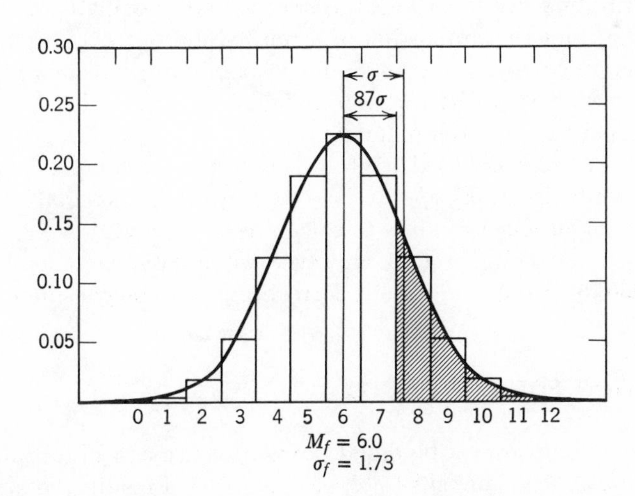

FIGURE 8.4 Normal approximation of a binomial distribution where $p = q$, and $n = 12$.

which will result if we obtain an indefinitely large number of tosses of 12 coins. It will not necessarily be the mean of any empirical sampling distribution.

The standard deviation of the frequency of events represented by p or q, in any binomial distribution, is given by $\sqrt{npq}$ where the symbols n, p, and q have the same meaning we have previously given them. Now, however, we must point to a change in terminology. The standard deviation of the *sampling distribution* of any sample statistic, such as a proportion, or mean, or correlation coefficient, is called the *standard error* of that statistic. We have a theoretical sampling distribution of this statistic (a frequency) so the standard deviation of the distribution of frequencies is called a standard error. The formula for this standard error is given below.

$$\sigma_f = \sqrt{npq}$$

FORMULA 8.2
Standard error of the frequency of events in a binomial distribution

In this example, where samples of 12 coins ($n = 12$) are tossed, the standard error of the theoretical sampling distribution is $\sigma_f = \sqrt{npq} = \sqrt{12 \times .5 \times .5} = 1.73$. For this binomial distribution, then, we have $M_f = 6.0$ and $\sigma_f = 1.73$. These parameters are shown on the normal curve approximating the histogram in Figure 8.4.

Using the parameters ($M_f = 6.0$, $\sigma_f = 1.73$) of this theoretical sampling distribution, we must find the z score that is equivalent to 8 tails, and then we find the proportion of a normal curve to the right of that point. This proportion will give us the probability of 8 or more tails, and should be very close to .194, the probability we obtained from summing the individual terms of the binomial.

Since this z score is based upon a distribution of sample frequencies, we shall symbolize it as $z_f = (X_f - M_f)/\sigma_f$. It seems that we finally have all of the necessary values to calculate z_f, since we have X_f given as 8 tails, $M_f = 6.0$ and $\sigma_f = 1.73$. However, we cannot use 8.0 as the appropriate value for X_f. An important correction has to be considered.

CORRECTION FOR CONTINUITY

When sampling from a binomial population we can obtain only *integral numbers* of heads or tails (1, 2, or 3, etc.); as a result, the abscissa of the sampling distribution represents a discrete or discontinuous function. The result is the saw-tooth appearance of the histogram in Figure 8.4.

The normal curve, by contrast, is based on a continuous scale of measurement consisting of a theoretically unlimited number of subdivisions between integers. A normal curve will, in theory, have a perfectly smooth outline. As a consequence of using a continuous normal distribution curve to *estimate* proportions of a discrete binomial distribution, we must make a correction in the value we use for X_f. This is called a correction for continuity.

When the binomial is approximated by a normal curve, we treat its abscissa *as if* it represented a continuous function. Thus, a truly discrete count of 8 tails is treated as if it occupied the *interval* 7.5 to 8.5 tails. The portion of the curve representing 8 or more tails is therefore assumed to begin at 7.5, and to include the right tail of the curve from that point. Consequently, the value we must substitute for X_f in the formula for z_f is not 8.0, but 7.5. (Had we wished to find the z_f equivalent for *more* than 8 tails the appropriate value of X_f would be 8.5, which is the upper limit of the 7.5 − 8.5 interval.) The value of z_f for our problem is $z_f = (X_f - M_f)/\sigma_f = (7.5 - 6.0)/1.73 = .87$. Using Table N we find that 30.78% of a normal distribution falls between M ($z = .00$) and $z = .87$. If we have 50.00% of a normal distribution above the mean, we shall find 50.00 − 30.78 or 19.22% in the upper tail of the distribution, the portion above $z = .87$. Dividing 19.22% by 100 to convert the percentage to a proportion, the normal approximation of the binomial yields a probability of .192 for 8 or more tails. This probability is represented by the shaded area in the right tail of Figure 8.4, and it is quite close to the value of .194 which we found earlier by the much more tedious process of summing the 9th through the 13th terms of the expansion of $(p + q)^{12}$.

In the last problem we could compare the accuracy of the normal approximation to a binomial distribution because the expansion of $(p + q)^{12}$, while tedious, was not prohibitive. In the next problem the only reasonable approach is to use the normal approximation.

Suppose we wish to find the probability of obtaining 28 or more questions right by chance on a 100-item multiple choice test where each item has five alternatives. This seems like a forbidding problem, as indeed it would be if we had to use the binomial expansion, but the solution is relatively easy if we use the normal approximation.

With five alternatives for each item, the probability of an incorrect choice is given by $q = 4/5$, and a correct choice by $p = 1/5$. Since there are 100 items, the appropriate binomial is $(q + p)^{100}$. We need the sum of the final 73 terms that represent the probabilities of 28 items correct through 100 items correct. (Of course, if a solution to the problem were actually attempted by expanding the binomial, it would be much easier to sum the first 28 terms and then subtract the result from 1.00.)

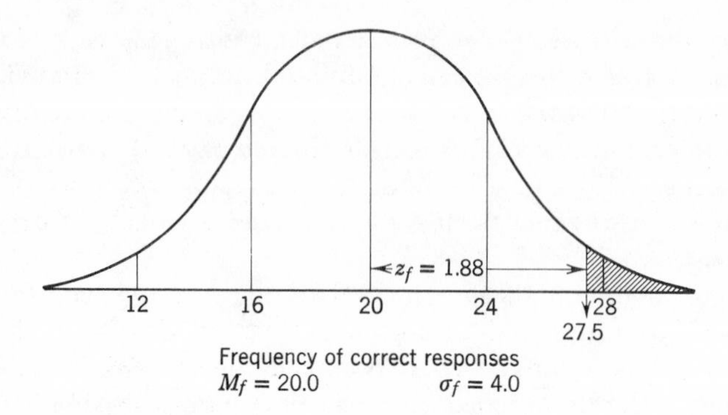

FIGURE 8.5 Normal curve approximating the binomial expansion of $(p + q)^{100}$, where $p = 4/5$ and $q = 1/5$.

Assume a theoretical sampling distribution based on samples of 100 items ($n = 100$). Our problem is to determine the proportion of samples that will contain 28 or more items answered correctly. A solution using the normal approximation is appropriate because although $p \neq q$, $n > 5$. In fact, since np is so large we can expect the approximation to be very good. Again, the first step is to find the parameters of the sampling distribution, M_f and σ_f, so that we can sketch the normal curve which approximates this binomial.

The mean number of correct responses in this theoretical sampling distribution is $M_f = np = 100 \times 1/5 = 20$, and the standard error is $\sigma_f = \sqrt{npq} = \sqrt{100 \times 1/5 \times 4/5} = 4.0$. At this point the investigator may avoid confusion if he sketches a normal curve with these parameters. Figure 8.5 shows such a curve.

We must find the proportion of this normal curve falling above a z score equivalent to 28 correct responses; this is actually 27.5, since we must correct for continuity. The z_f equivalent to a score of 27.5 is given by $z_f = (X_f - M_f)/\sigma_f = (27.5 - 20)/4 = 7.5/4 = 1.88$. The area from the mean to $z = 1.88$ is found from Table N to be 46.99%. Since 50% of the area of the curve is above the mean, the area in the portion of the tail above a score of 27.5 will be 50.00 − 46.99, or 3.01%. This is the portion of the curve in the shaded right tail of Figure 8.5. The percentage is converted to a proportion or probability value when it is divided by 100, so the probability of getting 28 or more items correct by chance under these circumstances is .03.

The rule of thumb which states that the normal approximation can be used when np (or nq, if nq is smaller than np) is equal to or greater than 5, is just a rule of thumb. If $p = 1/5$ and $n = 25$, we could presumably use the normal approximation because np would be equal to 5. However, even the slight skew in this sampling distribution would yield substantial error in our estimates of probability *if* we had to make use of the extreme tails of the distribution. For example, if we calculate the probability of getting all 25 items *wrong* on a five-alternative, 25-item multiple choice test, we shall need to evaluate the first term of the expansion $(q + p)^{25}$ where the probability of getting an item wrong by chance is $q = 4/5$. The value of $(4/5)^{25}$ is .004. If we solve the same problem with the normal approximation we shall find $M_f = np = 5$ and $\sigma_f = \sqrt{npq} = 2$. This will yield $z_f = (0.5 - 5.0)/2 = -4.5/2 = -2.25$. The probability of obtaining z values below -2.25 is .012. This probability, based on the normal approximation, is three times as large as the correct value obtained from the binomial expansion.

The reason for the error in this particular problem is not difficult to understand. We dealt with a very extreme tail of the normal approximation and, at that point, even a moderate skew in the binomial distribution produces a very poor fit with the precisely symmetrical normal curve. This proportion of error would be substantially reduced if the problem had required us to approximate the binomial distribution over its center section, or if the problem had required the use of a larger portion of the curve than the very small area below $z = -2.25$.

Actually, regardless of the values of p and q, the binomial distribution approaches the normal curve as a limit as the exponent of the binomial approaches infinity. For practical purposes, however, we must consider not only the value of np (or nq, if nq is smaller than np), but also the section of the normal curve used in the approximation. The rule of thumb that np must be equal to or greater than 5 is a rule which must be used with caution.

THE APPROXIMATION FOR PROPORTIONS

The problems discussed so far have all been concerned with the *frequency* of some event's occurrence in a sample. We determined the probability of obtaining samples with eight or more tails when 12 coins are thrown, and we found the probability of correctly guessing the answers to 28 or more items on a five-alternative, 100-item, multiple choice test.

In these and similar problems, the abscissa of the distribution represents the frequency or number of events occurring in each sample. There are other problems where it is more convenient for the abscissa to represent the proportion of events occurring in each sample. In Figure 8.1, which shows histograms of several different sampling distributions the abscissas are scaled for *both* frequencies and proportions. No new principle is involved here; it is only a change in an arbitrary scale for the abscissa.

However, if we are to fit a normal curve to a sampling distribution of proportions, we must find M_p and σ_p, the mean and standard error of the proportions of events occurring in each sample. Since frequencies are converted to proportions when they are divided by n, we can simply divide the formulas for M_f and σ_f by n to obtain formulas for M_p and σ_p. Thus we have two formulas.

$$M_p = \frac{M_f}{n} = \frac{np}{n} = p$$

FORMULA 8.4
Mean of the theoretical sampling distribution of proportions when samples of size n are drawn from any binomial population

and

$$\sigma_p = \frac{\sigma_f}{n} = \frac{\sqrt{npq}}{n} = \sqrt{\frac{pq}{n}}$$

FORMULA 8.5
Standard error of the proportion of events in a theoretical sampling distribution drawn from a binomial population

We shall use these formulas to determine the probability of obtaining .67 (8) or more tails when 12 coins are tossed. We solved the same problem earlier using M_f and σ_f, and we found the probability of 8 or more tails to be .192. Now we shall base our solution on M_p and σ_p as a check on the accuracy of our previous answer.

The first step is to determine M_p which is simply p, the proportion or probability of tails in the hypothetical population of tosses; $p = .5$, so $M_p = .5$. The value of σ_p will be $\sqrt{(pq)/n} = \sqrt{(.5 \times .5)/12} = .144$. We now have a mean and standard error for the proportion of tails in this binomial distribution. A normal curve which approximates this distribution is sketched in Figure 8.6.

The area in which we are interested falls to the right of X_p, a point representing .67 (8) or more tails out of 12 coins tossed. However, we must still correct for continuity, so the appropriate proportion for X_p will be 7.5/12, or .625.

We are now in a position to determine z_p, which is analogous to z_f, but uses M_p and σ_p instead of M_f and σ_f. We have: $z_p = (X_p - M_p)/\sigma_p = (.625 - .500)/.145 = .87$. The proportion of a normal curve falling above this point is .192, which is the same as the probability we obtained when the problem was solved on the basis of frequencies.

Up to this point we have confined our attention to artificial situations, such as coin tossing or dice throwing, in order to illustrate the principles

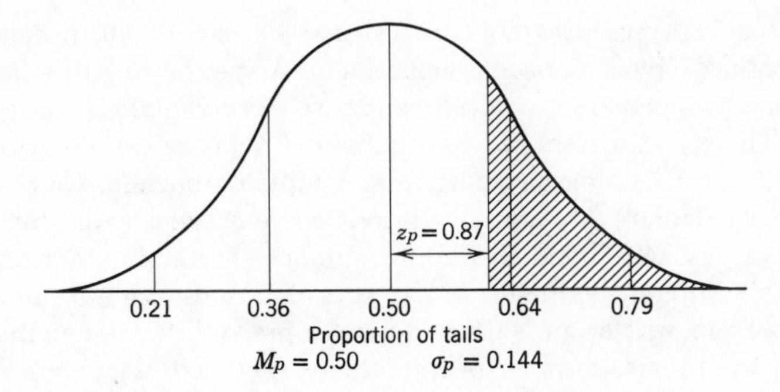

FIGURE 8.6 Normal approximation of a binomial distribution, where $p = q$, and $n = 12$.

involved. Now let us apply these principles to somewhat more realistic research situations.

HYPOTHESIS TESTING

Suppose we have the following problem: as Director of Alumni Affairs for a college we are asked to determine if more than 40% of the alumni who have graduated within the past ten years will support a special alumni scholarship program. We have a list of about 8000 alumni, but decide that contacting this entire population is much too expensive unless there is very good evidence that more than 40% of them favor the proposition. Keep in mind that we are looking for evidence that p, the proportion of alumni favoring the scholarship program, is greater than .40.

There are three mutually exclusive and exhaustive possibilities for this population proportion; $p = .40$, $p < .40$, and $p > .40$. If we can show that $p = .40$ and $p < .40$ are very unlikely, then the remaining possibility $p > .40$ will have strong support. The strategy is to obtain a sample from the population of alumni and find the proportion of alumni in this sample who favor the scholarship program. Then we determine the probability that such a sample proportion could have arisen by chance if $p = .40$ or $p < .40$. If this probability is very low, we reject these alternatives and conclude that the proportion of alumni favoring the program does indeed exceed .40.

You will note that we seem to have combined $p = .40$ and $p < .40$ into a single null hypothesis, $p \leqq .40$. This hypothesis will be rejected if we obtain a sufficiently improbable sample proportion. Actually, however, the only hypothesis we can test *directly* is $p = .40$, because only this hypothesis gives a specific value for p. A specific hypothesis for the population proportion is required before we can calculate a value for M_p and σ_p. These parameters must be known so that we can determine the probability of obtaining any particular sample proportion. Quite clearly, we cannot calculate M_p and σ_p without some specific value for p. The hypothesis $p < .40$ yields an infinite number of sampling distributions, each with a different value for M_p and σ_p. As a result, any obtained sample proportion will occur with a different probability, depending upon which of the infinite number of hypotheses ($p < .40$) we choose to test, and we obviously cannot test them all.

This difficulty is handled in a relatively straightforward way. We combine the hypothesis $p < .40$ and the hypothesis $p = .40$ into *one* hypothesis $p \leqq .40$, and then use $p = .40$ to set the parameters of the theoretical sampling distribution. For this hypothesis to be rejected, the sample proportion must fall into the *right* or *upper* tail of the theoretical sampling distribution. A sample proportion is evidence to reject $p \leqq .40$, at some fixed level of α, only if it is sufficiently larger than .40 to fall in the upper 5% or 1% of the distribution. If a sample proportion is in the upper 5% of the sampling distribution for $p = .40$, its occurrence will be even more improbable in any of the infinite number of sampling distributions which might be generated by $p < .40$.

Now that we have outlined the strategy, let us proceed with the calculations. Suppose we question a randomly selected sample of 100 alumni and find that 50 of them support the scholarship program. Does this sample proportion, $X_p = .50$, fall far enough into the upper tail of the theoretical sampling distribution for us to reject $p \leqq .40$ at the 5% level of significance? The theoretical sampling distribution is based on the expansion of $(q + p)^{100}$, where $p = .40$ and $q = .60$. Using the formulas for M_p and σ_p, we find $M_p = p = .40$, and

$$\sigma_p = \sqrt{(pq)/n} = \sqrt{(.60 \times .40)/100} = .049.$$

The normal curve which approximates this theoretical sampling distribution is sketched in Figure 8.7.

Now we can determine the probability of obtaining a sample with 50 or more favorable responses. The probability of this event is represented in Figure 8.7 by the area falling to the right of .495, the value of X_p, corrected for continuity, which is equivalent to 50 or more agreements in a sample of 100. Since sample proportions are not continuous even when $n = 100$, we should correct for continuity, although with such a large sample the correction makes little practical difference. In this problem we make the correction by using the lower limit of the interval repre-

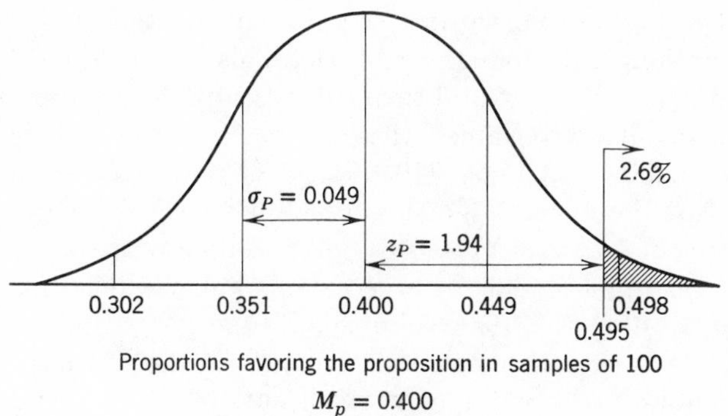

FIGURE 8.7 Normal approximation of a binomial distribution, where $p = .40$, and $n = 100$.

senting 50 agreements, which is 49.5 agreements. When this is converted to a proportion of 100 subjects we have $X_p = .495$. The z equivalent is

$$z_p = (X_p - M_p)/\sigma_p = (.495 - .400)/.049 = .095/.049 = 1.94.$$

The shaded area in the right tail of the distribution represents the percentage of sample proportions with 50 or more favorable responses when the proportion of favorable responses in the population is actually .40. From Table N we find that only 2.6% of the normal curve falls above $z = 1.94$. According to the criteria previously agreed on, we will reject the null hypothesis that $p \leqq .40$, accept the alternative hypothesis that $p > .40$, and continue the alumni scholarship project.

Instead of calculating the probability of obtaining a particular sample and then rejecting the null hypothesis if the *probability* is less than a previously specified value, we can use a slightly different approach. We can calculate the sample proportion which would just reach significance at the 5% level, and then any obtained sample proportion which reaches or exceeds this value will lead to the rejection of the null hypothesis. When this procedure is used we do not actually calculate the probability of obtaining a particular sample proportion; we simply establish a critical value for the sample proportion, and if an obtained proportion exceeds this value we know it will be improbable enough to reject the hypothesis that produced the theoretical sampling distribution.

We shall apply this procedure to the previous problem. We first find the value of z that cuts off the upper 5% of the normal curve. This can be determined from Table N and is $z = 1.645$. Second, we find the proportion of favorable responses in a sample that, for this theoretical sampling distribution, will produce $z = 1.645$. This can be found by solving $z_p = (X_p - M_p)/\sigma_p$ for X_p, and then substituting the appropriate values for M_p and σ_p. We determined these values earlier so we have $X_p = M_p + z_p\sigma_p = .400 + 1.645 \times .049$. Figure 8.8 shows the same curve as Figure 8.7, but the upper 5% of Figure 8.8 has been shaded to show the region of rejection above $X_p = .481$. Any sample proportion, corrected for continuity, which equals or exceeds this value will result in the rejection of $p \leqq .40$ at the 5% level of significance.

If we decide to find the region of rejection at the 1% level, we must find from Table N the value of z which cuts off the upper 1% of the normal curve. We substitute this value, $z = 2.33$, for $z = 1.645$, and then we proceed as before. Solving for X_p with $z = 2.33$, we have $X_p = M_p + z_p\sigma_p = .400 + 2.33 \times .049 = .514$. Any sample proportion, corrected for continuity which is at least .514, will result in the rejection of $p \leqq .40$ at the 1% level.

Let us review some of these principles with a different problem. Suppose we know that 20% is the normal survival rate among a strain of laboratory rats subjected to a certain environmental stressor. However, we have developed a special program that we feel provides a degree of immunity to this stressor. We shall test it by obtaining a sample of 400 rats, introducing the treatment and then subjecting them to the stressor. If the treatment is effective, then this sample of 400 S's has come from a population in which the survival rate exceeds 20%. We can accept this hypothesis if we can reject the hypotheses that $p = .20$ and $p < .20$. The null hypothesis, which actually provides the theoretical sampling distribution is $p = .20$, but of course we must also be able to reject

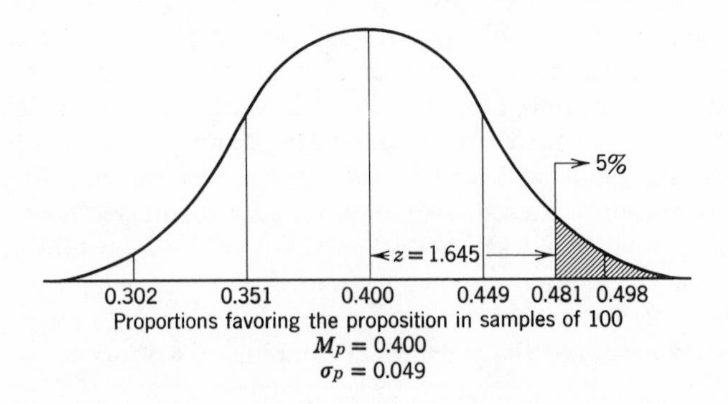

FIGURE 8.8 Normal approximation of a binomial distribution, where $p = .40$, and $n = 100$.

$p < .20$ before we can conclude that our treatment is effective. Only if the sample proportion surviving the stressor is well *above* .20 will we reject the hypothesis that $p \leqq .20$. In this problem too, only one tail of the curve holds a region of rejection.

What proportion of 400 rats must survive the stressor for us to reject the hypothesis that $p \leqq .20$ at the 1% level? The first step is to determine M_p and σ_p for the binomial distribution based on the expansion of $(q + p)^{400}$, where $p = .20$ and $q = .80$. We have $M_p = p = .20$, and $\sigma_p = \sqrt{(pq)/n} = \sqrt{(.80 \times .20)/400} = .020$. We must now find the z score that marks off the upper 1% of the normal curve. Table N gives this as $z = 2.33$. The critical region for sample proportions is given by: $X_p = M_p + z_p\sigma_p = .200 + 2.33 \times .020 = .247$. If the obtained sample proportion of surviving rats equals or exceeds .247, we will reject the hypothesis that $p \leqq .20$, accept the hypothesis that $p > .20$ and conclude that our treatment does provide a degree of immunity to stress.

ONE- AND TWO-TAILED TESTS

In each of the preceding problems evidence to reject the null hypothesis consisted of obtaining a sample proportion which fell into a previously specified tail of the sampling distribution. In the last problem, we could have rejected $p \leqq .20$ only if a sample proportion had been obtained that equaled or exceeded .247. In the first problem, we could have rejected $p \leqq .40$ only if a sample proportion had been obtained which equaled or exceeded .514. In these tests of significance, the sample proportion had to fall into a predetermined tail of the theoretical sampling distribution for the null hypothesis to be rejected. This, of course, was because the experimental hypothesis had specified the *population* proportion to be greater or smaller than some particular value. Tests of significance in which the region of rejection falls in only one tail or side of a distribution are called one-tailed, one-sided, or directional tests of significance.

There are also two-tailed, two-sided, or nondirectional tests of significance, and these, as the name implies, utilize both tails to establish a critical region for the rejection of a null hypothesis. Suppose we wish to show that some population proportion is *not* .60. Support for this hypothesis can be obtained by showing that $p = .60$ is highly improbable. If the null hypothesis specifies that the population proportion *is* .60 then sample deviations in *either direction* can result in its rejection. Under these circumstances the most deviant 5% of the curve consists of the upper *and* lower 2.5%, the areas above $z_p = 1.96$ and below $z_p = -1.96$.

We can reject this null hypothesis at the 5% level if the sample proportion falls in either of these regions of rejection. For significance at the 1% level we require the sample proportion to fall in the upper or lower $\frac{1}{2}$% of the curve, above $z_p = 2.58$ or below $z_p = -2.58$.

It is sometimes a statistical convention to refer to significance at a *level* when a two-tailed test is used, and significance at a *point* for a one-tailed test. You should be aware of the convention, but it has little to recommend it and we shall not follow it in this text.

We will have more to say about one- and two-tailed tests later. For the moment, let us consider an example of a two-tailed test. Suppose we have called for volunteers to serve in an experiment and we want to know if, on the basis of the number of male volunteers in the sample, we must reject the hypothesis that men and women are equally likely to volunteer. Let us assume that we have obtained 80 volunteers and 46 are male. Do we reject $p = .50$ at the 1% level of significance? Since the last two examples made use of theoretical sampling distributions of sample proportions, we shall work with sample frequencies in this problem. If the population of volunteers is evenly divided between males and females, then the mean frequency of males in samples of 80 is $M_f = np = 80 \times .50 = 40$. The standard error of this sampling distribution is

$$\sigma_f = \sqrt{npq} = \sqrt{80 \times .50 \times .50} = 4.47.$$

We have sketched a normal distribution, with $M_f = 40$ and $\sigma_f = 4.47$, in Figure 8.9.

The next step is to find the frequency of males, X_f, which will just permit the rejection of $p = .50$ at the 1% level. There will be two such values, one at the beginning of the upper $\frac{1}{2}$% of the curve and one at the

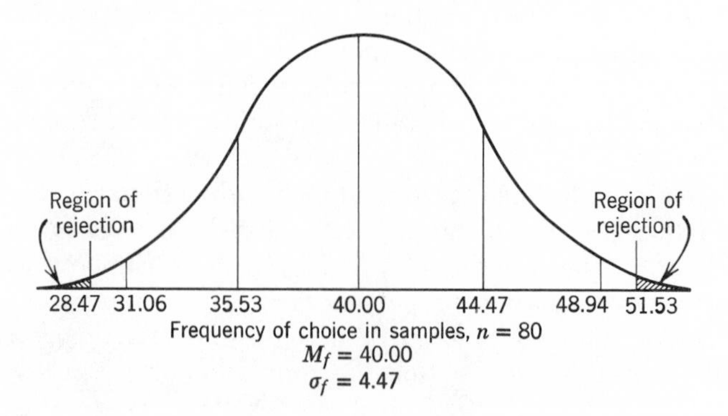

FIGURE 8.9 Normal approximation of a binomial distribution, where $p = q$, and $n = 80$.

beginning of the lower $\frac{1}{2}\%$. These values of X_f can be found by solving $z_f = (X_f - M_f)/\sigma_f$ for X_f. With $z_f = \pm 2.58$, and with M_f and σ_f taking the values already calculated, we have $X_f = M_f \pm z_f\sigma_f = 40 \pm 2.58 \times 4.47 = 28.47$ and 51.43. The sample frequency of 46 (45.5 when corrected for continuity) is well inside these limits; it does not provide satisfactory evidence against the hypothesis that men and women are equally likely to volunteer for the experiment.

CONFIDENCE INTERVALS

We have described the general procedure for testing a specific hypothesis about the proportion of events in a binomial population on the basis of the proportion of events found in a sample. The bulk of research in behavioral science involves hypothesis testing, and this topic will continue to be the focus for much of our discussion throughout this text. However, there are also situations in which the investigator may not have any preconceived hypothesis about the population proportion, but he may wish to estimate this parameter from the information provided by the sample. As an illustration, we may wish to estimate the proportion of people in some population who will vote "yes" on a particular issue. We can estimate this population proportion from a sample proportion. If we find that .60 of a randomly obtained sample favors the issue, then our best estimate of the population proportion favoring the issue will also be .60. This procedure, by which we use a sample to predict a specific value for a population parameter, is called point estimation.

Point estimation has the advantage of being a very specific prediction, but any particular point estimate also has a most unfortunate disadvantage. It is not very likely to be *exactly* right even though it is more likely to be right than any other point estimate. Let us illustrate. Suppose we have an urn containing thousands of pellets which are either black or red. We don't know the proportion of black pellets in this population, so we estimate it from a sample. We draw a random sample of 100 pellets and find that 54 are black and the remainder red. From this information, our best estimate of the proportion of black pellets in the population will also be .54. However, *if* the population proportion is in fact .50, it would be quite possible to obtain a sample proportion of .54. As a matter of fact, even if the population proportion were .54 we would not expect to obtain sample proportions of .54 very often, although this proportion would tend to occur more often than any other. You may recall a problem illustrating this point which we solved in the chapter on probability. There we found the probability of throwing exactly seven heads when tossing 14 unbiased coins. In this case, the probability that the sample proportion of heads will exactly match the population proportion of heads was only .209. Although this probability is not very great, it is greater than the probability of obtaining any *other* proportion of heads.

The investigator may not be satisfied with a point estimate. Instead, he may prefer to specify a range of values within which the parameter will have some specified probability of occurrence. For example, suppose we obtain a sample based on 200 cases in which the proportion of events is .60. Of course .60 will be our *best* estimate for the proportion of events in the population from which the sample comes, but it is also quite reasonable to expect that the population proportion could be .59 or .61, or even .58 or .62. We might prefer to specify a *range* of values within which the population proportion is quite likely to fall. This may be preferable to a specific point estimate with its rather uncertain reliability. If we want to be quite confident that the interval includes the population proportion, then it should be a rather wide interval extending a considerable distance on either side of the point estimate. If we are willing to reduce our confidence somewhat, we will specify narrower limits for the interval. Of course, we could set the confidence interval so that its probability of containing the population proportion will approach certainty, perhaps being as high as .9999, but if we do this, the absolute size of the interval may be so large it would provide us with very little useful information about the parameter. The convention is to determine confidence intervals which have a probability of either .95 or .99 of containing the parameter in question. These are called the 95% and 99% confidence intervals.

The procedure for determining the upper and lower limits of a confidence interval (called the fiducial limits) is not too difficult. One simply uses the point estimate as a base, then adds and subtracts 1.96 times the standard error for the 95% confidence interval, or 2.58 times the standard error for the 99% confidence interval. For this problem, where .60 of a sample of 200 favored an issue, the fiducial limits of the 95% confidence interval are given by the following calculations.

$$X_p \pm 1.96\sigma_p$$

$$.60 \pm 1.96\sqrt{\frac{.60 \times .40}{200}}$$

$$.60 \pm .0679$$

Thus the fiducial limits for the 95% confidence interval are .532 and .668. This means that the probability is .95 that the population proportion is between .532 and .668.

Suppose we have drawn a sample with $n = 100$ and we find that the sample proportion is .80. We can determine the fiducial limits of the 99% confidence interval by the following formula.

$$X_p \pm 2.58\sigma_p$$

$$.80 \pm 2.58\sqrt{\frac{.80 \times .20}{100}}$$

$$.80 \pm .103$$

Thus, for this problem, the fiducial limits of the 99% confidence interval are .697 and .903. This means that when such a sample has been obtained, the probability is .99 that it was drawn from a population in which the proportion of events was between .697 and .903.

SAMPLING FROM FINITE POPULATIONS

We have now illustrated several types of problems in which the binomial distribution (or a normal curve approximating it) served to provide tests of significance. A characteristic common to all of these problems and examples has been the very small ratio of sample size to population size. In most of the illustrations we have hypothesized an infinite binomial population, and when this has not been the case we have specified populations very much larger than the samples that have been drawn from them. You may have been surprised to find that we used the same formulas to test hypotheses about population proportions whether the population was infinite, or consisted of some comparatively small number of cases, as it did in the example involving a population of 8000 alumni. Actually, when we have sampled without replacement, as we ordinarily do in a research situation, a correction in the standard error becomes important only if the sample constitutes a substantial proportion of the population from which it was drawn. The correction term by which any standard error should be multiplied when the sample size is 5% or more of the population, is given below.

$$c = \frac{N - n}{N - 1}$$

FORMULA 8.6
Correction term for the standard error when sample size exceeds 5% of the population and sampling is without replacement

From Formula 8.6 we find that when samples of 25 are drawn from a population of 500 cases, the correction terms yields $c = .976$. Under these conditions, where the sample is 5% of the population, the standard error will not be modified very much. It should be obvious that when the sample is smaller than 5% of the population, the correction term becomes relatively insignificant. This is a very difficult thing for most laymen to understand because it means that a sample of 1000 will predict the proportion of events in a population of 100,000 very nearly as well as for a population of 100 million. The value of the correction term under these two circumstances will be left as an exercise for the student.

POWER AND THE TYPE II ERROR

In Chapter 6, "The Nature of Research," we commented at length on the Type I error. A Type I error is made when we reject the null hypothesis because of a deviant sample when, in fact, the null hypothesis is true. The probability of making such an error, symbolized by α, is set by the experimenter when he determines the level of significance required for the rejection of the null hypothesis.

A Type II error occurs if we fail to reject the null hypothesis and the hypothesis is in fact false. The probability of making this error, symbolized by β, is a function of sample size, the level of α set by the experimenter, the true value of the parameters, and the particular test of significance used by the investigator. Other things being equal, β becomes a measure of the sensitivity, or power, of a test of significance. If β is high, it means that we have little chance of rejecting a false null hypothesis. If β is low, it means that the test of significance will probably produce a "significant" difference *if* the null hypothesis is really false. The power of a test of significance is defined as $(1 - \beta)$, where β is the probability of a Type II error.

In a previous example we found that if 46 males occur in a sample of 80 we cannot reject the hypothesis that males and females are evenly divided in the population from which the sample was drawn. Let us determine the probability of having made a Type II error in this situation if the proportion of males in the population is, in fact, .60. We shall calculate β for this example on the assumption that the hypothesized value of p is not .50, but is actually .60. The question we are attempting to answer is this, "What is the probability of failing to reject $p = .50$, with a sample $n = 80$, when the true value is $p = .60$?" Figure 8.10 will help you to understand the principles involved. If we follow it carefully, we see that the normal curve A is simply a redrawing of Figure 8.9. It represents a smoothed theoretical expected sampling distribution derived from the *false* hypothesis $p = .50$. The critical sample frequencies required for

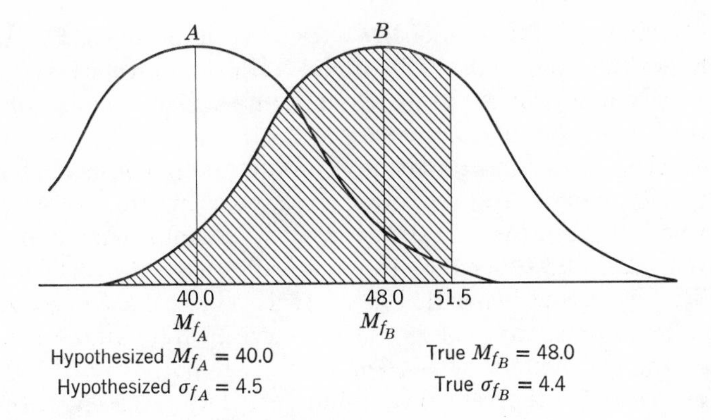

FIGURE 8.10 Normal approximations of binomial distributions, where $p = .50$ (A), and $p = .60$ (B), with $n = 80$.

the rejection of this hypothesis were determined in the last problem as 28.5 or less, and 51.5 or more. Sample frequencies, corrected for continuity, which fall within these limits will *not* result in the rejection of the hypothesis $p = .50$.

The normal curve B approximates the theoretical sampling distribution based on the *true* hypothesis, $p = .60$. For this curve we have $M_{f_B} = np = 80 \times .60 = 48$, and $\sigma_{f_B} = \sqrt{npq} = \sqrt{80 \times .60 \times .40} = 4.4$. Knowing M_{f_B} and σ_{f_B} we can find the proportion of the *true* curve which falls below 51.5. This represents the proportion of samples which will *not* result in the rejection of the false null hypothesis $p = .50$.

In order to find the proportion of curve B below 51.5, we first find the value of z_{f_B} which is equivalent to a frequency of 51.5. We have $z_{f_B} = (X_{f_B} - M_{f_B})/\sigma_{f_B} = (51.5 - 48.0)/4.4 = .80$. The proportion of curve B between M_{f_B} and $z_{f_B} = .80$ can be determined from Table N; it is 29%. Consequently, the proportion of the curve falling below 51.5 is .79. Since any sample frequency of less than 51.5 will not lead to the rejection of the null hypothesis and will, therefore, produce a Type II error, it follows that β, the probability of a Type II error, is .79.

Please pay careful attention to the variables which influence the size of β. Notice that if the true population proportion were .80, the true theoretical sampling distribution, B, would move much further to the right and a far smaller proportion of curve B would fall below 51.5. This would greatly reduce β and increase the power of the test against this alternative hypothesis. In addition, σ_f would be smaller if the true p were larger, and this would further reduce β. The power of a test of significance is obviously a function of the alternative hypothesis. For an alternative hypothesis, $p = .60$, the binomial test of significance with $n = 80$ has relatively little power, but if the alternative hypothesis is $p = .80$, the power is substantially increased.

Sample size is a second important consideration. We have shown the power of the binomial test to be relatively weak when $n = 80$, and the alternative hypothesis is $p = .60$. However, if sample size is increased, say to $n = 320$, and other factors remain the same, β is decreased substantially. Figure 8.11 helps illustrate how this happens. Curves A and B are both based on samples of $n = 320$. Curve A approximates the theoretical sampling distribution based on the false hypothesis $p = .50$. Curve B approximates the theoretical sampling distribution based on the alternative and presumably true hypothesis $p = .60$. The means and standard errors of these distributions have been calculated below.

$$M_{f_A} = np \qquad\qquad M_{f_B} = np$$
$$= 320 \times .50 \qquad\qquad = 320 \times .60$$
$$= 160 \qquad\qquad = 192$$
$$\sigma_{f_A} = \sqrt{npq} \qquad\qquad \sigma_{f_B} = \sqrt{npq}$$
$$= \sqrt{320 \times .5 \times .5} \qquad\qquad = \sqrt{320 \times .6 \times .4}$$
$$= 8.9 \qquad\qquad = 8.8$$

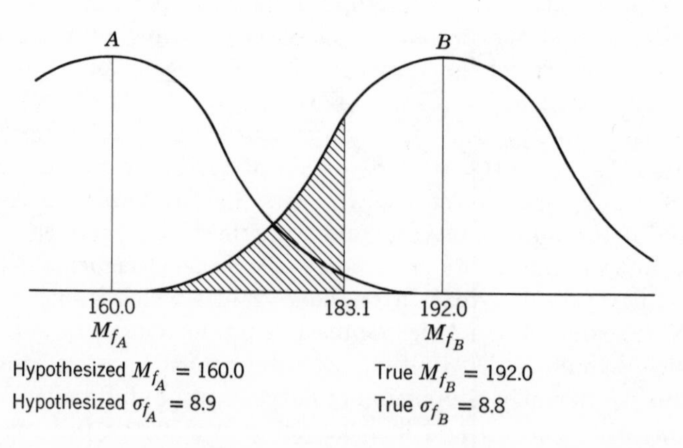

FIGURE 8.11 Normal approximations of binomial distributions where $p = .50$ (A), and $p = .60$ (B), with $n = 320$.

As we did in the preceding problem, we shall first determine from distribution A, the frequency of males in samples of 320 that would place a sample in the region of rejection, the upper or lower $\frac{1}{2}\%$ of the sampling distribution. The calculations below show that sample frequencies must be less than 137 or more than 183.

$$\begin{aligned} X_{f_A} &= M_{f_A} \pm 2.58\sigma_{f_A} \\ &= 160 \pm 2.58 \times 8.9 \\ &= 137 \text{ and } 183 \end{aligned}$$

The next task is to determine the z score equivalent to a frequency of 183 using the parameters of theoretical sampling distribution B, which is the *true* sampling distribution. Once we know this z score we can determine the proportion of samples with less than this frequency, and that proportion will be β, the probability of a Type II error. The required calculations appear below.

$$z_{f_B} = \frac{183 - 192}{8.8} = -1.02$$

Notice that the shaded portion of curve B now falls below $z_{f_B} = -1.02$. Only 15% of curve B falls below this point so the probability of a Type II error is reduced from .79 to .15 by our fourfold increase in sample size.

A third factor in determining β is the level of α, set by the experimenter. In the problem above we used the 1% level of significance and a two-tailed test. If we had used the 5% level, the value 183 would have become $M_{f_A} + 1.96\sigma_{f_A} = 160 + 1.96 \times 8.94 = 177.5$. The z_{f_B} equivalent of this value is $(177.5 - 192)/8.8 = -1.65$, and the percentage of curve B falling below $z_{f_B} = -1.65$ is given by Table N as 4.9%. If we had set $\alpha = .05$ and used an $n = 320$ instead of 80, we would have substantially increased the probability of rejecting $p = .50$ if the true population proportion had been .60.

A final and very important factor determining power is the test of significance itself. For a given n, level of α, and particular alternative hypothesis, some tests of significance will be much more likely than others to produce Type II errors. At this point in your study of statistics, when we have discussed only the binomial distribution as a test of significance, a comparison of the relative power of two tests would be premature. Later, when we have discussed a variety of tests, we shall discuss why some are more powerful than others. The investigator should always select the most powerful test of significance suitable for his data.

ASSUMPTIONS FOR THE USE OF THE NORMAL APPROXIMATION

In order to use the normal approximation of the binomial as a test of significance, we must be able to make certain assumptions about our data. If these assumptions cannot be met, then the shapes of the theoretical sampling distributions will not be approximately normal and consequently, probabilities derived from Table N will be in error.

As we mentioned earlier in the chapter, the lesser of the two values, np or nq, should be at least 5 before we can make use of the normal approximation. A more conservative, safer, minimum is sometimes suggested as 10, particularly when $p \neq q$ and a one-tailed test is to be used. This is occasioned by the fact that when $p \neq q$, and n is small, we have a good deal of skew in the binomial; but as n increases the binomial becomes progressively more "normal" even if $p \neq q$.

When the binomial $(p + q)^n$ is used to determine probabilities, the n refers to *independent* events. If one coin is thrown four times it is quite reasonable to assume that the probability of a head on toss 2 is not influenced by what happened to the same coin on toss 1. If four different coins are tossed, it is equally reasonable to assume that the probability of a head on coin 4 is uninfluenced by the outcome of coin 1. *Only* under these circumstances where the outcomes are independent does the binomial provide an appropriate probability model. The assumption of independence is so easily met in illustrations of coin and dice throwing that it is sometimes forgotten when the binomial is applied to behavioral data. Consider the following example. Suppose we wish to determine if games A and B are equally popular with a population of twelve-year old boys. If we select a sample of ten children and have each child choose game A or game B *without knowing the choices made by any of the others,* we shall have the $n = 10$ independent events required to use the expansion of $(p + q)^{10}$ as a probability model. If A and B are equally popular the probability of 10 children *independently* choosing A is only 1/1024.

On the other hand, suppose we allow the children to choose *while they are in a group,* so that each child knows which game the others have chosen before he makes his own choice. Under these circumstances we may find that the more dominant children choose first and the submissive children then tend to make the same choice. If we find that all 10 children choose game A, it may be because game A is, in fact, more popular among children making their choices individually; or it may be that the games are equally popular, but children choosing last are influenced by those choosing first. We have no way to decide which of these alternatives is the appropriate reason for rejecting the binomial model. We shall have more to say on the subject of independence when we examine Chi square, the test of significance to be discussed in the next chapter.

Exercises

1. What percentage of the normal curve falls above $z = 1.76, 2.12, -.71$, and $-.06$?
2. Suppose we have a program which we hypothesize will increase the proportion of students successfully completing an examination. The proportion presently completing it is .40. Evidence in favor of this program consists of rejecting the null hypothesis that $p \leqq .40$. We find that 31 of 60 students pass the examination. Can the null hypothesis be rejected at the 5% level of significance? What value of z_f is obtained? What value of z_f is required to reject the hypothesis?
3. We hypothesize that the proportion of events in a population is .50. We obtain a sample of 64 cases and find the proportion of events to be .375. Should the hypothesis be rejected at the 5% level of significance? What value of z_p is obtained? What value of z_p is required to reject the hypothesis? Remember to correct for continuity.
4. A theory of genetics suggests that when crossing a certain species of large- and small-flowered plants, we should expect 25% of the next generation to be small-flowered and the remaining 75% large-flowered. Actually, we find 20 out of 100 plants are small-flowered. (a) Give a statement of the null hypothesis. (b) Should this be a one- or two-tailed test of significance? (c) What value of z_f is obtained? (d) What value of z_f is required for significance at the 1% level?
5. Will failure to correct for continuity increase or decrease Type I errors? Explain your answer.
6. A deck of ESP cards consists of 5 suits with 5 identical cards in each suit. What number of cards must S call correctly before we can assume (1% level) that he has ESP?
7. In a sample of 200 voters, 119 prefer candidate A. Can we safely forecast victory (5% level) for this candidate in the next election?
8. We find 70% of paroled convicts meeting certain criteria will violate their parole within three years. A randomly selected group of 40 are given special assistance, and of these 19 violate their parole within a three-year period. Is this evidence (1% level) that the special assistance was effective?
9. We obtain a sample of 200 cases in which there are 30 type A events. What are the fiducial limits at the 5% level for the proportion of type A events in the population from which the sample was drawn?

CHAPTER 9

Chi Square

The binomial distribution and its normal approximation provide a test of significance for hypotheses about dichotomous data. It is an appropriate test to use when observations can be classified into one of two possible categories such as "yes–no," "male–female," or "correct–incorrect." When data can be classified in *more* than two categories, the binomial no longer provides a test of significance. For example, if a response can be classified as occurring "always," "often," "sometimes," or "never," the appropriate test of significance is called chi square (χ^2).

THE CHI SQUARE DISTRIBUTION

The chi square distribution differs in some important ways from the binomial and normal distributions that we have already discussed; however, no new principles are involved in chi square's use as a test of significance. Before discussing chi square, let us briefly review the normal approximation of the binomial as a test of significance. We begin with some hypothesis about the population proportion which, given the sample size, allows us to calculate the mean and standard error of a theoretical sampling distribution. With this information we can calculate the z equivalent to any obtained sample proportion. When it can be assumed that z is normally distributed, Table N gives the probability of obtaining a value of z as large or larger than that yielded by the sample. If this z falls into an extreme tail of the normal distribution, so that the probability of obtaining it is less than α, the hypothesis about the population proportion is rejected.

The use of the chi square distribution is quite similar. We calculate a statistic called chi square that is based on the discrepancy between frequencies in a sample and frequencies expected according to some hypothesis. There is an appropriate chi square distribution, and the value of χ^2 required for significance in this distribution is given by Table C of the Appendix. If the obtained χ^2 exceeds the tabled value for a given level of α, then the null hypothesis is rejected.

Let us consider the differences between the chi square and the normal distributions. Chi square is really a family of distributions, a family in which each member differs from the others according to the degrees of freedom available for the calculation of the chi square statistic. We won't define degrees of freedom (df) for the moment, but χ^2 distributions based on 1, 3, 5, and 10 df are shown in Figure 9.1. Since these distributions differ substantially, different values of χ^2 will be required to determine the 5% and 1% level of significance *in each distribution*. Note the markedly skewed χ^2 distribution for 1 df shown in Figure 9.1. A χ^2 of 3.84 marks off the upper 5%, and a χ^2 of 6.64 marks off the upper 1% of this distribution. On the other hand, in the more symmetrical distribution for 10 df, a χ^2 of 18.3 is needed for the 5% level and 23.2 for the 1% level. In *any* normal distribution a particular value of z always cuts off the same proportion of the curve; one percent falls beyond $z = 2.33$ and five percent falls beyond $z = 1.645$.

Chi square, like the normal curve and the binomial, is a theoretical

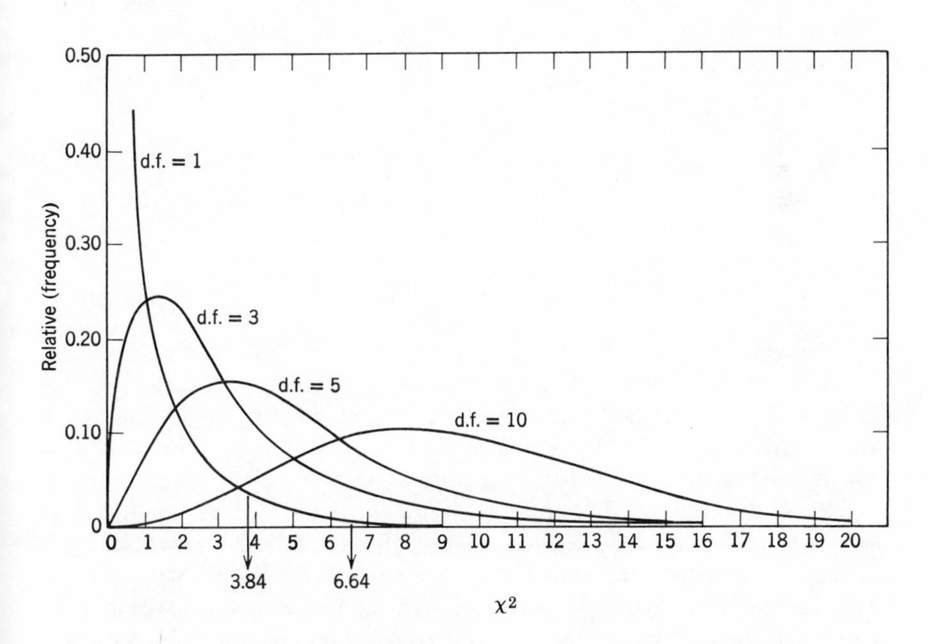

FIGURE 9.1 The relative frequency of χ^2 for different degrees of freedom.

sampling distribution. If one draws a very large number of random values of z from a normal distribution, squares these values and from the result constructs a frequency distribution, it will approximate the χ^2 distribution with 1 df. It follows that if we square the value of z which marks off any given portion of a normal curve we shall have the value of χ^2 which marks off the equivalent portion of the chi square distribution for 1 df. When the chi square distribution has 1 df, $\chi^2 = z^2$.

If we wished to approximate the chi square distribution with 2 df, we would randomly obtain two values of z from a normal distribution, square them and find their sum. The χ^2 distribution for 2 df will be approximated by the sampling distribution of these *sums*. In an analogous manner we could approximate chi square distributions with higher df. Degrees of freedom, then, refer to the components of χ^2 which are "free to vary randomly."

The values of χ^2 required to cut off specific areas of the right tail of chi square distributions with various df are given in Table C. Note that the tabled value of χ^2 for 1 df at the 5% level is 3.84, and the tabled value for the 1% level is 6.64. From Table C, you should find that a χ^2 equal to or greater than 9.21 is required for significance at the 1% level with df = 2. When df = 4, significance at the 5% level is obtained when χ^2 is at least 9.49, and a value of 13.28 is required for significance at the 1% level. Use Table C to determine if the probabilities associated with the following values of χ^2 are accurate.

$\chi^2 = 4.81$, df = 1, $p < .05$ $\qquad$ $\chi^2 = 7.94$, df = 1, $p < .01$

$\chi^2 = 6.11$, df = 2, $p < .05$ $\qquad$ $\chi^2 = 16.67$, df = 4, $p < .01$

$\chi^2 = 17.47$, df = 9, $p < .05$ $\qquad$ $\chi^2 = 4.10$, df = 2, $p > .05$

$$\chi^2 = \sum \frac{(|o - e| - .5)^2}{e}$$

FORMULA 9.1
χ^2 corrected for continuity

Formula 9.1 is used to calculate χ^2 for problems having one degree of freedom. In this formula e represents the expected, or theoretical, frequency of an event and o represents the obtained frequency. Note that the quantity $|o - e|$ is the *absolute* value of the difference between frequencies, and that .5 is subtracted from this value *before* it is squared.

The subtraction of .5 from $|o - e|$ is a correction for continuity. The calculation of χ^2 is based on frequencies, which are discrete data; but the chi square distribution is continuous. A correction for continuity is needed here just as it was when we approximated the binomial distribution with a normal curve. The correction is accomplished for chi square by reducing the absolute value of each $|o - e|$ term by .5 before squaring it. The correction should only be applied to problems with 1 df. Now consider the following illustrative problem. This is a problem which can be solved by using *either* chi square or the normal approximation of a binomial distribution. We will use chi square first.

Suppose we have a hypothesis that .50 of a population prefer candidate A and .50 prefer someone else. In a randomly obtained sample of 100 voters, except for the effects of sampling error, we shall expect to find 50 voters favoring A and 50 opposing A. However, suppose we find that 60 prefer A and only 40 oppose A. Should we reject the null hypothesis? It is customary to structure such a problem as it appears below.

	For A	Against A	Border Totals
Observed Frequency	60	40	100
Expected Frequency	50	50	100

$$\chi^2 = \sum \frac{(|o - e| - .5)^2}{e} = \frac{(9.5)^2}{50} + \frac{(9.5)^2}{50} = 3.61$$

This problem has one degree of freedom. Degrees of freedom in chi square problems are given by the number of $|o - e|$ components which are free to vary independently once the border totals have been specified. This is most readily determined by simply finding the number of "observed" cell entries that are free to vary independently. There are two "observed" cell entries for this problem, but only one of these can vary independently of the other once the border total of 100 observations has been determined. If there are 100 votes altogether, and 60 are for A, then the frequency in the "against A" cell is fixed and must be 40. (Notice that if the frequency in only one cell is free to vary randomly then only one of the two $|o - e|$ values is free to vary.) With 1 df the χ^2 of 3.61, which we have obtained for this problem is just short of significance at the 5% level. On the basis of this evidence one could *not* reject the hypothesis that the proportion of votes for candidate A is .50.

The problem can also be approached by using the normal approximation to a binomial distribution. We would proceed as follows to find the parameters.

$$M_f = np = 50$$

$$\sigma_f = \sqrt{npq} = 5$$

$$z_f = \frac{X_f - M_f}{\sigma_f} = \frac{(60 - 50) - .5}{5} = 1.90.$$

The z required for significance at the 5% level with a two-tailed test is ± 1.96. We have obtained 1.90, which, like the χ^2 of 3.61, is just short of significance at the 5% level.

ONE- AND TWO-TAILED TESTS USING CHI SQUARE

The example above illustrates a curious and apparently contradictory situation; chi square is a test of significance which permits a two-tailed test with only one tail of the sampling distribution. The sampling distribution which approximates χ^2 is composed of squared positive *and* negative values of z. As a result, improbably small deviations, positive and negative, contribute to the left tail of the chi square distribution and improbably large deviations, positive and negative, contribute to the right tail. Consequently, the area in the right tail alone is used to make a "two-tailed" test of significance.[1]

It is possible to use the χ^2 distribution to make a "one-tailed" or "directional" test. If the null hypothesis in the last problem had been that the proportion of people favoring A was equal to or less than .50 ($p \leqq .50$), we would have had a directional test of significance. With 60 votes for A in a sample of 100, the obtained $\chi^2 = 3.61$ would have been evaluated by using one-half the probability value for a two-tailed test given by the chi square table. As a result, instead of being just short of significance at the 5% level, this χ^2 would have been just short of significance at the 2.5% level. Of course, as we have pointed out before, if directional tests of significance are used they must be specified in advance of any data collection. Now again, let us direct your attention to the modification required to make a one-tailed test with χ^2. When χ^2 is used as a directional test one evaluates it on the basis of one-half the tabled *probability* values for a two-tailed test. This is *not* the same as saying that one-half the tabled value of χ^2 is required for significance. We mean, for example, that a chi square of 2.706, "significant" at only the 10% level in a nondirectional test with 1 df, will, in a directional test, be significant at the 5% level. Similarly, a chi square of 5.412, which is significant at the 2% level in a nondirectional test, will be significant at the 1% level in a directional test.

Even if we use only one tail of the χ^2 distribution as a test of significance, the distribution does have two tails. What is meant by a very

[1] This confusing terminology causes some authors to prefer the term "nondirectional" test of significance. Confusion, however, can only be eliminated by thoroughly understanding the way each theoretical distribution is used to provide a test of significance.

small value of χ^2, one which falls toward the end of the other tail? The appropriate conclusion is that we have obtained an improbably "good fit" to the χ^2 distribution, that our observed frequencies show an improbably *small* amount of sampling error. Of course, we should expect very small values of χ^2 occasionally, but if they occur very often we may have calculated χ^2 incorrectly, or perhaps used the distribution with inappropriate data.

The advantage of χ^2 over the binomial test is that χ^2 may be used when frequencies are divided among more than two alternatives. Let us consider an example: suppose a theory of genetics states that when long-flowered and short-flowered plants are crossed, the offspring will be long, medium, and short-flowered in a 1:2:1 ratio. Our observations of 200 such offspring show 64 long-flowered, 100 medium-flowered, and 36 short-flowered plants. The null hypothesis for this example is that the observed deviation from the expected frequencies of 50:100:50 is the result of sampling error. The data may be displayed in a table such as that below.

	L.-flowered	M.-flowered	S.-flowered	Border Totals
Observed	64	100	36	200
Expected	50	100	50	200

$$\chi^2 = \sum \frac{(o - e)^2}{e} = \frac{(14)^2}{50} + \frac{(0)^2}{100} + \frac{(14)^2}{50} = 7.84$$

This problem has 2 df. Once the border total of 200 observations has been established, only two of the observed cell entries are free to vary. Once two of them have been set, the third will be fixed. Since the problem has 2 df we do not correct for continuity. This χ^2 of 7.84 is significant at the 5% level. We should conclude that these results are rather improbable if the theory is true and the experiment has been carefully conducted.

When reporting a χ^2 value of this sort, the convention is to do it in the following way.

$$\chi^2 = 7.84, \text{df} = 2, p < .05$$

THE SIGNIFICANCE OF DIFFERENCES BETWEEN PROPORTIONS

We have illustrated chi square (and the binomial before it) with examples in which we have had some a priori hypothesis about a population proportion. Tests of significance have involved the difference between an a priori hypothesis about the population proportion and the proportion of events observed in a sample.

A much more common situation involves research of a type we mentioned in Chapter 6, in which a scientist found that only 46 of 1000 subjects who used a vaccine developed colds, while the disease occurred in 164 of 1000 untreated subjects. Here the investigator's problem requires him to find the probability of drawing random samples as different as these two from a common population. Notice that in this illustration we are testing the significance of a *difference between two sample proportions*, not the difference between a hypothesized population proportion and a single observed sample proportion.

Our best estimate, a point estimate, of the population proportion developing colds, assuming that the difference in sample proportions is entirely due to random error, is based on the pooled observations from the two samples, 210/2000, or .105. The question now becomes: given a population proportion of .105 people contracting colds, what is the probability of drawing two random samples of 1000 people, one of which contains 46 cold sufferers and the other 164 cold sufferers? The most convenient test of this hypothesis uses the χ^2 distribution. The procedure is as follows.

We have a total of 2000 subjects; 1000 in the vaccinated group, and 1000 in the control group. Of these 2000 subjects, 210 have developed colds and 1790 have not. These observations compose the border totals to the right of the 2×2 table below.

	Vaccine	No vaccine	Border Totals
Colds	46	164	210
No colds	954	836	1790
Border Totals	1000	1000	2000

Of the vaccinated subjects, 46 developed colds and 954 did not; of the subjects who were not vaccinated, 164 developed colds and 836 did not. These observations supply the observed frequencies *within* each cell of the table.

Our next problem is to obtain the expected frequencies for each cell. These are the frequencies expected if the "vaccine" and "no vaccine" groups have come from the same population. We have seen that the best estimate for the proportion of those developing colds in this population is found by pooling the frequencies in the two samples. By doing so we arrived at 210/2000, or .105, as the expected proportion of colds. Subtracting this proportion from 1.000, we have $1.000 - .105 = .895$, which is the expected proportion of subjects failing to develop colds. If there are 1000 subjects in each group then we should expect frequencies of 105 "colds" and 895 "no colds" in each group. Of course, these are the expected frequencies if the null hypothesis is true and the vaccine is ineffective. We will record these expected frequencies in the upper left corner of the appropriate cells in the 2×2 table below.

	Vaccine	No vaccine	Border Totals
Colds	[105] 46	[105] 164	210
No Colds	[895] 954	[895] 836	1790
Border Totals	1000	1000	2000

A general rule for finding the expected frequency for *any cell* of any chi square problem is to find the product of the border totals for the row and column containing that cell; then divide this product by the total number of observations. The expected frequency for the upper left cell above is thus $(210 \times 1000)/2000 = 105$. When one expected frequency is calculated for a 2×2 table, the other expected frequencies can be obtained by subtraction from the border totals. Notice that the sum of the expected frequencies and the sum of the obtained frequencies are equal for each column and for each row of the table. This must *always* be the case and provides a good check on the accuracy of computations.

The number of degrees of freedom for this chi square is determined by using the same kind of reasoning that was applied to the preceding problems. In the 2×2 table above only *one* of the observed cell entries is free to vary if the border totals are fixed. Once we have specified one cell entry, all of the rest can be determined by subtraction from the border totals. Consequently, this problem has just one degree of freedom.

With one degree of freedom we should correct our calculations for continuity although, for this particular problem, N is so large that the correction is of little consequence. We have the following calculations.

$$\chi^2 = \sum \frac{(|o - e| - .5)^2}{e} = \frac{(58.5)^2}{105} + \frac{(58.5)^2}{105} + \frac{(58.5)^2}{895} + \frac{(58.5)^2}{895} = 72.8$$

$$\chi^2 = 72.8, \text{ df} = 1, p < .001.$$

The director of research for the drug company would have little reason to question the rejection of the null hypothesis with these results. It is quite improbable that these samples have come from the same population. The vaccine is apparently effective.

Let us consider another type of problem which can be solved using χ^2. Suppose we want to know if a significant relationship exists between the proportion of students on probation and their membership in college fraternities. We discover that 614 men are in fraternities and 88 of them are on probation. We also find that 230 of 1818 nonfraternity men are on probation. This represents 14.3% of fraternity men and 12.6% of nonfraternity men in the same uncomfortable circumstance. The Dean of Men maintains that this is clear evidence for the failure of fraternities to support scholarship and the President of Interfraternity Council claims it is purely the result of sampling error; χ^2 is called for.

	Probation	No Probation	Total
Fraternity	80.3 88	533.7 526	614
Independent	237.7 230	1580.3 1588	1818
Total	318	2114	2432

The expected frequency for the upper left cell is $(614 \times 318)/2432 = 80.3$. The remaining expected frequencies can be calculated directly or obtained by subtraction from the border totals.

$$\chi^2 = \sum \frac{(|o - e| - .5)^2}{e} = \frac{(7.7 - .5)^2}{80.3} + \frac{(7.7 - .5)^2}{533.7} + \frac{(7.7 - .5)^2}{237.7} + \frac{(7.7 - .5)^2}{1580.3}$$

$$\chi^2 = .64 + .10 + .22 + .03$$

$$\chi^2 = .99, \text{ df} = 1, p > .05$$

Such a χ^2 could easily occur as a result of sampling error. To the extent that anyone can win an argument with a Dean of Men, the I. F. president would appear to have a good chance.

Notice, however, that there is an important difference between the last two examples. In the first, we assumed that the subjects were randomly assigned to treatment groups so that we had a proper independent variable, treatment with the cold vaccine. The finding of a significant χ^2 in that problem allowed the conclusion that the difference between groups was *an effect* produced by the vaccine.

On the other hand, even if we had found a significant χ^2 in the second problem, we could only have concluded that *a relationship* existed between fraternity membership and scholarship. We could *not* have concluded that fraternity membership caused poor scholarship, because whatever variables led a man to join a fraternity might also be the ones which affected his scholarship.

We shall now consider a χ^2 problem having 3 df. An educational psychologist wishes to investigate the relative effectiveness of instructional set on problem solving. He has four different types of instructions and he wishes to know if these differentially influence ability to solve a particular type of problem. Eighty students are randomly assigned to the four instructional groups, *A*, *B*, *C*, and *D* with the restriction that each group shall contain equal numbers of subjects. The numbers of students achieving and failing to achieve solution in each group are given below.

	A	*B*	*C*	*D*	Total
Solution	12.5 / 18	12.5 / 16	12.5 / 10	12.5 / 6	50
No Solution	7.5 / 2	7.5 / 4	7.5 / 10	7.5 / 14	30
Total	20	20	20	20	80

The null hypothesis in this problem specifies that the proportions achieving solution in each group show only random variation from some common population proportion. The best estimate of this common population proportion will be found by pooling the number of students in the "solution" category, obtaining their proportion of the total, and then doing the same for the students in the "no solution" category. Thus, we have $50/80 = .625$ achieving solution, and $30/80 = .375$ failing to achieve solution for all groups.

If the null hypothesis is true, we shall expect these proportions in each instructional group. Since there are 20 subjects in each group, the expected frequency of those achieving solution will be $20 \times .625$, or $(20 \times 50)/80$, which is 12.5. You should note that the calculations involved follow the rule previously given for determining expected frequencies; the row total multiplied by the column total divided by the total number of observations yields the expected frequency for the cell at the intersection of that column and row.

Once again we can deduce the df available by noting that if the border totals are all fixed, the entries in only 3 of the 8 cells are free to vary. In the next problem we shall provide a general formula for determining the df for any χ^2 table having fixed border totals.

The calculation of χ^2 for this problem is as follows.

$$\chi^2 = \sum \frac{(o-e)^2}{e} = \frac{(18-12.5)^2}{12.5} + \frac{(2-7.5)^2}{7.5} + \cdots + \frac{(6-12.5)^2}{12.5} + \frac{(14-7.5)^2}{7.5} = 19.40$$

$$\chi^2 = 19.40, \text{ df} = 3, p < .001$$

This χ^2 is highly significant for the 3 df of this problem, and we would therefore conclude that the proportions reaching solution for the four instructional groups were not the result of random sampling from a common population. These instructions apparently do affect, differentially, the proportions of individuals who can solve this kind of problem.

Suppose we wish to determine if college class is *related* to political conservatism. We have a test on which a score of 0–25 is defined as indicating liberal, 26–50 middle-of-the-road and 51–75 conservative. This test is given to 100 randomly selected members of each class, freshman through senior. The following data result.

	Liberal		Middle-Road		Conservative		Total
Senior	30	36	50	36	20	28	100
Junior	30	32	50	46	20	22	100
Sophomore	30	24	50	58	20	18	100
Freshman	30	28	50	60	20	12	100
Total		120		200		80	400

$$\chi^2 = \sum \frac{(o-e)^2}{e} = \frac{(6)^2}{30} + \frac{(2)^2}{30} + \cdots + \frac{(2)^2}{20} + \frac{(8)^2}{20} = 16.99$$

We shall now give a general formula for determining df in χ^2 problems. It is: df $= (r-1)(c-1)$, where r = the number of rows and c = the number of columns in the table. In this problem there are 4 rows and 3 columns so there are $(4-1)(3-1) = 6$ df. For 6 df a χ^2 of 16.99 is significant beyond the 1% level.

Since we have surveyed a random sample from each class, does the significant χ^2 mean that college experience *affects* political attitudes, or is it simply *related* to political attitudes? The answer is "only related." We have randomly selected subjects for questioning; we have not randomly *assigned* subjects to four, three, two, and one years of college. The significant value of chi square means that there is a *relationship* between year in college (at this college) and political attitudes (measured by *this* questionnaire).

RESTRICTIONS ON THE USE OF χ^2

In spite of its wide range of usefulness, χ^2 has some limitations. When using the normal approximation to the binomial we suggested that np and nq be $\geqq 5$. When we have a χ^2 problem with 1 df we must follow the same rule, no *expected* frequency may be less than 5. With more than 2 df a single cell may have an expected frequency less than 5, provided it is not less than one.

We must also be able to make the same assumptions about the independence of observations for chi square that we had to make in order to use the binomial. This means that the location of any observation in a particular cell of the table must not affect the cell in which any other observation is located. In the previous problem, the fact that Sam Smith was a conservative senior did not at all influence the classification of Joe Jones or *any other student.* This independence can best be assured by using random sampling to select observations in each category.

There are many instances in psychological research when we must deal with the differences between proportions of correlated rather than randomly selected groups. Chi square can still be used in these situations, but its use requires some special considerations because, if groups are based on correlated observations, a fundamental assumption of χ^2 has been violated. Correlated observations are obviously not independent!

CHI SQUARE FOR CORRELATED OBSERVATIONS

A psychologist, interested in the effect of some independent variable on his subjects' behavior, might give them a pretest, then expose them to the independent variable, and assess the effects of this variable with a posttest on the *same* subjects. In such an experiment, each subject serves as his own control, so we do not have a comparison between two randomly drawn samples. Since the samples were not independently obtained, a different technique for the use of chi square is required; this is the chi square test for correlated observations.

Let us assume we have the following problem: we have drawn a single sample of 100 subjects and find that 40 of them support a particular proposition. Two weeks later we question the same subjects and find that the number supporting the proposition has increased to 60. We wish to determine if this shift is too great to be expected on the basis of sampling error. From these data we can obtain the border totals (but not the cell entries) for the 2×2 table below.

		Test 1 (Original)		
		Favor	Opposed	Total
Test 2 (Two Weeks Later)	Opposed	*a* 30	*b* 10	40
	Favor	*c* 10	*d* 50	60
	Total	40	60	100

To obtain the cell entries we must classify each subject according to whether he is favorable or opposed to the proposition at each of the two inquiries. Each subject can respond in one of four ways: favorably at both inquiries; opposed at both; favorably at the first, but opposed at the second; and opposed at the first but favorably at the second. We could record the number of subjects making each of these four responses or, given the border totals, we could simply record the frequency in any cell and obtain the other by subtraction from the border totals.

Look at cells *a* and *d* from the 2×2 table; the *sum* of these two cells records the total *changes* of opinion, the changes in *both* directions. If the null hypothesis is true and nothing more than sampling error is operating, then these two cells should have equal or nearly equal frequencies; we should expect one direction of change to be as frequent as the other. In terms of the χ^2 procedure we described earlier when we tested the departure of an observed from an expected frequency, we assume that cells *a* and *d* should, within the limits of sampling error, have the same frequencies. The extent to which they do not reflects a significant *change* in the proportion of people holding a particular opinion. This line of reasoning yields the chi square table below.

	Favor_1 Opposed_2	Opposed_1 Favor_2	Total
Observed Changes	30 (*a*)	50 (*d*)	80
Expected Changes	40 $\frac{(a+d)}{2}$	40 $\frac{(a+d)}{2}$	80

$$\chi^2 = \sum \frac{(|o - e| - .5)^2}{e} = \frac{(9.5)^2}{40} + \frac{(9.5)^2}{40} = 4.51$$

The $\chi^2 = 4.51$ is significant at the 5% level for 1 df. We would conclude from these data that there is a significant shift in opinion between the times of the two surveys.

We shall consider another example requiring a test of significance for correlated proportions. Let us assume that in a randomly drawn sample of voters we find 66 favoring the Republican candidate and 34 favoring the Democratic candidate. A week following the initial poll we conduct another survey of the *same subjects*. Now we find that the subjects are evenly divided with 50 favoring the Democratic candidate and 50 favoring the Republican. We wish to know if this represents a significant change of opinion.

It is clear that more respondents have shifted their preference *to* the Democratic candidate than have shifted their preference *to* the Republican candidate, but whether or not this shift is significant depends on the total number of shifts that have taken place. Consequently we need to know the frequencies in the categories "shifted preference Republican to Democrat" and "shifted preference Democrat to Republican." Suppose these frequencies are 23 and 7 respectively. If there has been no change in the proportion of the *population* favoring the Republican candidate, then the change we have observed in our sample must be viewed as the result of sampling error. The question becomes: if there are 30 changes of opinion in our sample, what is the probability that 23 will occur in one direction and 7 in another *if* nothing but chance is operating? If the null hypothesis is true we should expect 15 changes of opinion in each direction. Thus, the chi square table will look like this:

	R to D	D to R
Observed	23	7
Expected	15	15

$$\chi^2 = \sum \frac{(|o - e| - .5)^2}{e} = \frac{(7.5)^2}{15} + \frac{(7.5)^2}{15} = 7.50$$

$$\chi^2 = 7.50, \text{ df} = 1, p < .01$$

It seems quite clear that the shift in opinion is significant.

χ^2 WITH PROPORTIONS AND PERCENTS

Chi square is used to determine the significance of the difference between observed and expected frequencies. When χ^2 is used as a test of the significance of differences between proportions or percents, the proportions are converted to frequencies, and it is the *frequency* which is actually entered in the cells of the chi square table. We cannot enter the proportion or percent of events directly into the cells of a chi square table because proportions and percents are independent of sample size and the value of chi square is considerably influenced by sample size.

In the illustration below, we shall show how chi square increases with sample size even though the *proportions* of events in each class remain constant. Suppose we have 70% of a sample favoring some proportion and we wish to determine if this is significantly different from 50%. It can be seen that the value of χ^2 will be quite different for the three situations below, depending upon sample size, even though the *proportions* are the same in each sample.

	Favor	Opposed
Observed	7	3
Expected	5	5

$N = 10$

$\chi^2 = .90$

	Favor	Opposed
Observed	70	30
Expected	50	50

$N = 100$

$\chi^2 = 15.21$

	Favor	Opposed
Observed	700	300
Expected	500	500

$N = 1000$

$\chi^2 = 159.20$

It is necessary to bear in mind that χ^2 is ultimately based upon frequencies. When proportions or percents are given, they must be converted to frequencies before χ^2 is calculated.

USING χ^2 FOR TESTS OF CONTINUOUS DATA: THE MEDIAN TEST

Up to this point our illustrations of chi square have been based on frequencies or counts; that is, on discontinuous scales whether orderable, as in college class, or nonorderable, as in types of instruction. In the next few examples we shall show how chi square may be used as a test of significance if essentially continuous data are treated as if they were discontinuous. This test of significance is called the median test even though it is really an application of chi square.

Suppose we extinguish a maze-running response in one group of animals under condition A and in another group under condition B. The time it takes each subject to run the maze on the fifth day is recorded below. We want to know if there is a significant difference between the running times for the two groups. Notice that running time is a continuous scale. Nevertheless, we can record the number of *animals* in each group which fall above and below the median running time for the *combined* groups, and in that rather roundabout way make use of χ^2 as a test of significance. We shall work from the data recorded below.

A. 14, 22, 24, 16, 12, 10, 13, 18, 16, 22, 30, 26, 26, 27, 20, 20

B. 15, 11, 9, 8, 6, 12, 18, 19, 13, 26, 12, 13, 11, 23, 11, 9

The median for the pooled data is 15.5. If these two samples came from populations with the same median, then, except for sampling error, the proportion of subjects in group A who are above the median and the proportion of subjects in group B who are above the median should be the same. However, we find 12 of 16 S's in group A are above the combined groups' median of 15.5, and only 4 of 16 S's in group B are above this point. The data can be cast into the 2×2 table below.

	A	B	Total
Above median	8 12	8 4	16
Below median	8 4	8 12	16
	16	16	32

The value of χ^2 is 6.12, df $= 1$, $p < .02$. The obtained value of χ^2 is significant at the .02 level.

If samples A and B have been drawn from populations having the same median, then the proportion of sample A falling above the median of the combined groups should be, within the limits of sampling error, the same as the proportion of sample B which falls above this point. The sampling error of the difference between two proportions can be tested with χ^2, as we have shown in earlier illustrations. The same procedures have been applied here. Problems requiring the median test should have $n_A + n_B = 20$ because of the restrictions on the smallest expected cell frequency discussed earlier.

Let us apply the median test to another set of data in which $n_A \neq n_B$.

The following scores on a mechanical comprehension test were obtained for a group of college men and women.

32 men: 46, 67, 39, 26, 43, 38, 52, 61, 63, 52, 56, 38, 37, 31, 42, 41, 43, 44, 56, 52, 58, 54.

12 women: 26, 19, 35, 44, 53, 21, 39, 27, 26, 48, 51, 28.

The null hypothesis in this problem is that these two groups vary no more than might be expected on the basis of random sampling from populations with the same median. We must obtain the proportion of subjects in each group whose scores fall above the median of the combined groups. If these proportions are significantly different by χ^2, we can reject the null hypothesis.

	Men	Women	Total
Above median	10.4 / 12	5.6 / 4	16
Not above median	11.6 / 10	6.4 / 8	18
Total	22	12	34

(Note the nature of the dichotomy. "Not above the median" includes scores *at* the median and accounts for the inequality between rows.)

$$\chi^2 = \sum \frac{(|o - e| - .5)^2}{e} = \frac{(1.1)^2}{10.4} + \frac{(1.1)^2}{11.6} + \frac{(1.1)^2}{5.6} + \frac{(1.1)^2}{6.4} = .62$$

$$\chi^2 = .62, \text{ df} = 1, p > .05$$

There seems to be no evidence from these data that there is a significant difference between the mechanical comprehension of men and women college students.

The failure to reject the null hypothesis in this study should alert an intelligent investigator to the possibility of a Type II error. It has been fairly well established that college men have better mechanical comprehension than college women. The data from this experiment, while in that direction, does not permit the rejection of the null hypothesis. If a difference in mechanical comprehension exists between the populations, but the difference between the samples is not significant, we have a Type II error.

Chi square is actually not a very powerful test of significance. For that matter, neither is the binomial. Since the median test is based on chi square, the median test also has little power. In the next chapter we shall discuss much more powerful tests of significance for the kind of data we are subjecting to the median test. One of the reasons why these tests are more powerful than chi square is that they use more of the information available in the data. We shall explore this more fully.

The low power of the median test (or χ^2) is the result of its failure to be influenced by the *magnitude* of a score's departure from the median. In the median test *only the frequency* of observations above and below the median contribute to the value of χ^2. If a measurement is one point above the median it will contribute as much to χ^2 as a measurement thirty points above the median.

Suppose we conduct two experiments, each requiring the comparison of two sets of data. We shall use the median test in both experiments. The data from the two experiments appear below.

Experiment 1		*Experiment 2*	
Experimental	*Control*	*Experimental*	*Control*
50	21	21	21
50	21	21	21
50	21	21	21
50	21	21	21
Mdn. = 20.5		Mdn. = 20.5	
20	1	20	20
20	1	20	20
20	1	20	20
20	1	20	20

It is apparent that there is a considerable difference between the magnitudes of the experimental and control group scores in Experiment 1, while the scores for these groups in Experiment 2 are identical. We might, therefore, expect a larger χ^2 from the data in Experiment 1 than from the data in Experiment 2. In fact, both yield the same χ^2; they are both zero. This is the result of having half the experimental group and half the control group above the median in each experiment. Under these circumstances the χ^2 for both experiments will be zero, even though in Experiment 1 the experimental group scores above the median are far above, and the control group scores below the median are far below. The *distance* from the median does not affect the value of χ^2; all that contributes to the χ^2 is the proportion (frequency) of scores above or below the median. Since this proportion is the same in both experiments, the χ^2's are identical.

When orderable data are subjected to a median test, we disregard some of the information contained in the data. Actually, all of the scores could be reduced to 0's and 1's (if we let 0 stand for all scores below the combined median and 1 stand for all scores above the combined median), and the resultant χ^2 would be unaffected. There are tests of significance which make use of *all* the information contained in the data. These tests are more powerful than the median test and should always be used when their assumptions can be met. These more powerful tests will be discussed in subsequent chapters.

OVERVIEW

This is the third distribution we have studied. We have discussed the binomial distribution, the normal distribution, and now the chi square distribution. The use of all of these distributions in tests of statistical significance is quite similar. The distributions provide us with a theoretical relative frequency of events; for the binomial it is the relative frequency, or probability, of obtaining any proportion of events in a sample of size n, given the proportion of events in the population from which the sample was randomly drawn; for the normal distribution it is the relative frequency, or probability, of obtaining samples yielding values of z as deviant as those listed in Table N; for the chi square distribution with various df it is the probability of obtaining χ^2 values as large or larger than those listed in Table C.

In each case, when we select an appropriate test of significance, we assume that if the null hypothesis is true, our data should conform to that theoretical sampling distribution. When the test is significant, it means that on the basis of the hypothesized sampling distribution, the results are quite improbable. However, before we can reject hypotheses about the population parameters, it is quite important that the remaining assumptions about the distribution have been met, for example, that observations are randomly obtained and that we have the proper df. If we have not met these assumptions, we shall be dealing with an unknown distribution and obtaining meaningless levels of significance.

For example, if a test of significance is carried out and reveals a χ^2 of 6.74, df $= 1$, $p < .01$, the conclusion is to reject the null hypothesis, since the sampling distribution of chi square for 1 df contains less than 1% of its values above 6.68. If we have made an error, and the proper sampling distribution for this statistic is the chi square distribution for 4 df, we will have rejected the null hypothesis when such a rejection was inappropriate.

The assumptions required for using these sampling distributions are rather easy to meet, but the tests of significance are usually restricted to comparisons of frequencies and proportions. In the next section we will deal with distributions designed for tests of significance of means, variances and correlation coefficients. The principle is the same, obtaining a probability estimate for a deviation of a given magnitude, but the distributions are a bit more complex in their assumptions. Happily, they are also much more powerful, more likely to demonstrate that a true difference is significant.

Exercises

1. A research investigator has 100 seeds randomly selected from each of the brands marketed by 6 different companies. The number of seeds germinating are: A, 65; B, 60; C, 50; D, 55; E, 55; F, 45. He wishes to know if there is a significant difference among the proportions germinating. Find the value of χ^2 and interpret it.
2. In Exercise 1, another investigator selects the brands having the highest and lowest proportions of germinating seeds and performs a χ^2 test on just these two proportions. The result is significant, but no one is the least surprised. What is the value of χ^2? What assumption has been violated?
3. An investigator wishes to determine if there is a relationship between birth order and election to office among members of certain campus organizations. He finds that among the organizations there are 26 officers and 100 who are not officers for a total of 126 members. Of the 26 officers 16 are either first born or only children while 36 of the 100 non-officeholders fall in this category. The remaining 10 officers and 64 non-officeholders have older siblings. Determine the value of χ^2 to test the significance of the association.
4. Scores on a statistics test for students with at least one year of college math are as follows: 26, 37, 28, 27, 29, 31, 36, 42, 41, 29, 36, 38, 39, 42, 53, 56, 34, 38, 42, 48, and 47. Students with no college math had scores of: 18, 26, 31, 18, 23, 36, 45, 51, 23, 19, 32, 28, 27, 31, 45, 34, and 32. Is it reasonable to assume both samples were drawn from populations with the same median ($p < .05$)?
5. Calculate χ^2 for the following data.

	Yes	?	No	
Men	30	40	130	200
Women	20	30	50	100
	50	70	180	300

Interpret the meaning of a significant χ^2 for these data.

There is a "hitch" to the calculation or evaluation of χ^2 in most or perhaps all of the examples below. In some cases assumptions have been violated, the wrong procedure has been used, or some information necessary to the calculation has been omitted. See if you can spot the trouble, if any, in each example.

6. An investigator, upon examining the relationship between scholarship and birth date, discovers an abnormally large difference in the proportion of honor students born in March and September.

	March	September	Total
Honor Student	26	12	38
Not Honor Student	135	172	307
	161	184	345

He subjects these frequencies to a χ^2 analysis and will assume that month of birth is related to scholarship if χ^2 is significant.

7. In a mock election .54 of the Sophomore class favor a proposition and .46 oppose it. What is the probability of obtaining such a deviant sample if the population proportion is .50 favoring and .50 opposing?

8. One hundred rats are given two trials in a maze. On the first trial 46 turn right; on the second trial 58 turn right. Is this a significant difference?

9. At least a 2/3 majority is needed to approve a proposition. We have drawn a sample of 200 in which .60 favor the proposition and .40 oppose it. We wish to determine the probability of obtaining a sample proportion this small if the proportion in the pouplation is .67. The χ^2 is 4.4. This value of chi square means that if the population proportion is .67 or more, then the probability of obtaining such a low sample proportion is less than 1/20.

10. We wish to assess the difference between men and women drivers regarding their tendency to make left hand turns without signaling. We observe several dozen women until they make a total of 192 left hand turns. We find that they signal on 147 of them. About 20 men are observed until they make 110 left hand turns. We find that they signal on 96 of these. Is there a significant difference between the proportions of left hand turns signalled by men and women?

11.

	Yes	?	No	
Men	8 14	14 14	8 2	30
Women	8 2	14 14	8 14	30
	16	28	16	60

$$\chi^2 = \frac{(5.5)^2}{8} + \frac{(5.5)^2}{8} + 0 + 0 + \frac{(5.5)^2}{8} + \frac{(5.5)^2}{8} = 15.1$$

(a) Assuming that two randomly drawn samples of 30 men and 30 women have been obtained and their responses to a question recorded above, comment on the calculation of χ^2.

(b) Suppose the investigator, after noting the results, discards the "?" category and, retaining the same χ^2, reduces the df required for evaluating it.

12. Suppose we wish to determine if the first 5 questions at the end of this chapter are of a different level of difficulty than the second 5 questions. The number of questions in the two groups answered correctly by each class member appears below. How could you use χ^2 to evaluate the significance of the difference?

	A	B	C	D	E	F	G	H	I	J	K	L	M	N	O	P	Q	R
1st.	5	4	3	2	5	1	3	4	3	4	4	3	3	2	3	5	4	4
2nd.	4	2	1	3	5	0	3	2	1	2	2	1	1	3	3	4	5	3

Do not use the median test, but see if you can develop a technique which takes advantage of the "repeated observations" aspect of the problem.

CHAPTER 10

The *t* Distribution

In previous chapters we have discussed methods for determining the probability of obtaining any particular proportion of events in a sample, given the sample size and the proportion of events in the population. We also discussed methods for determining the probability that two or more randomly drawn, or two correlated sample proportions came from a common population. In each case we defined two or more mutually exclusive categories, counted, or enumerated, the observations classified in these categories and then used tests of significance based on these enumerations.

When true measurement is possible, the statistician has considerably more powerful tests of significance at his disposal than those we previously discussed. These are more powerful tests, in part because they make use of the greater amounts of information provided when we measure the characteristics of objects rather than count objects that possess certain characteristics. For example, we can count the number of children finishing a problem in less than 5 seconds and report that .40 of the group met this standard, or we can measure the time taken by each child and report a group mean of 7.18 seconds. We obviously receive more information if we know that a child took 8.17 seconds to solve a problem than if we know he required "more than 5 seconds." This increase in information permits more powerful tests of significance than those we discussed in previous chapters.

HYPOTHESES ABOUT THE POPULATION MEAN

Suppose we have an infinitely large population of measurements and we want to know if it is reasonable to assume that the population mean (M_H) is 100. Since the population is infinitely large, we cannot compute the population mean but we can draw a sample of n cases and compute a sample mean ($\bar{X}$) with the following equation.

$$\bar{X} = \frac{\Sigma X}{n}$$

If we continue to select samples and calculate sample means, we can construct a sampling distribution of sample means. This process was discussed briefly in Chapter 6, and Figure 6.1 illustrated such a sampling distribution.

There are two important facts about sampling distributions of sample means: first, the sampling distribution will tend to be normal, and if n is reasonably large this will be true even if the samples come from a decidely skewed population. Second, the mean of the sampling distribution of sample means will, in the long run, equal M, the population mean. This second property of the sample mean permits it to be called an unbiased estimator of its parameter, the population mean.

Let us consider some implications of the first property of the sample mean, that its sampling distribution tends to be generally normal in form. This is an extremely helpful situation because it allows us to use z, the familiar normal deviate, to test *hypotheses* about the population mean. When z is used for this purpose we shall have

$$z = \frac{\bar{X} - M_H}{\sigma_{\bar{X}}}$$

where $\bar{X}$ is an obtained sample mean, M_H is an *hypothesized* value for the *unknown* population mean, and $\sigma_{\bar{X}}$ is the standard deviation of the theoretical sampling distribution of sample means, called the standard error of the mean. This z can be evaluated by reference to Table N, the table of the normal curve. The null hypothesis in such a problem will state that the difference between the obtained sample mean and the hypothesized population mean is the result of sampling error. For a two-tailed test, z must equal at least ± 2.58 for rejection at the 1% level. You should recognize this as essentially the same procedure we applied in Chapter 8 to testing hypotheses about population proportions; only the labels have changed.

The procedure seems straightforward enough until we actually try to calculate the standard error of the mean ($\sigma_{\bar{X}}$), the standard deviation of the theoretical sampling distribution of sample means. If the population is very large, the number of possible samples which compose the theoretical sampling distribution will also be very large, and will, for all practical purposes, pose an insurmountable computational problem. However, this parameter can be calculated from knowledge of the population standard deviation, σ, and the sample size, n. The formula is given below.

$$\sigma_{\bar{X}} = \frac{\sigma}{\sqrt{n}}$$

FORMULA 10.1

The standard error of the mean calculated from the population standard deviation and sample size

Let us examine this formula a moment. First, remember that $\sigma_{\bar{X}}$ is a measure of the variability of sample *means*, just as σ is a measure of the variability of individual measurements. Notice from Formula 10.1 that $\sigma_{\bar{X}}$ is directly proportional to σ, the variability of the individual measurements in the population from which the sample means have come, and inversely proportional to $\sqrt{n}$, the square root of sample size. When the population of measurements is quite variable, the sampling distribution of sample means is also quite variable; however, sampling distributions based on very large samples will be less variable than sampling distributions based on small samples. In the limiting case when $n = 1$, the formula quite reasonably shows that the standard error of the mean will be the same as the standard deviation of measurements in the population.

Consider this illustration: suppose that in some normally distributed population of measurements we have $M = 50$ and $\sigma = 10$. We know that in any such population $M \pm 2.58\sigma$ will include 99% of the measurements; so if $\sigma = 10$ for this population, we should expect to find 99% of the individual measurements falling between 24.2 and 75.8, or within a range of about 52 units. However, if we have a distribution of sample *means* drawn from this population, and each sample is based on $n = 100$ measurements, we will find the sample means only $(1/\sqrt{n})$ 1/10 as variable as the individual measurements in the parent population. Thus, we shall find 99% of the distribution of sample *means* falling between 47.4 and 52.6, a range of only 5.2 units.

Unfortunately, Formula 10.1, which gives the standard error of the mean in terms of σ and n, is not very useful because it requires σ, a population parameter. If we could calculate σ we could also calculate M, and there would be no need of any hypothesis testing. Populations are usually much too large for the direct calculation of any parameters. Since parameters are not available, we might consider using a *sample* standard deviation as an *estimate* of the population standard deviation, just as we can use the sample mean, $\bar{X}$, as an estimate of the population mean, M. This apparently reasonable approach would work perfectly well, except that the sample standard deviation, and its square, the sample variance, are *biased* estimators of their respective population parameters.

In the next few paragraphs we explain why this bias occurs and what can be done about it.

We mentioned that $\overline{X}$, the sample mean, is an unbiased estimator of M, the population mean. However, the sample variance and its square root, the sample standard deviation, are both *biased* estimators of their equivalent population parameters. The mean of a sampling distribution of sample variances will *not*, on the average, equal the population variance. The reason for this situation may be clarified somewhat if you compare the formula for a population variance $\Sigma(X - M)^2/N$ with the formula for a sample variance $\Sigma(X - \overline{X})^2/n$. Notice that the difference involves taking the sum of squared deviations from the population mean to determine the *population* variance, and taking the sum of squared deviations from a *sample* mean to determine the sample variance. The bias is in the "sum of squares" term, the sum of squared deviations from the mean which forms the numerator of both formulas.

The *sample* sum of squares, $\Sigma(X - \overline{X})^2$, consistently *underestimates* the population sum of squares, $\Sigma(X - M)^2$. It can be shown that for any distribution the sum of squares taken from the mean will be less than the sum of squares taken from *any* other point in that distribution. Since the *sample* mean will not ordinarily coincide *exactly* with the population mean, the sum of squared deviations taken from the sample mean will be less than it would have been had it been taken from the population mean. Thus, the sample variance will underestimate the population variance. When samples are small (and sample means tend to be very variable), the amount of bias will tend to be large and the population variance may be severely underestimated.

Fortunately, the bias in the sample variance as an estimator of the population variance can be corrected quite easily if we divide the sample sum of squares by $n - 1$, the number of degrees of freedom, instead of by n, the number of observations. We have $n - 1$ degrees of freedom for the sum of squares because, once the sample mean has been determined, only $n - 1$ of the observations are free to vary randomly. For example, if $\overline{X} = 7$ and $n = 4$, there are only 3 df since only three of the four numbers which compose the sample can vary randomly. Once the values X_1, X_2 and X_3 have been obtained the value of X_4 is fixed. We have already discussed the concept of df in the chapter on chi square; it is simply being applied here in a slightly different context.

It can be shown that when the sample sum of squares is divided by df, $(n - 1)$, instead of by n, the bias in the estimate of the population variance is eliminated. Unfortunately, the square root of this unbiased estimate of population variance is not quite an unbiased estimate of the population standard deviation. Nevertheless, we shall use it as our best estimate of the parameter. Thus, the formula for an *estimate* of the population standard deviation, which will be symbolized by S, is as follows.

$$S = \sqrt{\frac{\Sigma(X - \overline{X})^2}{n - 1}}.$$

FORMULA 10.2
An estimate of the population standard deviation obtained from a sample of n measures

Formula 10.2 is somewhat unwieldy for computational purposes, so an equivalent formula which permits the calculation of S directly from raw scores is given below.

$$S = \sqrt{\frac{\Sigma X^2 - \frac{(\Sigma X)^2}{n}}{n - 1}}$$

FORMULA 10.3
Computational equivalent for Formula 10.2

Although the division of the sum of squares by $n - 1$ provides an *estimate* of σ from a sample, this does not mean that the formula for σ given in Chapter 4 has been changed. That formula, in its various forms was,

$$\sigma = \sqrt{\frac{\Sigma(X - M)^2}{N}} = \sqrt{\frac{\Sigma x^2}{N}} = \sqrt{\frac{\Sigma X^2 - \frac{(\Sigma X)^2}{N}}{N}}$$

and is, the proper formula for the standard deviation of any population. If the interest is in the standard deviation of a specific set of 1000 scores defined as a population, the formula for σ is used. If the 1000 scores are a *sample* from a much larger population, and an *estimate* of σ in that larger population is required, then we use the formula for S.

The problems posed by the estimation of σ have caused a considerable digression from our original concern, testing hypotheses about the population mean. Let us return to that problem. You may remember that σ was required in order to calculate $\sigma_{\overline{X}}$ which, in turn, was needed as the denominator of the z ratio $(\overline{X} - M_H)/\sigma_{\overline{X}}$. The probability of obtaining such a z ratio would determine whether or not we could reject M_H as a population mean. We had already explained that the unknown parameter, σ, can be estimated from a sample by calculating the statistic S. We can also estimate the unknown parameter $\sigma_{\overline{X}}$ by substituting S for σ in Formula 10.1. If we let $S_{\overline{X}}$ represent an estimate of $\sigma_{\overline{X}}$ we have:

$$S_{\overline{X}} = \frac{S}{\sqrt{n}}.$$

FORMULA 10.4
An estimate of the standard error of the mean

This formula is not very convenient for computing $S_{\bar{X}}$, but a few judicious substitutions and a little algebra produce Formula 10.5 below which permits the calculation of $S_{\bar{X}}$ directly from raw scores.

$$S_{\bar{X}} = \sqrt{\frac{\Sigma X^2 - \frac{(\Sigma X)^2}{n}}{n(n-1)}}$$

FORMULA 10.5
Computing formula for an estimate of the standard error of the mean

We finally have all of the formulas necessary to calculate the normal deviate z. The new equation is as follows.

$$z^1 = \frac{\bar{X} - M_H}{S_{\bar{X}}}$$

We can now try an example problem using these ideas. Suppose a sample has been drawn for which $n = 100$, $\bar{X} = 97.1$, and $\Sigma X^2 - (\Sigma X)^2/n = 9{,}900$. A theory specifies that M_H, the population mean, is 100. Can this hypothesis be rejected at the 1% level? We have the following calculation.

$$z = \frac{\bar{X} - M_H}{\sqrt{\frac{\Sigma X^2 - \frac{(\Sigma X)^2}{n}}{n(n-1)}}} = \frac{-2.90}{1} = -2.90 \cdot$$

A z of -2.90 is considerably more deviant than the ± 2.58 required for significance at the 1% level for a two-tailed test, and we therefore reject the hypothesis that $\bar{X} = 97.1$ was the result of random sampling from a population having $M_H = 100$.

This is probably a good time to review some symbols and formulas because we have discussed so many in such a short time. Subsequent material will be much easier to understand if you know what each symbol means.

M A population mean, $M = \dfrac{\Sigma X}{N}\cdot$

$\bar{X}$ A sample mean (which is also an unbiased estimate of the population mean), $\bar{X} = \dfrac{\Sigma X}{n}\cdot$

[1] The student will shortly find that this ratio does not have a precisely normal distribution, although the correspondence is good when $n = 50$.

- x The deviation of a variable from the mean of its distribution, $x = X - M$ or $x = X - \overline{X}$, depending upon whether we are dealing with a sample or a population.
- N Total number of observations in a population.
- n A subset of observations constituting a sample.
- σ^2 The population variance, $\sigma^2 = \dfrac{\Sigma(X - M)^2}{N}$.
- S^2 The unbiased estimate of the population variance derived from a sample of n observations, $S^2 = \dfrac{\Sigma(X - \overline{X})^2}{n - 1}$.
- σ The standard deviation of a population, $\sigma = \sqrt{\dfrac{\Sigma(X - M)^2}{N}}$.
- S An estimate of the population standard deviation derived from a sample of n observations, $S = \sqrt{\dfrac{\Sigma(X - \overline{X})^2}{n - 1}}$.
- $\sigma_{\overline{X}}$ The standard error of the mean, which is the standard deviation of the theoretical sampling distribution of sample means, $\sigma_{\overline{X}} = \sigma/\sqrt{n}$.
- $S_{\overline{X}}$ An estimate of the standard error of the mean derived from S, $S_{\overline{X}} = S/\sqrt{n}$.

REVIEW OF THE NORMAL DEVIATE z

Notice the applications of the normal deviate z. We began by defining a "z score" as $z = (X - M)/\sigma$. The numerator of this ratio is the difference between a score and the mean of the population from which it comes; the denominator is the standard deviation of that population. When data are normally distributed, any z score is evaluated by reference to the table of the normal curve from which an appropriate centile or probability value can be determined.

When tests of significance involved the frequency of events in a sample, a very similar formula, $z_f = (X_f - M_f)/\sigma_f$ was used. The numerator of this ratio is the difference between the frequency of events *observed* in some sample and the hypothesized mean of the frequencies in some theoretical sampling distribution. The denominator is the standard deviation, or standard error, of the theoretical sampling distribution of sample frequencies. When conditions are such that z is normally distributed, the probability that the sample proportion was the result of sampling error is given by the proportion of the normal curve falling beyond z.

In the case of proportions, the normal deviate is $z_p = (X_p - M_p)/\sigma_p$. The numerator is the difference between the proportion of events *observed* in a sample and the mean of the proportions hypothesized for the theoretical sampling distribution. The denominator is the standard deviation (standard error) of this hypothesized theoretical sampling distribution of proportions. The normal deviate z is interpreted as it was in the previous example.

Testing hypotheses about the population mean, M, involves a continuation of the same theme. For large samples, the approximately normal deviate, $z = (\bar{X} - M_H)/S_{\bar{X}}$, is also the ratio of a difference to a standard deviation (standard error). The only change is the use of $S_{\bar{X}}$ to estimate $\sigma_{\bar{X}}$. When z is significant, that is, larger than ± 2.58 for the 1% level, or ± 1.96 for the 5% level, the hypothesis is rejected that $\bar{X}$ was obtained by a random selection of cases from a population with mean M_H.

When the normal curve is used to interpret these relative deviates, the score, frequencies, proportions, or sample means *must* be normally distributed. However, the sampling distribution of the ratio $(\bar{X} - M_H)/S_{\bar{X}}$ is only approximately normal even when n is large. When n is less than 30, which is often the case in behavioral research, the sampling distribution of this ratio begins to depart substantially from normal, and the departure becomes more pronounced as n decreases. Procedures for coping with this situation are described in the next section.

THE t DISTRIBUTION AND HYPOTHESES ABOUT THE POPULATION MEAN

The nature of bias in the sampling distribution of sample standard deviations and the fact that this bias is largely corrected when the sum of squared deviations from the mean is divided by $n - 1$, instead of n, has already been discussed. Another, and rather different kind of problem with this sampling distribution will now concern us. Although the mean value of the sampling distribution of S comes quite close to the parameter σ, the sampling distribution of S is skewed when small samples are used, and becomes progressively more skewed as the sample size decreases. The skew is positive, which means that a randomly obtained value of S is more likely to be below σ than above it, even though the mean of the sampling distribution of S comes very close to σ as the number of samples increases. If S typically underestimates σ, then ratios based on the use of σ in the denominator will tend to be larger when σ is estimated than when σ is known; thus, $(\bar{X} - M_H)/S_{\bar{X}}$ will tend to be larger than

$(\bar{X} - M_H)/\sigma_{\bar{X}}$. Moreover, this discrepancy will *increase* as the number of degrees of freedom used in the estimation of σ *decreases*.

We have already seen that the ratio $(\bar{X} - M_H)/\sigma_{\bar{X}}$ is distributed normally and, therefore, we would expect this ratio to exceed ± 2.58 just 1% of the time as a result of sampling error. The ratio $(\bar{X} - M_H)/S_{\bar{X}}$ is *not normally distributed*, although its distribution approaches normal form as the number of degrees of freedom for the sum of squares term in S increases. However, when df is as low as 6, the ratio $(\bar{X} - M_H)/S_{\bar{X}}$ will exceed ± 2.58 two and one-half times as often as the normally distributed ratio $(\bar{X} - M_H)/\sigma_{\bar{X}}$.

When the denominator of this ratio consists of the *estimate* of the standard error, the ratio follows a distribution known as t rather than z, and the ratio itself is referred to as a t ratio. For example, $t = (\bar{X} - M_H)/S_{\bar{X}}$, while $z = (\bar{X} - M_H)/\sigma_{\bar{X}}$. Like χ^2, the t distribution takes a variety of forms depending on the number of degrees of freedom available for the problem. If there are only a few degrees of freedom, the t distribution has much fatter tails than the normal distribution. As df increases, these tails become progressively more like those of a normal distribution and, as a result, the t required for significance approaches z. Figure 10.1 shows a normal distribution and a t distribution for df = 10. You can see from the figure that the t ratio required to mark off the extreme 1% of the t distribution will be larger than the ± 2.58 required to designate the extreme 1% of a normal curve.

The values of t required for different levels of significance with various degrees of freedom are given in Table T of the appendix. Both t and z can be either positive or negative, but, as with the normal curve table, we have only recorded positive values of t. The table is designed for use with either one- or two-tailed tests. (Use Table T to verify that a $t = 3.169$ is required for significance at the 1% level when 10 df are available and a two-tailed test is in order.) As with z the obtained t must equal or exceed the tabled value before we can assume the result is significant at the

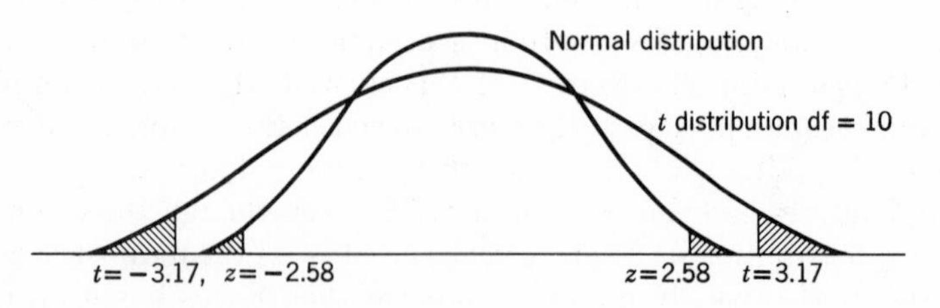

FIGURE 10.1 Areas of significance at the 1% level for a normal distribution and a t distribution with df = 10.

indicated level. Remember that unless a directional test is indicated we are not concerned with the sign of t.

Consider the following problem which requires the use of the t distribution. Suppose we have a sample, $n = 20$, in which $\bar{X} = 42.5$ and $\Sigma(X - \bar{X})^2 = 1680$. We wish to know if it is reasonable ($\alpha = .05$) to assume that this sample mean was obtained by random sampling from a population in which $M_H = 48.0$. We shall make use of the formula below.

$$t = \frac{\bar{X} - M_H}{S_{\bar{X}}}$$

FORMULA 10.6
t ratio for the difference between a sample mean ($\bar{X}$) and a hypothesized population mean (M_H)

$$t = \frac{\bar{X} - M_H}{S_{\bar{X}}} = \frac{\bar{X} - M_H}{\sqrt{\dfrac{\Sigma(X - \bar{X})^2}{n(n - 1)}}} = \frac{42.5 - 48.0}{\sqrt{\dfrac{1680}{20(20 - 1)}}} = -2.62.$$

A t test of an hypothesis about a population mean will have $n - 1$ degrees of freedom where n is the number of cases in the sample. Thus, for the problem above, we have $t = 2.62$, df $= 19$, $p < .05$. Since a $t = 2.093$ is required at the .05 level for 19 df, and we have obtained a t of 2.62, the hypothesis that this sample mean came from a population in which $M_H = 48.0$ can be rejected at the 5% level. Had the normal curve been used instead of the t distribution, the null hypothesis would have been rejected at an incorrect level of significance.

In this problem the t distribution was used to test a specific hypothesis about the population mean, that $M_H = 48$. However, instead of testing one specific hypothesis after another, it might be more economical to establish the limits within which any hypothesis about the population mean is tenable; we might establish the fiducial limits for the population mean. The logic for this procedure is exactly the same as that which led to the development of fiducial limits for the population proportion, a topic discussed in Chapter 8. We shall review these ideas briefly as they apply to the population mean.

In the previous problem $M_H = 48.0$ was rejected, but other values for M_H could have been chosen which would not have been rejected. Obviously, since the sample mean is the best estimate of the population mean, any value for M_H that is quite close to $\bar{X}$ cannot be rejected. We can, in fact, determine a *range* of possible values for M_H, both above and below $\bar{X}$, which will not be rejected at the 5% level. Any hypothesis which places the population mean within these limits, called fiducial limits, will not be rejected at the 5% level. Conversely, any hypothesis which places

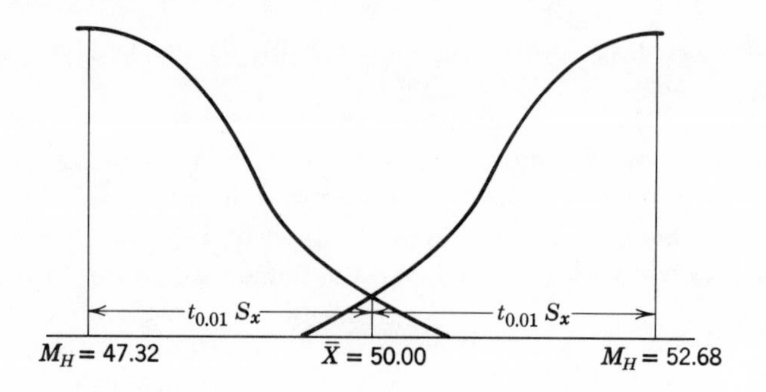

FIGURE 10.2 The fiducial limits of the mean.

the population mean outside these limits will be rejected at the 5% level. In short, the probability is .95 that the population mean will fall within these limits. As a result, the limits define what is called the 95% confidence interval. Of course, we can also construct the 99% confidence interval, or for that matter an appropriate interval for any level of confidence. The fiducial limits define the confidence interval.

In order to determine the fiducial limits of the mean we must solve the equation $t_{.05} = (\overline{X} - M_H)/S_{\overline{X}}$ for M_H. Since t can take both positive and negative values, we shall have for the lower fiducial limit $M_H = \overline{X} - t_{.05}S_{\overline{X}}$ and for the upper fiducial limit $M_H = \overline{X} + t_{.05}S_{\overline{X}}$. The fiducial limits for the data from the problem above are 38.1 to 46.9. Any hypothesis which places M_H *within* these limits cannot be rejected at the 5% level. They are the fiducial limits of the 95% confidence interval.

Suppose $n = 51$, $\overline{X} = 50.00$, and $\Sigma(X - \overline{X})^2 = 2550$. What is the 99% confidence interval for the population mean? The fiducial limits are found again from the equation $M_H = \overline{X} \pm t_{.01}S_{\overline{X}}$. The sample mean, $\overline{X}$, is given as 50.00. The value of t required for significance at the 1% level for 50 df $= \pm 2.68$, and $S_{\overline{X}} = \sqrt{\Sigma(X - \overline{X})^2/[n(n - 1)]} = 1.00$. Therefore, $M_H = \overline{X} \pm t_{.01}S_{\overline{X}} = 50 \pm 2.68$, and the fiducial limits are 47.32 and 52.68. When the data from the sample corresponds to that given above, the probability that the population mean falls within this confidence interval is .99.

Figure 10.2 illustrates the confidence interval just calculated. If $\overline{X} = 50$, $S_{\overline{X}} = 1.00$, and $t_{.01} = 2.68$, then sampling from a population with $M = 52.68$ will produce sample means below 50.00 just $\frac{1}{2}$% of the time. Similarly, sampling from a population with $M = 47.32$ will produce sample means above 50.00 just $\frac{1}{2}$% of the time. These values of M_H constitute the fiducial limits of the population mean at the 1% level and define the 99% confidence interval.

THE SIGNIFICANCE OF A DIFFERENCE BETWEEN MEANS

Situations which require us to determine the probability that a sample mean, $\bar{X}$, came from a population with mean M, are rather rare in applied research; few psychological theories predict population means. We are more commonly required to determine the probability that sampling error was responsible for a *difference* between two sample means. We may have taught two classes by different teaching procedures. If their mean achievement test scores are 28 and 33 we cannot immediately conclude that teaching methods have produced the difference, because we would expect some difference due to sampling error even if the teaching procedures were equally effective. The question we must be able to answer is this: How probable is the obtained sample mean difference on the basis of sampling error alone, when no mean difference exists between the populations being sampled? If the sample mean difference is highly improbable on the basis of sampling error we can reject the null hypothesis that sampling error alone was responsible. This permits us to accept the alternative hypothesis that the difference between the sample means was the result of sampling from populations with *different* means, a difference that presumably resulted from the different conditions of instruction.

Imagine that there are two infinite populations with means M_1 and M_2. Suppose a sample of n_1 cases is drawn from the first population, and the sample mean, $\bar{X}_1$, is calculated. Then a second sample of n_2 cases is drawn from the second population and that sample mean, $\bar{X}_2$, is calculated. Now the second sample mean is subtracted from the first, and the difference is recorded. If this procedure is continued for a large number of pairs of samples, we shall have the data to construct a sampling distribution of mean differences, $\bar{X}_1 - \bar{X}_2$. When the samples are composed of reasonably large n's, and an infinite number of pairs of sample means have been obtained, the sampling distribution of mean differences will have certain characteristics: its mean will coincide with $M_1 - M_2$, the difference between the means of the populations from which the samples came; and it will be normally distributed.

The concept of a sampling distribution of mean differences was discussed in Chapter 6. Table 6.3 (reproduced here as Table 10.1), shows a frequency distribution of differences between sample means when each sample has been drawn from the *same* population. Of course, if each sample comes from the same population, then $M_1 = M_2$ and $M_1 - M_2 = 0$. If we have drawn a large number of samples, and if $M_1 - M_2 = 0$, we should expect the mean of the sampling distribution of mean differences to be rather close to zero, and the form of the sampling distribution to be approximately normal. The problem facing the investigator is to determine the probability that a *particular* sample mean difference, $\bar{X}_1 - \bar{X}_2$, will occur given some hypothesis about $M_1 - M_2$, the difference between the means of the populations from which these samples have presumably come.

TABLE 10.1 A Sampling Distribution of 100 Mean Differences Where Samples ($n = 10$) Are Drawn from a Table of Random Numbers

2.5 to 3.4	|||
1.5 to 2.4	𝍸 𝍸 |
.5 to 1.4	𝍸 𝍸 𝍸 𝍸 ||
−.5 to .4	𝍸 𝍸 𝍸 𝍸 𝍸 |
−1.5 to −.6	𝍸 𝍸 𝍸 𝍸 ||||
−2.5 to −1.6	𝍸 𝍸 |
−3.5 to −2.6	|
−4.5 to −3.6	|

Since the theoretical sampling distribution of mean differences tends to be normal, the probability of obtaining any specific sample mean difference should be calculable if we can determine the standard deviation of sample mean differences. This statistic is called the standard error of the difference between means; it is symbolized by $\sigma_{\bar{X}_1 - \bar{X}_2}$. The standard error of the difference between means cannot normally be calculated directly, but it can be estimated, just as S provided an estimate of σ and $S_{\bar{X}}$ provided an estimate of $\sigma_{\bar{X}}$. The formula for this *estimate*, which we will symbolize as $S_{\bar{X}_1 - \bar{X}_2}$, is given by

$$S_{\bar{X}_1 - \bar{X}_2} = \sqrt{\frac{S_1^2}{n_1} + \frac{S_2^2}{n_2}}$$

FORMULA 10.7
The estimate of the standard error of the difference between two independent means

where S_1^2 is the unbiased estimate of variance for the population yielding $\bar{X}_1$, and S_2^2 is the unbiased estimate of variance for the population yielding $\bar{X}_2$.

When we can assume that the two samples have come from populations with the *same* variance, we can pool the sums of squares and the degrees of freedom from each sample to arrive at a single estimate of the population variance. This common estimate is given by

$$S^2 = \frac{\Sigma(X - \bar{X}_1)^2 + \Sigma(X - \bar{X}_2)^2}{n_1 + n_2 - 2}.$$

When this single estimate is substituted for each of the separate sample estimates in Formula 10.7 that formula reduces to

$$S_{\bar{X}_1 - \bar{X}_2} = \sqrt{S^2 \cdot \frac{n_1 + n_2}{n_1 n_2}}.$$

If we now use the raw score formula $\Sigma X^2 - (\Sigma X)^2/n$ for each of the sums of squares that are pooled to yield S^2, we shall have a computing formula for $S_{\bar{X}_1 - \bar{X}_2}$ (which may look rather unwieldy, but it will actually be very convenient when real calculations are required). This computing formula is

$$S_{\bar{X}_1 - \bar{X}_2} = \sqrt{\frac{\Sigma X_1^2 - \frac{(\Sigma X_1)^2}{n_1} + \Sigma X_2^2 - \frac{(\Sigma X_2)^2}{n_2}}{n_1 + n_2 - 2}\left(\frac{n_1 + n_2}{n_1 n_2}\right)}$$

FORMULA 10.8
A computing formula, based on raw scores, for the estimated standard error of the difference between two means

Once we know the standard error of the differences between means we may divide it into the difference between the obtained sample mean difference and the hypothesized population mean difference. The result is a t ratio whose distribution has already been discussed.

$$t = \frac{(\bar{X}_1 - \bar{X}_2) - (M_{H_1} - M_{H_2})}{S_{\bar{X}_1 - \bar{X}_2}}$$

FORMULA 10.9
t ratio of a difference between hypothesized and obtained differences divided by the estimated standard error of the differences

Notice the parallel between Formulas 10.6 and 10.9. Formula 10.6 gives t as the ratio of the difference between a sample mean and a hypothesized population mean divided by the estimated standard error of the sample mean. Formula 10.9 gives t as the ratio of the difference between sample *differences* and hypothesized population mean differences, divided by the estimated standard error of the sample mean *differences.*

Formula 10.9 is not very convenient for the direct computation of t. First, we are ordinarily interested in testing whether or not the difference between two sample means is significantly different from zero. That is, we normally hypothesize that $M_{H_1} - M_{H_2} = 0$. When this is the situation and the computational formula is substituted for $S_{\bar{X}_1-\bar{X}_2}$ we have Formula 10.10 which is essentially a computational version of Formula 10.9 with $M_{H_1} = M_{H_2}$.

$$t = \frac{(\bar{X}_1 - \bar{X}_2) - 0}{S_{\bar{X}_1-\bar{X}_2}} = \frac{\bar{X}_1 - \bar{X}_2}{\sqrt{\dfrac{\Sigma X_1^2 - \dfrac{(\Sigma X_1)^2}{n_1} + \Sigma X_2^2 - \dfrac{(\Sigma X_2)^2}{n_2}}{n_1 + n_2 - 2}\left(\dfrac{n_1 + n_2}{n_1 n_2}\right)}}$$

FORMULA 10.10
Computational formula for 10.9 when $M_{H_1} = M_{H_2}$

The df for the t given by 10.9 or its computational equivalent, 10.10 is $n_1 + n_2 - 2$. This is the result of using two estimates of the population variance each having $n - 1$ degrees of freedom. The total degrees of freedom are then $n_1 + n_2 - 2$.

Keep in mind that Formula 10.9 can be used to determine if a difference between sample means is the result of sampling error when the hypothesized difference between population means is *not* zero. Suppose a hypothesis specifies that the mean score on a test of mechanical ability will be 24 points higher for college men than for college women. Thus, $M_{H_M} - M_{H_W} = 24$. If we find a sample mean of 116 for men and a sample mean of 100 for women we have $\bar{X}_M - \bar{X}_W = 16$ and the numerator of our t ratio for this problem will then be $(\bar{X}_M - \bar{X}_W) - (M_{H_M} - M_{H_W}) = -8$. Remember that we are testing the significance of the departure of a sample mean difference (16 points) from a hypothesized population mean difference (24 points). If we wished to test the hypothesis that the populations had the *same* mean, then the numerator of the t ratio would be 16. In that situation the hypothesized population mean difference would be zero.

We shall illustrate the use of the t test with the following example. Suppose we have a group of 32 children who have been randomly assigned to one of two instructional programs designed to improve their vocabularies. At the termination of the experiment the children in Program A have a mean vocabulary test score of 37.56, and those in Program B have a mean vocabulary test score of 30.12. The investigator must determine whether these means are significantly different. The raw data and the t test are given below.

Group A		*Group B*	
36	31	26	29
42	35	25	27
31	47	36	31
39	43	37	33
47	46	31	29
38	36	24	30
36	33	38	22
29	32	36	28

$$\Sigma X_A^2 = 23{,}101 \qquad \Sigma X_B^2 = 14{,}872$$

$$\Sigma X_A = 601 \qquad \Sigma X_B = 482$$

$$n_A = 16 \qquad n_B = 16$$

$$\overline{X}_A = 37.56 \qquad \overline{X}_B = 30.12$$

$$t = \frac{\overline{X}_A - \overline{X}_B}{\sqrt{\dfrac{\Sigma X_A^2 - \dfrac{(\Sigma X_A)^2}{n_A} + \Sigma X_B^2 - \dfrac{(\Sigma X_B)^2}{n_B}}{n_A + n_B - 2}\left(\dfrac{n_A + n_B}{n_A n_B}\right)}}$$

$$= \frac{37.56 - 30.12}{\sqrt{\dfrac{23{,}101 - \dfrac{(601)^2}{16} + 14{,}872 - \dfrac{(482)^2}{16}}{16 + 16 - 2}\left(\dfrac{16 + 16}{16 \cdot 16}\right)}} = \frac{7.44}{1.91} = 3.90$$

$t = 3.90$, df $= 30$, $p < .01$

Before going on to the next section, conduct a t test to determine if the mean of an experimental group based on scores of 10, 10, 9, 8, and 7 is significantly different ($\alpha = .05$) from the mean of a control group based on scores of 7, 7, 6, and 5. You should find $t = 3.27$, df $= 7$, and $p < .05$.

THE t TEST FOR PAIRED OBSERVATIONS

The procedure we described for obtaining a theoretical sampling distribution of mean differences assumed that the assignment of sample elements was entirely random, that each pair of samples was composed of $n_1 + n_2$ independent observations. The form of the t test we described in the previous section should be applied only when this assumption can be met. In fact, this is sometimes called an independent groups t test to stress this particular assumption. When subjects have been randomly assigned to two treatment groups, as they were in the illustrative problem, the assumption of independence is certainly reasonable. However, there are other research situations where we may wish to assume the equivalence of experimental and control groups by matching or equating subjects on some variable highly correlated with the dependent variable. For example, if we are investigating methods of instruction using two groups of school children, we might wish to match a child in one group with a child of equivalent intelligence in another group. If pairs of children with equivalent intelligence test scores are assigned so that one member of each pair is in the experimental group and the other member is in the control group, we would certainly expect to find a relationship between these pairs on the dependent variable at the conclusion of the experiment. Under these circumstances we do not have $n_1 + n_2$ independent observations, but only n independent observations where n is the number of pairs. Consider another very similar situation. Suppose, instead of matching pairs of subjects, we test each subject twice, once under the experimental condition and once under the control condition. If the experimental design permits, this is the best possible match; each subject serves as his own control. However, if each subject is tested twice, we certainly cannot assume that the two sets of data are composed of independent observations. When the data consist of such *paired* observations, a modification of the t test is used called the matched pairs t test.

We have discussed the sampling distribution of the difference between randomly obtained sample means. Now we shall consider a somewhat different population of differences. Imagine two populations of individual measurements or scores such that each measurement in one population is paired with some measurement in the other population. If the second drawn measurement of each pair is subtracted from the first, $X_1 - X_2$, we shall have a population of differences (D), where $X_1 - X_2 = D$. The mean of these differences (M_D) will be equal to the difference between the means of the two parent populations ($M_1 - M_2$).

$$M_D = \frac{\Sigma D}{N} = \frac{\Sigma(X_1 - X_2)}{N} = \frac{\Sigma X_1}{N} - \frac{\Sigma X_2}{N} = M_1 - M_2$$

Assume that we have drawn a sample of paired scores from this population and that by subtracting the second score from the first we can obtain a difference score, D. This sample of difference scores will have a sample mean, $\bar{D} = \Sigma D/n$, and can also be used to provide an estimate of the standard deviation in the population of difference scores,

$$S_D = \sqrt{\frac{\Sigma D^2 - (\Sigma D)^2/n}{n - 1}}.$$

If we were to continue drawing such samples and calculating the mean of the differences, $\bar{D}$, for each sample, we could construct a sampling distribution of these mean differences. The mean of a sampling distribution of mean differences will, on the average, be equal to the mean difference for the population of paired measurements from which the sample came. The estimated standard error of this sampling distribution is given by Formula 10.11.

$$S_{\bar{D}} = \frac{S_D}{\sqrt{n}} = \frac{\sqrt{\dfrac{\Sigma D^2 - \dfrac{(\Sigma D)^2}{n}}{n - 1}}}{\sqrt{n}}$$

FORMULA 10.11
The estimated standard error of the mean of paired difference scores

The t ratio which tests the significance of a difference between a sample mean difference and a hypothesized population mean difference is given by

$$t = \frac{\bar{D} - M_{D_H}}{S_{\bar{D}}}.$$

When the hypothesized population mean difference is zero, as it usually is, a more appropriate computing formula is

$$t = \frac{\bar{D}}{S_{\bar{D}}} = \frac{\bar{D} - 0}{\sqrt{\dfrac{\Sigma D^2 - (\Sigma D)^2/n}{n(n - 1)}}}$$

FORMULA 10.12
t for paired observations when the hypothesized population mean difference is zero

For the evaluation of this t, we have $(n - 1)$ df, where n is the number of *pairs* of measurements.

We shall now illustrate a problem which would require the use of the matched pairs t test. Assume that we have used the same subjects in a pretest posttest situation, and we want to know if the pretest and posttest means are significantly different. Such paired observations *do not* consist of two randomly obtained, independent sets of scores, so we shall make use of a matched pairs t test. The data and computations are given in Table 10.2. Using the data from Table 10.2 we can make the following calculations.

TABLE 10.2 A Matched Pairs t Test Conducted for a Hypothetical Pre-Post Test

Person	Pretest	Posttest	Difference	D^2
A	16	9	7	49
B	18	12	6	36
C	22	16	6	36
D	14	16	−2	4
E	16	14	2	4
F	12	13	−1	1
G	19	16	3	9
H	22	17	5	25
I	22	19	3	9
J	17	8	9	81
	$\bar{X}_1 = 17.8$	$\bar{X}_2 = 14.0$	$\Sigma D = 38.0$	$\Sigma D^2 = 254$
			$\bar{D} = 3.8$	

$$S_D = \sqrt{\frac{\Sigma D^2 - (\Sigma D)^2/n}{n - 1}} = \sqrt{\frac{254 - \frac{(38)^2}{10}}{9}} = 3.49$$

$$S_{\bar{D}} = \frac{S_D}{\sqrt{n}} = \frac{3.49}{3.16} = 1.10$$

$$t = \frac{\bar{D}}{S_{\bar{D}}} = \frac{3.80}{1.10} = 3.45$$

$t = 3.45$, df $= 9$, $p < .01$ ($r = .50$ between pretest and posttest measurements)

The difference between these sample means is significant at the 1% level.

We have already mentioned that the matched pairs t test is not confined to situations in which the same subjects are tested twice; that it can also be used when subjects have been matched into pairs on some third variable believed to be correlated with the dependent variable. For example, we might wish to match pairs of subjects on mathematics aptitude prior to their instruction in mathematics and then give them instruction under two different classroom teaching procedures. We might select 12 subjects for one group who could be matched on mathematics *aptitude* test scores with 12 subjects in the other group. Then, after subjecting the two groups to different instructional methods, we would treat the resulting differences between each pair of mathematics *achievement* test scores exactly as we did the data in the last example. The data from such an experiment might look like that illustrated in Table 10.3. Using the data from Table 10.3 we can make the following calculations.

TABLE 10.3 Mathematics Achievement Test Scores of Matched Pairs Subjects Having Different Methods of Instruction

Group I			Group II				
	Matched on						
Person	Aptitude	Achievement	Person	Aptitude	Achievement	D*	D^2
A	160	27	G	158	21	6	36
I	149	22	D	152	20	2	4
F	146	18	H	145	16	2	4
V	131	26	E	126	19	7	49
L	124	19	K	124	16	3	9
P	122	18	W	122	16	2	4
X	120	24	M	121	18	6	36
O	118	16	N	117	12	4	16
B	115	14	C	114	10	4	16
J	113	13	T	110	16	−3	9
Q	112	13	S	111	12	1	1
U	109	12	R	107	10	2	4
$\bar{D} = 3.00$						$\Sigma D = 36$	
							$\Sigma D^2 = 188$

* D, the difference between the pairs of achievement scores.

$$S_D = \sqrt{\frac{\Sigma D^2 - \frac{(\Sigma D)^2}{n}}{n - 1}} = \sqrt{\frac{188 - 108}{11}} = 2.70$$

$$S_{\bar{D}} = \frac{S_D}{\sqrt{n}} = \frac{2.70}{3.46} = .78$$

$$t = \frac{\bar{D}}{S_{\bar{D}}} = \frac{3.00}{.78} = 3.85$$

$$t = 3.85, \text{ df} = 11, p < .01$$

Notice that D, the difference in which we are interested, is between the pairs of achievement test scores. Achievement is the dependent variable in this experiment. Since our value of t is significant at the 1% level, we can reject the hypothesis that these two populations, defined by the difference in instruction, have the same mean achievement in mathematics.

When some reasonable basis exists for matching subjects, it will usually be advantageous to do so and to make use of the matched pairs t test. This is because matching subjects, or using the same subject as his own control as in a pretest and posttest design, usually results in a fairly high correlation between the two sets of scores on the dependent variable which reduces the variability of the difference scores (D's) and consequently reduces $S_{\bar{D}}$. The reduction in $S_{\bar{D}}$ means that the distribution of sample mean differences will cluster more tightly about the population mean difference and this *increases* the power of the t test. We can illustrate the effect of correlation with the two sets of data given below.

Correlated			*Uncorrelated*		
X_1	X_2	D	X_1	X_2	D
10	9	1	10	24	−14
19	17	2	19	9	10
6	8	−2	6	8	−2
26	24	2	26	17	9

Notice that we have exactly the same observations with the same difference between means. The data from one set consist of highly correlated observations, such as might be obtained if each subject served as his own control; the data in the other set consist of uncorrelated observations, such as might result from random assignment. For the differences among the correlated observations, S_D is 1.9; for the differences among the uncorrelated observations, consisting of the same scores, S_D is 11.2. Since S_D is smaller for the correlated group, the $S_{\bar{D}}$ will also be smaller, $S_{\bar{D}} = S_D/\sqrt{n}$. The reduction in S_D will yield a larger t for the correlated group, even though $\bar{D}$ is the *same* in both situations.

The concept of the sampling error of a difference between means, $S_{\bar{X}_1-\bar{X}_2}$, is the same as the concept of the sampling error of a mean difference, $S_{\bar{D}}$. The formula for $S_{\bar{D}}$ differs from that for $S_{\bar{X}_1-\bar{X}_2}$, because $S_{\bar{D}}$ takes the correlation into account, although $S_{\bar{X}_1-\bar{X}_2}$ does not. When there is no correlation between the pairs of measures on the dependent variable, the two formulas will give identical results. One might suppose that since this is the case we should always match on some variable; there seems to be nothing to lose and $S_{\bar{D}}$ might be lower than $S_{\bar{X}_1-\bar{X}_2}$, however we do lose degrees of freedom by attempting to match. The t test for independent groups, using $S_{\bar{X}_1-\bar{X}_2}$, has $(n_1 + n_2 - 2)$ df while the matched pairs t test, using $S_{\bar{D}}$, has only $(n - 1)$ df where n is the number of pairs. For the matched pairs procedure to be more powerful, we *must* have enough correlation between the pairs of measurements on the dependent variable to offset the loss in degrees of freedom. Matching is always preferred when it leads to a high correlation between the pairs on the dependent variable.

There is another test of significance, called the sign test, which also can be applied to matched data. We shall use it here to show what happens when a less powerful test is used in place of a more powerful one. You may recall that the sign test was presented before; it was given as a solution to Exercise 12 in Chapter 9. Let us apply the sign test to the data in Table 10.2 that provided us with our first illustration of the matched pairs t test. There are 10 pairs of observations; in 8 of them the pretest score is higher than the posttest score giving a + sign to the difference; in 2 of them the pretest score is lower than the posttest score giving a − sign to the difference. If the populations are normally distributed, and if the difference between their means is zero, we will expect an equal number of positive and negative differences; any departures from equality should occur only as a result of random sampling. Under the null hypothesis, for the problem at hand, we expect 5 positive signs and 5 negative signs in the 10 pairs of observations. Since the expected frequencies are at least 5, the problem will yield the following chi square.

	Positive	Negative	
Observed	8	2	10
Expected	5	5	10

$$\chi^2 = \sum \frac{(|o - e| - .05)^2}{e} = \frac{(2.5)^2}{5} + \frac{(2.5)^2}{5} = 2.5$$

$$\chi^2 = 2.5,\ \text{df} = 1,\ p > .10$$

A χ^2 of 2.5 for 1 df is not significant at the 10% level, yet for a matched pairs t test the *same data* led to a t of 3.45 which, for 9 df, was significant at the 1% level. This is not an error; it simply illustrates the fact that the matched pairs t test is much more powerful than the sign test. As we mentioned in Chapter 9, the median test, and its companion for matched data, the sign test, do not take the magnitudes of differences into account. While the sign test is totally insensitive to the magnitudes of differences which lie behind the signs, the t test is decidely influenced by them. This makes the t test much more powerful than the sign test and generally preferable to it.

One might be inclined to ask why bother with such tests as the median test, or the sign test, if the t test for independent groups and t test for matched pairs are available and are so much more powerful. The answer is that there are certain assumptions about the data which *must* be met before we can legitimately use the t test, but which are not required by the median test and the sign test. When the t test is used and these assumptions are not met, the t's we calculate will not be distributed as the tabled values of t, and the conclusions we draw may be inaccurate. In the next section we shall review some of the more important assumptions for the t test, but first we shall consider some additional example problems illustrating the concepts we have just covered.

Suppose we wish to know if the conditions A and B have any differential influence on the trials to learn some particular task. We might randomly assign 10 S's to each group and record their trials to reach some criterion of performance on the task. If there is a significant difference between mean trials for the two groups we would conclude that the conditions differentially affect learning. The hypothetical data are recorded in Table 10.4. This is the t test for independent groups, so we shall use Formula 10.10 to determine t.

TABLE 10.4 Number of Trials for a Learning Task Under Two Different Conditions (Independent Groups Design)

X_A	X_A^2	X_B	X_B^2
3	9	9	81
4	16	6	36
8	64	5	25
9	81	8	64
3	9	7	49
5	25	6	36
2	4	8	64
4	16	7	49
3	9	2	4
1	1	8	64
$\Sigma X_A = 42$	$\Sigma X_A^2 = 234$	$\Sigma X_B = 66$	$\Sigma X_B^2 = 472$
$\overline{X}_A = 4.2$		$\overline{X}_B = 6.6$	

The difference between sample means is significant at the 5% level.

$$t = \frac{\bar{X}_A - \bar{X}_B}{\sqrt{\dfrac{\Sigma X_A{}^2 - \dfrac{(\Sigma X_A)^2}{n_A} + \Sigma X_B{}^2 - \dfrac{(\Sigma X_B)^2}{n_B}}{n_A + n_B - 2}\left(\dfrac{n_A + n_B}{n_A n_B}\right)}}$$

$$= \frac{4.20 - 6.60}{\sqrt{\dfrac{234 - \dfrac{(42)^2}{10} + 472 - \dfrac{(66)^2}{10}}{10 + 10 - 2} \times \dfrac{10 + 10}{10 \cdot 10}}}$$

$$t = \frac{-2.40}{\sqrt{1.044}} = -2.35$$

$$t = 2.35,\ \text{df} = 18,\ p < .05$$

The same study could have been done by matching subjects on the basis of their performance on some *similar* learning task before assigning them to the treatment groups A and B. If the same 20 subjects were given a pretest, the following scores might have been obtained

Performance on the Matching Pretest

C	20	B	18	S	14	F	10	H	7
L	20	A	16	O	14	R	10	D	7
K	20	I	15	Q	13	P	9	M	6
J	19	C	15	G	10	E	8	N	6

After listing the subjects by the rank order of their pretest scores, as we have done above, we form matched pairs by assigning one member of the first pair (subjects C and L) to treatment group A, and the other to treatment group B. This process is continued through all pairs following the rank order down through pair M and N. The result is two groups of subjects where each individual in one group is paired with someone of equivalent ability in the other group. One group is then subjected to condition A, the other to condition B. The data are recorded in Table 10.5. Using these data and Formula 10.12 we have calculated the matched pairs t which appears on page 194. The difference between these sample means is significant at the 1% level.

We wish to stress again that the higher t in this second example, where S's are matched into pairs, was produced by arranging for a correlation between the pairs of measures on the dependent variable. The high correlation decreased $S_{\bar{D}}$ and, thereby, increased the power of the t test. Correlations between the pairs of measures on the dependent variable result from careful matching, or pairing, of subjects on the basis of some third variable which is known to correlate highly with the depen-

TABLE 10.5 Number of Trials for a Learning Task under Two Different Conditions (Matched Pairs Design)

A		B		Difference	D^2
C	9	L	9	0	0
K	8	J	8	0	0
A	5	B	8	−3	9
C	4	I	7	−3	9
S	4	O	8	−4	16
G	3	Q	5	−2	4
F	3	R	6	−3	9
P	3	E	7	−4	16
H	1	D	6	−5	25
N	2	M	2	0	0
				$\Sigma D = -24$	$\Sigma D^2 = 88$

$$\bar{D} = \frac{\Sigma D}{n} = -2.4$$

$$t = \frac{\bar{D}}{\sqrt{\dfrac{\Sigma D^2 - \dfrac{(\Sigma D)^2}{n}}{n(n-1)}}} = \frac{-2.4}{\sqrt{\dfrac{88 - \dfrac{(-24)^2}{10}}{10(10-1)}}} = \frac{-2.4}{.580} = -4.13$$

$t = 4.13$, df $= 9$, $p < .01$

dent variable. This tends to produce some correlation between scores on the dependent variable. When it is practical, the most effective matching is accomplished by pairing each subject with himself, by testing each subject twice, once in each of the experimental conditions. This is called the pre-post test design, and was used in an earlier illustration of the matched pairs t test.

While the pre-post test design is extremely powerful because each subject serves as his own control, this very fact introduces some problems for its use in certain kinds of experiments. Suppose a counseling psychologist decides to institute a program to help probationary students improve their grades. He selects 10 subjects at random from the group placed on probation at the end of the previous term, and for three months devotes himself to intensive therapy sessions with these students. When the next term's grades are released he is gratified to find that a matched pairs t test comparing pretherapy and posttherapy grades reveals a significant improvement for his group of counselees and he concludes that the counseling has had a beneficial influence. Possibly it has, but it cannot be demonstrated in such a study. Other factors such as parental pressure and the draft were probably also at work, and so the counselor's efforts constitute a confounded variable.

Such a study should have a control group, but this does not mean that we must forego a matched pairs t test. For the proper conduct of this study we need two groups of subjects matched on the basis of grades at the beginning of "therapy." One group is subjected to therapy and the other is not. We then determine pre-post test differences for each group. Our interest lies in the *difference between these differences.* Consider the data recorded in Table 10.6.

The columns labeled Pre_1 and Pre_2 are the end-of-term grades on the basis of which the S's are matched into two groups. The Pre_2 and $Post_2$ improvement (D_2) demonstrates the facilitative effect of parents, deans and miscellany. This is the control group. The Pre_1-$Post_1$ improvement (D_1) demonstrates the facilitative effect of these influences *and* the counselor's efforts. The experimental question involves the *difference* between these differences ($D_1 - D_2$). The null hypothesis would stipulate

TABLE 10.6 The Effect of Counseling on Student Achievement, Hypothetical Data for a Matched Pairs t Test

	Experimental				Control				
Subject	Pre_1	$Post_1$	D_1	Subject	Pre_2	$Post_2$	D_2	$D_1 - D_2$	$(D_1 - D_2)^2$
A	20	28	8	D	19	25	6	2	4
B	24	30	6	E	25	30	5	1	1
C	14	21	7	I	16	22	6	1	1
K	20	26	6	J	21	26	5	1	1
Q	27	34	7	F	26	31	5	2	4
P	15	19	4	N	19	23	4	0	0
G	12	18	6	S	15	20	5	1	1
H	19	27	8	R	12	18	6	2	4
T	13	20	7	O	13	18	5	2	4
L	18	25	7	M	17	23	6	1	1
								$\overline{D_1 - D_2} = \overline{1.3}$	
									$\Sigma(D_1 - D_2)^2 = 21$

that this mean $\overline{(D_1 - D_2)}$ differs from zero only as a result of sampling error. The t test of this hypothesis would proceed as follows.

$$t = \frac{\overline{D_1 - D_2}}{\sqrt{\dfrac{\Sigma(D_1 - D_2)^2 - \dfrac{[\Sigma(D_1 - D_2)]^2}{n}}{n(n-1)}}} = \frac{1.3}{\sqrt{\dfrac{21 - 16.9}{90}}} = 6.10$$

$$t = 6.10, \text{ df} = 9, p < .001$$

The counseled group has shown significantly more improvement than the uncounseled group.

The same research could have been conducted without matching pairs on the basis of grades at the beginning of the term. We would still be comparing D_1 and D_2 but since the subjects were not paired we would treat these difference scores as two sets of independent measures and conduct an independent groups t test of the difference between their means. If we used the independent groups procedure we would proceed as follows, this time using the data in Table 10.7.

TABLE 10.7 The Data from Table 10.6 When Subjects Are Unmatched on the Pretest

Experimental				Control			
Pre_1	$Post_1$	X_1	X_1^2	Pre_2	$Post_2$	X_2	X_2^2
20	28	8	64	19	25	6	36
13	20	7	49	25	30	5	25
18	25	7	49	16	22	6	36
14	21	7	49	21	26	5	25
20	26	6	36	26	31	5	25
19	27	8	64	19	23	4	16
15	19	4	16	15	20	5	25
12	18	6	36	12	18	6	36
27	34	7	49	13	18	5	25
24	30	6	36	17	23	6	36
		$\Sigma X_1 = 66$				$\Sigma X_2 = 53$	
			$\Sigma X_1^2 = 448$				$\Sigma X_2^2 = 285$

$$t = \frac{\bar{X}_1 - \bar{X}_2}{\sqrt{\dfrac{\Sigma X_1^2 - \dfrac{(\Sigma X_1)^2}{n_1} + \Sigma X_2^2 - \dfrac{(\Sigma X_2)^2}{n_2}}{n_1 + n_2 - 2}\left(\dfrac{n_1 + n_2}{n_1 n_2}\right)}}$$

$$t = \frac{6.6 - 5.3}{\sqrt{\dfrac{448 - \dfrac{(66)^2}{10} + 285 - \dfrac{(53)^2}{10}}{10 + 10 - 2} \times \dfrac{20}{100}}} = 3.04$$

$$t = 3.04, \text{ df} = 18, p < .01$$

These examples should illustrate the fact that the t test is not a technique to be applied arbitrarily whenever two means are presented. The experimenter must constantly bear in mind the nature of his experimental design. He must know the kind of question he wishes to ask, and he must be clever enough to select the data which will answer it. Tests of significance are essential to most such enterprises, but they provide only ambiguous answers unless they are intelligently used.

ASSUMPTIONS FOR t

The t distribution is based upon the assumption that samples have been drawn from normally distributed populations, and that these populations have the same variance. In fact, the t test is fairly robust. This means that even if the proper assumptions cannot be met exactly, the probability of obtaining significant t's under the null hypothesis will still be quite close to the α levels given in Table T for the specified df.

When the data appear to have come from skewed populations, it is advisable to avoid the use of small samples and the use of one-tailed tests. With two-tailed tests based on reasonably large n's, the tabled values of t will still be fairly accurate even when the parent populations show considerable skewness.

In order to use the t distribution we should also be able to assume that the populations from which the samples come have equal variances. You may recall that we made use of this assumption when we pooled the sum of squares and degrees of freedom from each sample into one common estimate of the population variance. This estimate was then used in Formula 10.8 to determine $S_{\bar{X}_1 - \bar{X}_2}$, the estimated standard error of the difference between means. If the populations seem to have unequal variances, and there are techniques which will be discussed in the next chapter for determining the significance of the difference between sample variances, then there are some special modifications of the t test which should be used.

First, if the population variances are unequal we should not pool the sums of squares and degrees of freedom from both samples to arrive at a common estimate of the population variance. Under these circumstances, we should calculate $S_{\bar{X}_1-\bar{X}_2}$ from Formula 10.7 rather than Formula 10.8.

Second, there is a change in the procedure for determining the *t required* for any given level of significance. The required t cannot be obtained directly from a t table; it must be calculated from the following formula which is an approximation developed by Cochran and Cox.[2] When the population variances are assumed to be unequal, the *t needed for significance* is given by Formula 10.13 below.

$$t = \frac{t_1 \dfrac{S_1^2}{n_1} + t_2 \dfrac{S_2^2}{n_2}}{\dfrac{S_1^2}{n_1} + \dfrac{S_2^2}{n_2}}$$

FORMULA 10.13
Value of t required for significance when populations have unequal variances

The elements of this formula should already be familiar, except for t_1 and t_2. These are taken directly from the t table and are the values of t required for significance at the level set by the experimenter; t_1 is the t required for $(n_1 - 1)$ df, and t_2 is the t required for $(n_2 - 1)$ df. The formula will not seem so complicated after we have illustrated its use with an example problem.

We shall consider an illustrative problem in which the sample variances are sufficiently different to warrant the use of the modified t test. Remember, however, that there is a test for the significance of the difference between sample variances; it is described in the next chapter. One would ordinarily use this test to determine if the sample variances are significantly different before resorting to this modification of the t test. Now for the problem: suppose we have obtained the following data and we wish to determine if there is a significant difference between the sample means. We shall specify a two-tailed test and require the 1% level of significance before rejecting the null hypothesis.

$$\bar{X}_1 = 1348.76 \qquad \bar{X}_2 = 1265.41$$

$$S_1^2 = 15697 \qquad S_2^2 = 3108$$

$$n_1 = 20 \qquad n_2 = 14$$

[2] Cochran, W. G. and Cox, G. M. *Experimental Designs*, New York: Wiley, 1950.

The first step is to obtain the estimated standard error of the difference between means. We should use Formula 10.7 which yields the following.

$$S_{\bar{X}_1-\bar{X}_2} = \sqrt{\frac{S_1^2}{n_1} + \frac{S_2^2}{n_2}} = \sqrt{\frac{15697}{20} + \frac{3108}{14}} = 31.73$$

Once the estimated standard error of the difference between means has been found it can be divided into the difference between the sample means to arrive at t. This yields

$$t = \frac{\bar{X}_1 - \bar{X}_2}{S_{\bar{X}_1-\bar{X}_2}} = \frac{1348.76 - 1265.41}{31.73} = 2.63.$$

Now that t has been obtained, we must determine if it is significant. The *t required for significance* when population variances are different is given by Formula 10.13.

$$t = \frac{t_1 \dfrac{S_1^2}{n_1} + t_2 \dfrac{S_2^2}{n_2}}{\dfrac{S_1^2}{n_1} + \dfrac{S_2^2}{n_2}} = \frac{2.861 \dfrac{15697}{20} + 3.012 \dfrac{3108}{14}}{\dfrac{15697}{20} + \dfrac{3108}{14}} = 2.89$$

Since Formula 10.13 shows that a t of 2.89 is required for significance at the 1% level, and since the t we have obtained is 2.63, we cannot reject the hypothesis that the two samples have come from populations with the same mean.

In spite of the fact that the t test assumes equality of population variances, the t distribution is *not* severely affected by unequal variances *unless* the samples are also small and unequal in size. There is evidence that sample variances that stand in a ratio of 4:1 will not unduly influence t, provided sample n's are equal and each sample consists of at least 15 cases. Of course, when variances are quite different and sample n's are small and unequal, we can resort to the modified t test given above.

TRANSFORMATIONS OF THE SCALE

The assumption of normally distributed populations is relatively unimportant when the t test is based on large samples, but it becomes critically important when samples are small. Unfortunately, the typical procedures for measuring certain psychological and educational variables tend to produce dedidedly non-normal distributions. For example, the running time of animals in a maze usually shows an extreme positive skew. When counts are made of behavior per unit time, or unit area, such as the distribution of words typed per minute or errors per page, the Poisson distribution results. This distribution looks like a lopsided normal curve and has a mean equal to its variance. When these or other variables yielding non-normal data are investigated with small samples, certain procedures are available to normalize the measures so that the t test can be used.

When distributions are not normal, or when the sample variances are quite disparate, it is sometimes possible to correct one or both conditions by transforming the scale of measurement. For example, running times in a maze are often converted to speed scores by taking the reciprocal of time; thus, speed $= 1/X$ where X is time. This conversion has some logical advantages pointed out by Crespi,[3] and may also tend to equalize disparate variances. If extreme skew exists, recording $\log_{10} X$ will often normalize the data. When high correlation exists between the means and variances of several subgroups, indicating Poisson distributions, we may be able to eliminate the correlation and normalize the data by substituting $\sqrt{X + .5}$ for the original dependent variable, X.

We can also normalize measures by converting them to normalized z scores, a procedure described in Chapter 4. You may remember that if a group of test scores is skewed, the resulting z scores will also be skewed, but the distribution can be normalized by finding the centile equivalent of each score, determining the z for that centile from Table N, and assigning the z score equivalent of that centile to the raw score. These are called normalized z scores and they will be normally distributed.

The scale of measurement used to describe any bit of behavior is arbitrary. There is nothing underhanded about transforming the scale so that the mathematics of a particular distribution applies. Of course, the conclusions that result will not necessarily apply to the variable if it is measured by some scale other than that on which the test of significance is based.

TESTS OF SIGNIFICANCE FOR CORRELATION COEFFICIENTS

The concept of a sampling distribution should be quite familiar by now. We have discussed sampling distributions of sample proportions, sample means, and sample mean differences. In this section we shall extend the concept to sampling distributions of correlation coefficients.

Suppose we have a population of *pairs* of measurements in which the

[3] Crespi, L. P. "Quantitative variation of incentive and performance in the white rat." *American Journal of Psychology*, 1942, **55**, 467–517.

correlation is zero. Nevertheless, if samples of 20 pairs of measurements are randomly obtained from this population we will find that the sample correlation coefficients range on either side of zero simply as a result of sampling error. This sampling distribution, like others we have studied, is a function of sample size. Large samples with high correlations are very unlikely if the population correlation is zero. Whether a sample correlation coefficient is significantly different from zero can be tested by t. For any sample of n pairs of measurements Formula 10.14 gives the relationship of r to t.

$$t = \frac{r\sqrt{n-2}}{\sqrt{1-r^2}}$$

FORMULA 10.14
t equivalent for sample r when the hypothesis is that the population correlation is zero

This t is evaluated with $(n - 2)$ df.

It is not actually necessary to determine t if we only wish to know if a sample r is significantly different from zero. Table R of the appendix contains a list of the critical values of r for a variety of degrees of freedom. This list specifies the sample r's which must be obtained in order to reject the hypothesis that the population correlation is zero. Table R is entered with $n - 2$ degrees of freedom.

Table R and Formula 10.14 can *only* be used when we are testing the hypothesis that the correlation in the population is zero. When the population correlation coefficient is not zero (particularly when the correlation is large, positive or negative), and samples are small, a special problem is encountered. We shall describe why this difficulty arises and what to do about it.

Assume we have a population of pairs of measurements with a correlation of .80 for the population. If we draw samples of 20 pairs of measurements, a correlation coefficient can be calculated for each sample and we can construct a sampling distribution of such correlation coefficients. Unfortunately, the sampling distribution will be markedly skewed, particularly when it is based on small samples drawn from populations in which r is high (positive or negative).

Figure 10.3 illustrates the sampling distribution of a correlation coefficient where $n = 8$, and the correlation in the population is .80. The skew is quite apparent and the reason is easy enough to understand. We have already seen that the variability of sample statistics is inversely related to sample size. As with the standard error of other sample statistics, the smaller the sample the larger the standard error of the correlation coefficient. However, regardless of the sample variability, the correlation coefficient cannot exceed the limits -1.00 to $+1.00$. If the samples are small

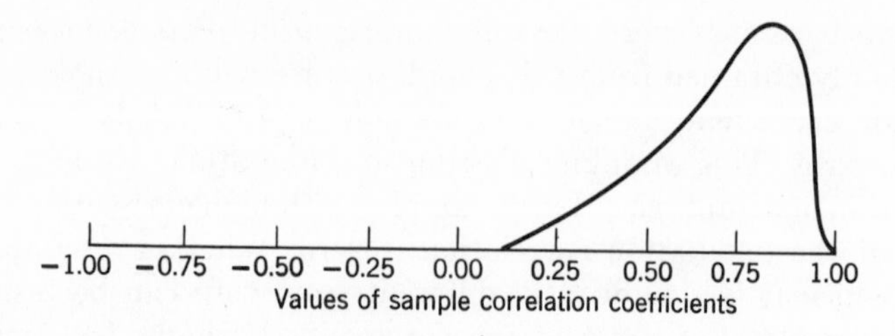

FIGURE 10.3 Distribution of sample correlation coefficients when n is small and the correlation in the population is .80.

and the correlation in the population is high, either positive or negative, these limits on the size of r will restrict one end of the sampling distribution more than the other, and this will produce the skewed distribution of sample correlation coefficients seen in Figure 10.3.

Since the sampling distribution of r is badly skewed when the population correlation is high and samples are small, the first step will be to normalize the distribution of sample r's by the use of a transformation called Fisher's Z' using the following equation.

$$Z' = \tfrac{1}{2}[\log_e (1 + r) - \log_e (1 - r)]$$

(Do not confuse Z', a transformed measure of r, with z the normal deviate.) It happens that Z' is distributed almost normally and has a known standard error.

$$\sigma_{Z'} = \frac{1}{\sqrt{n - 3}}$$

FORMULA 10.15
Standard error of Z'

Since Z' is distributed in an approximately normal fashion, we may set up a conventional normal deviate z from the following equation.

$$z = \frac{Z'_{\text{sample}} - Z'_{\text{population}}}{\sigma_{Z'}}$$

FORMULA 10.16
Relative deviation of a sample r from a hypothesized population r

Suppose we have a personality test which has been given to thousands of college sophomores and the correlation between the scores on this test and college grades has been found to be .30. Now imagine that we have shortened this test and administered it to a sample of 228 college sophomores. The correlation between their personality test scores and college grades is .39. We wish to know if this deviation is reasonable as a result of sampling from a population in which $r = .30$. In short, is the sample $r = .39$, with $n = 228$, sufficient evidence to reject the hypothesis that the sample has been drawn from a population in which the correlation is .30. We shall use a two-tailed test and the 5% level.

We proceed by finding the Z' equivalents for $r = .30$ and $r = .39$. In order to avoid the tedium of calculation, we have listed the Z' equivalents for most values of r in Table Z of the Appendix. Notice from this table that when $r = .30$, $Z' = .310$, and when $r = .39$, $Z' = .412$. Once Z' equivalents have been obtained we can use Formulas 10.15 and 10.16 to determine the *normal* deviate z. From Formula 10.15 we have the following calculation.

$$\sigma_{Z'} = \frac{1}{\sqrt{n-3}} = .067$$

With Formula 10.16, we obtain z.

$$z = \frac{Z'_{\text{sample}} - Z'_{\text{population}}}{\sigma_{Z'}} = \frac{.102}{.067} = 1.52$$

Since a z of ± 1.96 is needed to reject the hypothesis at the 5% level, and the obtained z for this problem is less than that, we cannot reject the hypothesis that sampling error accounts for our observations.

Now let us consider a different kind of problem. Suppose we obtain a sample of 250 cases in which the correlation between test A and some criterion variable, G, is .36. Then, in another equivalent sample of 250 cases we find the correlation between test B and variable G is .49. We want to determine if the difference between these correlation coefficients is significant. Can we reject the null hypothesis that the correlation coefficients could have arisen as a result of random sampling from some common population?

We can solve this problem by finding the standard error of the difference between randomly obtained Z' values, and then divide this standard error into the difference between the Z' equivalents of these sample r's. This will yield the normal deviate z, which may then be interpreted with reference to Table N of the normal curve. The standard error of the difference between randomly obtained Z' values is given by Formula 10.17 shown below.

$$\sigma_{Z'_1 - Z'_2} = \sqrt{\frac{1}{n_1 - 3} + \frac{1}{n_2 - 3}}$$

FORMULA 10.17
Standard error of the difference between two Z' values

When the Z' equivalents for the sample r's have been obtained from Table Z, we can find the normal deviate z from Formula 10.18.

$$z = \frac{Z'_1 - Z'_2}{\sigma_{Z'_1 - Z'_2}}$$

FORMULA 10.18
Normal deviate based on the standard error of the difference between Z' values

We can apply these formulas in the preceding problem. Suppose we have developed two tests and wish to determine if the correlation between each of them and some criterion differs significantly. The null hypothesis will be that the correlations differ only within the limits of sampling error. Two independent samples of 250 cases each are obtained. Sample 1 is given test *A* and the criterion test. Sample 2 is given test *B* and the criterion test. The following data are then available.

$$n_1 = 250, \qquad r_1 = .49$$

$$n_2 = 250, \qquad r_2 = .36$$

$$Z'_1 \text{ (from Table Z)} = .536$$

$$Z'_2 \text{ (from Table Z)} = .377$$

$$\sigma_{Z'_1 - Z'_2} = \sqrt{\tfrac{1}{247} + \tfrac{1}{247}} = .090$$

$$z = \frac{.159}{.090} = 1.77$$

A z of 1.77 is not significant at the .05 level for a two-tailed test, so we do not have adequate evidence to reject the null hypothesis. We cannot assume that tests *A* and *B* differ significantly in their correlation with the criterion.

It is apparent from these data that we must work with fairly substantial samples, or have very large differences between sample r's, if we hope to demonstrate statistically significant differences between correlation coefficients when samples are independent.

The Z' transformation procedure we have just described for testing the significance of differences between sample r's applies only to situations in which the samples are independent, as would normally be the case when the samples are composed of different sets of subjects. However, it is quite common for the research worker to compare two correlation coefficients which are *both* calculated from measurements of the *same* set of subjects. This is a much more efficient procedure, but the test of significance is now based upon the t distribution.

Suppose we contrast these two procedures by comparing our previous example, in which the groups were independent, with a slight modification which will yield correlated observations. Watch what happens to the level of significance when the more efficient procedure is used.

Imagine that we used two tests of academic ability in our previous example and that the criterion measure with which they were correlated was college grades. If we have 500 freshmen, we might randomly assign 250 of them to take test *A* and 250 to take test *B*, and at the end of the freshman year when grades have been announced, we might determine that $r_{AG} = .36$ and $r_{BG} = .49$. The significance of the difference between these two correlation coefficients was determined in the previous example with the use of Formulas 10.17 and 10.18. These formulas are appropriate here because the samples are independent. On the other hand, suppose we give all 500 freshmen both tests A and B, and then determine the correlation of each test with the criterion. The correlation coefficients are no longer independent because both have been obtained from measurements of the same subjects. In this situation we should make use of a modification of the t test where

$$t = \frac{(r_{BG} - r_{AG})\sqrt{(n-3)(1+r_{AB})}}{\sqrt{2(1 - r_{AG}^2 - r_{BG}^2 - r_{AB}^2 + 2r_{AG}r_{BG}r_{AB})}}$$

FORMULA 10.19
A t test for the significance of difference between correlation coefficients when samples are not independent

The elements of this formula should all be familiar except, perhaps, r_{AB}. This is the correlation between the two tests A and B for the 500 subjects in our sample. We shall give this intercorrelation a value of .70. For our example problem we have

$$t = \frac{(.49 - .36)\sqrt{497 \times 1.70}}{\sqrt{2[1 - .36^2 - .49^2 - .70^2 + 2(.36)(.49)(.70)]}} = 4.26.$$

$$t = 4.26, \text{ df} = 497, p < .001$$

This t should be evaluated with $n - 3$ degrees of freedom, where n is the number of subjects. With this procedure, the difference between these two correlation coefficients is significant at the .1% level.

Our discussion of significance tests for correlation coefficients has so far been limited to the Pearson product moment correlation coefficient. There are, of course, a wide variety of correlation coefficients and there are appropriate tests of significance for many of them. One useful correlation coefficient for enumerated data is the ϕ coefficient that we discussed in Chapter 5. We can test the significance of the sample ϕ by using the following relationship.

$$n\phi^2 \approx \chi^2$$

(read $\approx$ approximately equal to)

FORMULA 10.20
χ^2 test for the significance of a ϕ coefficient

This value of χ^2 has 1 df, and it is interpreted in the same way as any other value of χ^2. Suppose we have the following data.

	Freshman	Sophomore	
Auto on Campus	*a* 15	*b* 45	60
No Auto on Campus	*c* 45	*d* 15	60
	60	60	120

$$\phi = \frac{bc - ad}{\sqrt{(a + b)(a + c)(b + d)(c + d)}} = .50$$

$$\chi^2 \approx n\phi^2 \approx 30.0$$
$$\chi^2 \approx 30.0, \text{ df } = 1, p < .01$$

This value of χ^2, calculated directly from $\phi^2 n$, is not corrected for continuity. The following is a more exact formula that corrects for continuity.

$$\chi^2 = \frac{n[(bc - ad) - n/2]^2}{(a + b)(a + c)(b + d)(c + d)}$$

Although this formula looks fearsome, most of its elements must be calculated anyway to arrive at ϕ. Of course, if n is quite large and the approximation of χ^2 reveals it to be clearly significant (as in the example) or clearly insignificant, the decision to reject or not to reject the null hypothesis will not be changed by computing the corrected value of χ^2.

Remember that the degree of correlation is measured by ϕ, and its statistical significance is determined by χ^2, the two statistics provide quite different kinds of information.

SUMMARY

In previous chapters we discussed methods for determining the probability that sample proportions were randomly obtained from some defined population. Chapter 10 discusses methods for determining the probability that a sample mean, $\overline{X}$, has been drawn from a population with mean M, and for determining the probability that two independent or matched pair sample means have come from populations with any specified mean difference.

The concept of biased and unbiased statistics is introduced, and it is noted that while the sample mean is an unbiased estimator of the population mean, the sample variance is a biased estimator of the population variance. This bias is the result of taking the sum of squares from the sample mean, thus underestimating the value that would have been obtained had deviations been taken from the population mean. This bias may be eliminated if the sum of squares is divided by $n - 1$ instead of by n.

While sample means are normally distributed, S, the estimator of σ, is not normally distributed but is positively skewed. This means that S will underestimate σ more often than it will overestimate σ. Consequently, a mean difference divided by $S_{\bar{X}_1 - \bar{X}_2}$ will tend to give more high ratios than the normal curve predicts. This ratio is, therefore, evaluated by using a new distribution, t, which changes with df and whose use is required when sample sizes are not very large, unless, of course, σ is known and not estimated. When σ is *known*, the ratio of a difference to its standard error is normally distributed regardless of sample size. The t test for independent groups and matched pairs is discussed. The use of a matched pairs technique is suggested if matching results in a correlation between measures of the dependent variable. If no correlation is present the advantage is lost because of the reduction in df available for evaluating t.

The assumptions required for using t are that the variances of the separate samples be approximately equal and that the samples are large enough, or from a normally distributed population, so that the sample means will be normally distributed. These assumptions need not be met exactly, since t is a robust statistic and follows the tabled distribution in spite of modest departures from assumptions. Nevertheless, a t test has been discussed which should be used when populations appear to have unequal variances. Transformations of the scale are mentioned as a method of satisfying the assumptions of normality and the Z' transformation, and its use in evaluating the significance of differences between independently obtained correlation coefficients is illustrated. A t test for the significance of the difference between correlation coefficients for correlated observations is also given, as is a χ^2 test of the significance of ϕ.

Exercises

1. Use an independent groups t test to determine if the data below have been drawn randomly from populations with a common mean.
   Group A: 7, 9, 6, 11, 12, 16, 14, 21, 22, 14, 16, 17, 13, 17
   Group B: 8, 12, 14, 12, 11, 9, 10, 11, 16, 7, 6, 13, 19, 14
   Determine t = ______, df = ______, p______.
2. For a sample of 101 students we find $\bar{X} = 38$ and $\Sigma x^2 = 15{,}000$. Is it reasonable to assume that this sample mean was obtained by random sampling from a population in which $M = 43$ ($\alpha = .01$)?
   Determine t = ______, df = ______, p______.

3. (a) The following individuals have been paired on the basis of a hearing acuity test and then, after exposing one group to a high intensity noise, both are tested on tonal memory. Conduct a matched pairs t test. Determine t = ______, df = ______, p______.

No Noise		Noise	
Person	*Tonal Memory Test*	*Person*	*Tonal Memory Test*
A	16	K	18
B	17	L	14
C	24	M	16
D	25	N	19
E	19	O	15
F	15	P	20
G	28	Q	22
H	25	R	16
I	22	S	18
J	22	T	17

(b) Conduct an unmatched t test for the above data. Determine t = ______, df = ______, p______.

4. Conduct an independent groups t test on the following data.
Group A: 12, 8, 7, 7, 9, 6, 5, 4, 4, 3
Group B: 16, 4, 6, 9, 9, 8, 5, 10, 10, 12
Determine t = ______, df = ______, p______.

5. Assume that the groups in Exercise 4 consist of matched pairs, and conduct a matched pairs t test on the data.
Determine t = ______, df = ______, p______.

6. Test the significance of the difference between the following two correlation coefficients.

$$n_1 = 103 \qquad r_1 = .65$$

$$n_2 = 103 \qquad r_2 = .35$$

7. Find ϕ, and determine if it is significant.

Modern Art	Classical Music: Dislike	Classical Music: Like
Like	5	10
Dislike	10	5

8. We have recorded running times for two groups of animals. Convert these running times to speed scores (100/time = speed) and conduct a t test on the transformed scores.
   A: 2, 3, 5, 15, 4, 12, 3, 8, 4, 25.
   B: 3, 5, 4, 20, 2, 14, 2, 3, 5, 9, 6, 7.
   Determine t = ______, df = ______, p______.
9. T—F (and explain your answer). The standard error of any mean difference will be less if some way can be found to reduce the variance in the populations from which the means are drawn.
10. T—F (and explain your answer). If the sampling distribution of sample means is based on small samples, the means will not be normally distributed, and the appropriate distribution for statistical tests is t rather than z.
11. Conduct a t test of the significance of the difference between means for the two groups, A and B, assuming that they are independent samples.

A		B	
10	16	31	19
16	10	26	12
10	9	24	14
22	7	14	26
24	12	13	29
29	14	16	19
17	13		
20	16		

12. (a) Assume we have psychomotor tests A and B, each given to two groups of 200 subjects. Test A correlates .47 with a criterion of job performance, and test B correlates .36 with that criterion. Is this difference significant at the 1% level?
    (b) Assume that we have given both of these tests to all 400 subjects, that the same correlation with the criterion has been obtained, and that the intercorrelation between tests A and B is .70. Is the difference between r's significant at the 1% level with this procedure?

CHAPTER 11

Analysis of Variance I

When experiments involve two groups, such as an experimental and a control group, the t test provides an appropriate test of significance. However, some experiments involve more than two groups. When the investigator tests the hypothesis that several sample means vary among themselves more than should be expected on the basis of random sampling, the t test becomes clumsy and its interpretation ambiguous. For example, if an experiment consists of eight samples, and we wish to determine if all eight have come from the same population, we will have to compare each mean with every other and this will require 28 t tests. The amount of work involved is obvious.

Exactly what constitutes a significant t also raises some problems. Assume that the null hypothesis is true and that all eight means have come from the same population. If we compare two means selected at random and traditional definitions apply, we would expect a t significant at the 1% level to occur 1% of the time. But, if 28 t tests are conducted, the probability that one of them will be significant is considerably greater than .01. If the t's are independent and the additive rule can be applied we will find that the probability of obtaining a t significant at the 1% level is .28. This is obviously a contradiction in terms. We cannot reject the hypothesis that the k means have come from the same population if just *one* of the 28 comparisons yields a significant t. Should we then require a higher level of significance for t; should we require more comparisons to be significant at the 1% level, or should we require both? The solution is found by abandoning the t test altogether and using a different procedure called the analysis of variance. This is a very widely used and powerful statistical procedure which may be applied to the problem of comparing any number of sample means. It can be used when comparing only two sample means, but in that situation the t test is a more conventional treatment. The question to be answered by a simple analysis of variance is this: What is the probability that a group of sample means have occurred as a result of random selection from a common population? Such a question is appropriate if we wish to investigate the effectiveness of several different methods of instruction on achievement, several different degrees of deprivation on speed of running in a maze, or several different levels of illumination on the legibility of printed material.

The analysis of variance is based upon comparisons of unbiased estimates of population variance derived from different sources. Two methods of estimating population variance are already familiar because they were discussed in Chapter 10. One of these estimates is obtained from the variation of individual measures within a sample about the sample mean. The formula for that estimate of variance is $S^2 = \Sigma(X - \overline{X})/(n - 1)$. The second estimate is obtained from the variation of sample means about the grand mean of all samples. Recall that an estimate of the standard error of the mean, a measure of the variability of sample means, is given by Formula 10.4 as $S_{\overline{X}} = S/\sqrt{n}$. When this expression is squared and solved for S^2 we have $S^2 = nS_{\overline{X}}^2$, which gives an estimate of population variance in terms of variation among sample means and the sample size. It should be apparent then that we can estimate the population variance in *two distinctly different ways;* from the variation of measurements *within* samples about the individual sample means, and from the variation *among sample means* about a grand mean of all samples.

When there are k samples, *each sample* provides an estimate of population variance from the individual measurements *within* each sample varying about their sample mean. We can estimate the population variance from any one of these samples by $[\Sigma(X - \overline{X})^2]/(n - 1)$. The mean of such estimates for all k samples will then be

$$S_w^2 = \frac{\sum_k \sum_n (X - \overline{X})^2}{k(n - 1)}$$

FORMULA 11.1

The within samples estimate of the population variance. An estimate of the population variance obtained from the variation of individual measurements about their respective sample means

In this Formula, X's are individual measurements, $\overline{X}$ is the mean of a sample based on n measurements, and there are a total of k samples contributing to the mean of the estimates. (You will find that subscripts for Σ reduce ambiguity about what is being summed. The notation $\sum_n (X - \overline{X})^2$ tells us to sum the squared difference between each score and the sample mean for the n scores in that sample. The symbol $\sum_k$ directs us to sum these sums for all k samples.) Formula 11.1 is composed of two parts. The numerator is the sum of squared deviations within each sample summed over all k samples. This is referred to as the within sum of squares (SS_w). When this sum of squares is divided by $k(n - 1)$, the available df, we have S_w^2, the within samples estimate of variance.

There are $k(n - 1)$ df available for this estimate of variance, because *each* of the samples is based on $(n - 1)$ df, since only $n - 1$ measurements in each sample are free to vary randomly once the sample mean has been determined. If there are $(n - 1)$ df in each sample and there are k samples, then there are $k(n - 1)$ df available for S_w^2. Note that $k(n - 1)$ forms the denominator of Formula 11.1.

A second estimate of population variance can be obtained from the k sample means varying about the grand mean of all samples. We calculate $S_{\bar{X}}^2$, the estimated variance of sample means, directly from the sample means themselves and multiply the result by n, the sample size. The estimated variance of sample means is given by $[\Sigma(\bar{X} - \bar{\bar{X}})^2]/k - 1$, where $\bar{X}$'s are sample means, $\bar{\bar{X}}$ is the grand mean, or mean of sample means, and k is the number of sample means. When this quantity is multiplied by n, the sample size, we have

$$S_b^2 = nS_{\bar{X}}^2 = \frac{n \sum_k (\bar{X} - \bar{\bar{X}})^2}{k - 1}$$

FORMULA 11.2

The between samples estimate of the population variance. An estimate of the population variance obtained from the variation of sample means about the grand mean of all samples

Formula 11.2 is also composed of two parts. The numerator is the sample size, n, multiplied by the sum of squared deviations of sample means from the grand mean. This is called the between groups sum of squares (SS_b), and when it is divided by $k - 1$, the available df, we have S_b^2 which is the between samples estimate of variance. We have $k - 1$ df for this estimate because, once the grand mean is known, only $k - 1$ of the sample means can show random variation.

We have now shown that the population variance can be estimated in two ways, from the variation between samples, and from the variation within samples. We shall refer to these estimates as the between variance (S_b^2), and the within variance (S_w^2). These two estimates of the population variance are independent of each other, a fact we shall presently demonstrate.

There is however, a third way to estimate the population variance from several samples. We can simply pool all samples and consider the pooled data as one large sample. An estimate of population variance can be obtained by taking the sum of each measurement's squared deviation from the mean of the pooled data. This sum of squares (SS_t) is then divided by the total number of measurements less one, and yields an estimate of population variance which we shall symbolize by S_t^2. The formula for S_t^2 is

$$S_t^2 = \frac{\sum_k \sum_n (X - \overline{\overline{X}})^2}{nk - 1}$$

FORMULA 11.3
An estimate of the population variance based upon the variation of each measurement about the grand mean of all measurements

Here X's are individual measurements, $\overline{\overline{X}}$ is the grand mean of all samples, n is the number of observations in each sample, and k is the number of samples. Notice that the total number of observations will equal the n observations in each sample multiplied by the k samples. Thus, $N = kn$, so that $N - 1 = kn - 1$.

When our estimate of variance is based on the pooled measurements from all samples varying about the grand mean of all samples we will have a total of kn measurements of which $kn - 1$ are free to vary randomly once the grand mean has been obtained. This means that S_t^2 will have $kn - 1$ df, and this is the denominator of Formula 11.3.

We have now discussed three different estimates of population variance, each based on a sum of squares divided by the appropriate degrees of freedom. In the next section we will try to show some important relationships between these sums of squares, and between their degrees of freedom.

RELATIONSHIPS BETWEEN ESTIMATES OF THE POPULATION VARIANCE

Each of the three estimates of variance which we discussed in the previous section is derived from a sum of squares term (SS) divided by its appropriate degrees of freedom. The sum of squares *within* each of the k samples (SS_w), divided by $k(n - 1)$, yields S_w^2. The sum of squares *between* sample means (SS_b), divided by $k - 1$, yields S_b^2. When we obtained S_t^2, the numerator was a sum of squares term that resulted from pooling all k samples to arrive at a combined mean, and then taking the sum of squared deviations of individual measurements from that mean. This is called the *total* sum of squares (SS_t) because we can show that it is the sum of SS_b and SS_w. We can also show that the degrees of freedom for this sum of squares (df_t) is the sum of the degrees of freedom for SS_w and the degrees of freedom for SS_b.

First we shall consider the total sum of squares, given by $\sum_k \sum_n (X - \overline{\overline{X}})^2$. We can rewrite this expression as $\sum_k \sum_n [(X - \overline{X}) + (\overline{X} - \overline{\overline{X}})]^2$ because the deviation of a score from the grand mean, $(X - \overline{\overline{X}})$ is equivalent to the deviation of the score from the mean of its sample $(X - \overline{X})$ plus the deviation of that sample mean from the grand mean $(\overline{X} - \overline{\overline{X}})$. Therefore, we can write:

$$\sum_k \sum_n (X - \overline{\overline{X}})^2 = \sum_k \sum_n [(X - \overline{X}) + (\overline{X} - \overline{\overline{X}})]^2.$$

Expanding this expression we have

$$\sum_k \sum_n (X - \overline{\overline{X}})^2 = \sum_k \sum_n (X - \overline{X})^2 + 2 \sum_k \sum_n (X - \overline{X})(\overline{X} - \overline{\overline{X}}) + \sum_k \sum_n (\overline{X} - \overline{\overline{X}})^2$$

but since $\sum_n (X - \overline{X}) = 0$, the entire second term becomes zero, and we have

$$\sum_k \sum_n (X - \overline{\overline{X}})^2 = \sum_k \sum_n (X - \overline{X})^2 + n \sum_k (\overline{X} - \overline{\overline{X}})^2.$$

You may recognize $\sum_k \sum_n (X - \overline{X})^2$ as the within sum of squares from Formula 11.1, and $n \sum_k (\overline{X} - \overline{\overline{X}})^2$ as the between sum of squares used in Formula 11.2. The total sum of squares is thus shown to consist of two components, a between groups sum of squares and a within groups sum of squares. Using our symbols, $SS_t = SS_b + SS_w$.

The variance estimate, sometimes referred to as a mean square, is the sum of squares divided by degrees of freedom. We can also show that the total df of $kn - 1$ is the sum of the between groups df of $k - 1$ and the within groups df of $k(n - 1)$. Rewriting $k - 1 + k(n - 1)$ as $k - 1 + kn - k$, and collecting terms, we have $kn - 1$. This means that the value of S_t^2 can be obtained from sums of squares and df components that are also used to calculate S_b^2 and S_w^2, or it can be calculated directly from Formula 11.3.

THE F RATIO

Suppose we have drawn three samples from a common population, and we know that the differences among the sample means are due entirely to sampling error. Under these circumstances we would expect that the population variance estimated from the variation of these sample means about their own grand mean would be about the same as an estimate based on the variation of individual measurements within the samples about their respective sample means. If nothing more than random variation occurs among the sample means and if random variation is also the only source of variation within samples, then, within the limits of sampling error, we should expect to find S_b^2 about the same as S_w^2.

On the other hand, suppose that we have drawn a sample from each of three populations, and these populations have quite different means. Under these circumstances the variation among the sample means will be the result of random variation *and* variation produced by real differences among the population means. Since this is the case we would expect that, on the average, the population variance estimated from the variation among sample means will *exceed* the population variance estimated from variation within samples. If experimental treatments are effective and have produced differences among the means of the three populations, we would expect S_b^2 to be larger than S_w^2.

This means that the difference between S_b^2 and S_w^2 can be used as a test of the null hypothesis. If the null hypothesis is true, and the experimental treatments are without effect, S_b^2 will be about equal to S_w^2. If the null hypothesis is false, the different experimental treatments for each group will *increase* the variability among the means, and this can affect only S_b^2 making it larger than S_w^2.

We need some way to assess the significance of the difference between S_b^2 and S_w^2. If S_b^2 is divided by S_w^2, ratios in excess of 1.00 would be evidence that there is more than random variation among the sample means. If our null hypothesis states that there is no difference among the means of the populations from which the samples have come, then ratios of S_b^2 to S_w^2 that are much larger than 1.00 would constitute evidence that the null hypothesis should be rejected. On the other hand, if the population means are the same, then the ratio of S_b^2 to S_w^2 should be about 1.00 and would provide no evidence against the null hypothesis. The ratio of one variance to another is called the F ratio, and the values of F necessary for significance at the 1% and 5% levels are recorded in Table F. To use the table we must first obtain the F ratio, which is S_b^2/S_w^2, and then we must determine the degrees of freedom upon which each of the variance estimates is based. Tables of the F ratio include only the upper half of the F distribution, since only values of F which are larger than 1.00 can lead to rejection of the null hypothesis. To use the F table we must find the column associated with the between variance df and the row associated with the within variance df. The F ratio, or S_b^2/S_w^2, must be as large or larger than the tabled value of F for the result to be significant at the indicated level.

INDEPENDENCE OF S_b^2 AND S_w^2 IN A SAMPLE ANALYSIS

We will now conduct an analysis of variance for the data presented below.

Sample I	*Sample II*	*Sample III*	
2	4	6	$\bar{\bar{X}} = 6$
4	6	8	$n_1 = 3$
6	8	10	$n_2 = 3$
			$n_3 = 3$
$\bar{X} = 4$	$\bar{X} = 6$	$\bar{X} = 8$	$k = 3$

Our data consist of three samples of three cases each. If the data were genuine, the null hypothesis would state that the means of these samples had arisen by random sampling from a common population. The data have been highly simplified for illustrative purposes. Please be sure that you understand how the calculations below follow from the formulas.

$$S_w^2 = \frac{\sum_k \sum_n (X - \bar{X})^2}{k(n-1)} = \frac{SS_w}{df_w} \qquad \text{FORMULA 11.1}$$

$$= \frac{[(2-4)^2 + (4-4)^2 + (6-4)^2] + [(4-6)^2 + (6-6)^2 + (8-6)^2] + [(6-8)^2 + (8-8)^2 + (10-8)^2]}{6}$$

$$= \tfrac{24}{6} = 4$$

$$S_b^2 = \frac{n\sum_k (\bar{X} - \bar{\bar{X}})^2}{k - 1} = \frac{SS_b}{df_b} \qquad \text{FORMULA 11.2}$$

$$= \frac{3[(4-6)^2 + (6-6)^2 + (8-6)^2]}{2}$$

$$= \tfrac{24}{2} = 12$$

$$S_t^2 = \frac{\sum_k \sum_n (X - \bar{\bar{X}})^2}{kn - 1} = \frac{SS_t}{df_t} \qquad \text{FORMULA 11.3}$$

$$= \frac{[(2-6)^2 + (4-6)^2 + (6-6)^2] + [(4-6)^2 + (6-6)^2 + (8-6)^2] + [(6-6)^2 + (8-6)^2 + (10-6)^2]}{8}$$

$$= \tfrac{48}{8} = 6$$

There are several relationships to be noted here: first, $(SS_b = 24) + (SS_w = 24) = (SS_t = 48)$. This is the relationship between the component sums of squares which we developed earlier by algebra. Second, the df associated with these sums of squares also show that

$$(\text{df}_b = 2) + (\text{df}_w = 6) = (\text{df}_t = 8).$$

The analysis of variance is usually reported in a table similar to Table 11.1. The F ratio for these data was obtained by finding S_b^2/S_w^2, which is $12/4 = 3.00$. Now we must find the F ratio required for significance when, as in this example, the numerator of the ratio has 2 df and the denominator has 6 df. The F ratio required for significance will be located at the intersection of the 2 df column and the 6 df row of Table F. An $F = 5.14$

TABLE 11.1 Results of an Analysis of Variance

Source of Variation	Sum of Squares	df	Estimate of Variance (Mean Square)	F
Between groups	24	2	12	3.00
Within groups	24	6	4	
	48	8		

is required at the 5% level, so we do not have adequate evidence to reject the null hypothesis for these data.

The calculation of SS_t and df_t serve only as a matter of bookkeeping. We can check our arithmetic by showing that $SS_b + SS_w = SS_t$ and that $df_b + df_w = df_t$. The components of S_t^2 are calculated as a check on arithmetic, but S_t^2 itself doesn't enter the problem.

We pointed out in our initial discussion of S_b^2 and S_w^2 that these estimates of the population variance were independent. This fact is crucial to the logic of using the F ratio as a test of the null hypothesis. In this section we will demonstrate that S_b^2 and S_w^2 can vary independently of each other. We shall begin by using the data from our first example problem, but we shall modify them slightly so that all three sample means are equal. This equality can be achieved by adding two points to each measurement in Sample I and subtracting two points from each measurement in Sample III. The modified data and the estimates of variance calculated from them appear below.

Sample I	*Sample II*	*Sample III*	
4	4	4	$\overline{\overline{X}} = 6$
6	6	6	$k = 3$
8	8	8	
$\overline{X} = 6$	$\overline{X} = 6$	$\overline{X} = 6$	

$$S_b^2 = \frac{n\sum_k (\overline{X} - \overline{\overline{X}})^2}{k-1} = \frac{SS_b}{df_b} = \frac{3[(6-6)^2 + (6-6)^2 + (6-6)^2]}{2}$$

$$= \tfrac{3(0)}{2} = 0$$

$$S_w^2 = \frac{\sum_k \sum_n (X - \overline{X})^2}{k(n-1)} = \frac{SS_w}{df_w}$$

$$= \frac{[(4-6)^2 + (6-6)^2 + (8-6)^2] + [(4-6)^2 + (6-6)^2 + (8-6)^2] + [(4-6)^2 + (6-6)^2 + (8-6)^2]}{6}$$

$$= \tfrac{24}{6} = 4$$

$$S_t^2 = \frac{\sum_k \sum_n (X - \bar{\bar{X}})^2}{kn - 1} = \frac{SS_t}{\text{df}_t}$$

$$= \frac{[(4-6)^2 + (6-6)^2 + (8-6)^2] + [(4-6)^2 + (6-6)^2 + (8-6)^2] + [(4-6)^2 + (6-6)^2 + (8-6)^2]}{8}$$

$$= \tfrac{24}{8} = 3$$

By equalizing the sample means we have reduced SS_b to zero. That sum of squares is based on the variation of sample means about a grand mean, and these sample means no longer vary. Of course, if SS_b is zero, we shall have reduced S_b^2 to zero too.

Equalizing the sample means, however, has had no effect on SS_w. This sum of squares is based on the variation of individual measurements about their respective sample means and this remains unchanged. If SS_w remains unchanged, and df_w remains as it was, we shall *not* have altered S_w^2. Since $SS_t = SS_b + SS_w$, and we have reduced SS_b to zero while keeping SS_w as it was, we shall have reduced SS_t. The reduction in SS_t reflects the fact that our manipulations have reduced the total variation in the system. If we assume that the df available remains unchanged, then S_t^2 as a measure of variation per unit df will also be reduced.

The really critical point, however, is that the variation among sample means can be manipulated without altering the variation within samples. In the next example we shall modify the data so that the variation among sample means remains as it was in the first illustration, but we shall eliminate the variation *within* samples. This can be accomplished by adding two points to the lowest measurement and subtracting two points from the highest measurement in each sample. These modified data and the variance estimates based upon them appear below.

Sample I	*Sample II*	*Sample III*	
4	6	8	$\bar{\bar{X}} = 6$
4	6	8	$k = 3$
4	6	8	
$\bar{X} = 4$	$\bar{X} = 6$	$\bar{X} = 8$	

$$S_b^2 = \frac{n\sum_k (\bar{X} - \bar{\bar{X}})^2}{k-1} = \frac{SS_b}{\text{df}_b} = \frac{3[(4-6)^2 + (6-6)^2 + (8-6)^2]}{2}$$

$$= \tfrac{24}{2} = 12$$

$$S_w^2 = \frac{\sum_k \sum_n (X - \bar{X})^2}{k(n-1)} = \frac{SS_w}{\text{df}_w}$$

$$= \frac{[(4-4)^2 + (4-4)^2 + (4-4)^2] + [(6-6)^2 + (6-6)^2 + (6-6)^2] + [(8-8)^2 + (8-8)^2 + (8-8)^2]}{6}$$

$$= \tfrac{0}{6} = 0$$

$$S_t^2 = \frac{\sum_k \sum_n (X - \bar{\bar{X}})^2}{kn-1} = \frac{SS_t}{\text{df}_t}$$

$$= \frac{[(4-6)^2 + (4-6)^2 + (4-6)^2] + [(6-6)^2 + (6-6)^2 + (6-6)^2] + [(8-6)^2 + (8-6)^2 + (8-6)^2]}{8}$$

$$= \tfrac{24}{8} = 3$$

Now that we have removed the variation within samples, SS_w and S_w^2 become zero; but since the variation among sample means remains as it was originally, SS_b and S_b^2 are unchanged. With SS_w now zero, SS_t is reduced since $SS_t = SS_b + 0$.

The ability to isolate sources of variation in experimental data is an immensely valuable research tool. If we have several different experimental treatments, the variation among subjects *within* each treatment group is usually referred to as the "error" sum of squares, or "error" variance. The variation among subjects within samples cannot normally be produced by the independent variable because all subjects in a sample are treated alike. We assume then that the differences among subjects *within* samples are the result of sampling error or other uncontrolled factors. Consequently, the names "error" variance and "error" sum of squares are sometimes given to these terms.

The variation among the treatment means however, is composed of two different sources of variation; one of these is the result of different treatments and the other is the result of sampling error. (Even if the treatments exert no influence, we would hardly expect to obtain three identical sample means since sampling error will still play its role.) The between sum of squares and the between estimate of variance are measures of variation due to the combined effects of treatments *and* error. The F ratio may then be conceptualized as

$$F = \frac{\text{Variance due to treatments and error}}{\text{Variance due to error}}$$

If the treatments' effects are zero we would expect $F = 1.00$ except for the influence of sampling error.

COMPUTING FORMULAS AND EXAMPLES

The formulas previously given for different estimates of variance are not very efficient for real data. Computational formulas for the various component sums of squares are given below.

$$SS_t = \sum_N X^2 - \frac{\left(\sum_N X\right)^2}{N}$$

FORMULA 11.4
Computational formula for the total sum of squares

Here N is the total number of measurements ($N = nk$).

$$SS_b = \frac{\left(\sum_{n_1} X\right)^2}{n_1} + \frac{\left(\sum_{n_2} X\right)^2}{n_2} + \cdots + \frac{\left(\sum_{n_k} X\right)^2}{n_k} - \frac{\left(\sum_N X\right)^2}{N}$$

FORMULA 11.5
Computational formula for the sum of squares between groups

Here n_1, n_2, etc., are the number of measurements in each of the k treatment groups, and the final term, $\frac{\left(\sum_N X\right)^2}{N}$, is the same final term used in Formula 11.4.

$$SS_w = \sum_N X^2 - \left[\frac{\left(\sum_{n_1} X\right)^2}{n_1} + \frac{\left(\sum_{n_2} X\right)^2}{n_2} + \cdots + \frac{\left(\sum_{n_k} X\right)^2}{n_k}\right]$$

FORMULA 11.6
Computational formula for the sum of squares within groups

This formula for SS_w uses a combination of the terms from formulas for SS_t and SS_b. Of course, SS_w can be found most directly by subtracting SS_b from SS_t, since $SS_t = SS_b + SS_w$.

These computational formulas look rather unwieldy, but they will be quite indispensable when real data are analyzed. Our earlier conceptual formulas for the various sums of squares required the sum of squared deviation scores, and these deviation scores were obtained by subtracting a measurement from a mean, or by subtracting a sample mean from a grand mean. In our simplified examples, the means were all whole numbers and the samples were unrealistically small. When real data are analyzed, the samples will be much larger, the means will rarely be whole numbers, and the subtraction process will be quite tedious. The computational formulas allow us to work directly with the squares of raw scores, and then subtract a final correction term. This saves a great deal of time. In subsequent examples we will use numbers more like those which might be found in real data, and we shall make use of the computing formulas for the analyses.

Suppose we have a problem in which we have varied the intralist similarity of a group of nonsense syllables, and we wish to know if this variable has any significant effect on the trials required to learn the list. Assume that we have randomly assigned 10 subjects to each of the three different conditions of within-list similarity given below. The trials for each subject to reach a criterion of two errorless performances have been recorded.

Low Similarity	*Medium Similarity*	*High Similarity*
10	13	21
9	9	17
14	17	15
15	21	14
8	17	25
12	15	25
14	9	27
17	16	18
20	19	21
13	23	22
$\Sigma X = 132$	$\Sigma X = 159$	$\Sigma X = 205$
$\overline{X} = 13.2$	$\overline{X} = 15.9$	$\overline{X} = 20.5$

The first step in the analysis is to find the component sums of squares.

$$SS_t = \sum_N X^2 - \frac{\left(\sum_N X\right)^2}{N} \qquad \text{FORMULA 11.4}$$

$$= 8964 - \frac{(496)^2}{30} = 763.5$$

$$SS_b = \frac{\left(\sum_{n_1} X\right)^2}{n_1} + \frac{\left(\sum_{n_2} X\right)^2}{n_2} + \frac{\left(\sum_{n_3} X\right)^2}{n_3} - \frac{\left(\sum_{N} X\right)^2}{N} \qquad \text{Formula 11.5}$$

$$= \frac{(132)^2}{10} + \frac{(159)^2}{10} + \frac{(205)^2}{10} - \frac{(496)^2}{30}$$

$$= 272.5$$

$$SS_w = \sum_N X^2 - \left[\frac{\left(\sum_{n_1} X\right)^2}{n_1} + \frac{\left(\sum_{n_2} X\right)^2}{n_2} + \frac{\left(\sum_{n_3} X\right)^2}{n_3}\right] \qquad \text{Formula 11.6}$$

$$= 8964 - \left[\frac{(132)^2}{10} + \frac{(159)^2}{10} + \frac{(205)^2}{10}\right]$$

$$= 491.0$$

Once the component sums of squares have been obtained we can record the results in tabular form.

Source of Variation	*Sum of Squares*	df	*Estimate of Variance (Mean Square)*	*F*
Between groups	272.5	2	136.25	7.49**[1]
Within groups	491.0	27	18.18	
Total	763.5	29		

For 2 and 27 df, Table F shows that an F of 5.49 is required for significance at the 1% level. Our obtained F exceeds this value, so we can reject the null hypothesis at the 1% level. The sample means show more variation about the grand mean than can reasonably be attributed to sampling error. This is conventionally reported as $F = 7.49$, df $= 2/27$, and $p < .01$.

[1] Double asterisks (**) are usually used to denote an F ratio significant at the .01 level, while a single asterisk (*) denotes significance at the .05 level. We shall follow this style in tables throughout this book.

COMPARISONS FOLLOWING AN *F* TEST

The significant F in the last example tells us *only* that the variation among sample means cannot reasonably be attributed to chance. However, an inspection of the data shows that the high similarity list required the greatest number of trials to learn, medium similarity was next, and low similarity was easiest to learn. While this is an interesting and provocative relationship, it would be a mistake to assume that the significant F confirms the significance of this particular *ordering* of means. This statement may seem more reasonable if you consider that this particular ordering of difficulties is only one of six which could occur. (Three things may be ordered in six ways.) Any one of the six possible orderings of the sample means would have produced the *same* F ratio, so the obtained F cannot be cited as confirming the significance of this particular arrangement of means.

We have already pointed out the problems inherent in conducting t tests between pairs of sample means when the analysis involves more than two samples. These problems remain even when F is significant. Yet the finding that we have a significant between groups variance does not tell us which groups are responsible. In the example above, it appears that there may be no difference between the difficulty of low and medium similarity lists. Most of the between groups variance contributing to the significant F could well be the result of differences between these two means and the mean of the high similarity group. If this is true, it could lead us to suggest that difficulty is related to intra-list similarity, but only after a certain level of within-list similarity has been achieved. Such a question cannot be answered by a test of overall differences among the sample means.

There are a variety of statistical procedures available for treating the problem of comparisons between specific means following analysis of variance. Tuckey[2] has described procedures for discovering significant gaps in adjacently arranged means, for determining stragglers, and for finding if excessive variability exists in the groups not separated by the other two procedures. A process called "trend analysis," or "orthogonal polynomials," can be used to determine if a series of means, ordered according to size, can best be described by a straight line, (linear function) or a curved line (quadratic, or higher function). In our example, the sample means are probably best described by a curved line; they do not seem to fit a straight line. This is not determined by casual observation, but by appropriate statistical procedures.

[2] Tukey, J. W. "Comparing individual means in analysis of variance." *Biometrics*, 1949, **5**, 99–114.

Some experiments require that a control group be compared with several different experimental groups. Dunnett's[3] test permits us to assess the significance of comparisons such as these. In fact, a procedure described by Scheffe[4] can be used to test all possible comparisons between means. It is unlikely that the beginning student will be able to follow the logic and mathematics of these tests as they appear in the statistical journals. However, Edwards'[5] description should be within the grasp of the serious student. In the next section we shall discuss Duncan's Multiple Range Test, a very useful statistical test for comparisons following the analysis of variance.

DUNCAN'S MULTIPLE RANGE TEST

In the preceeding experiment, where we investigated the effect of intralist similarity on trials to learn, we found a significant F ratio, and we concluded that the variation among the three sample means was greater than should be expected on the basis of sampling error alone. Duncan's Test will enable us to determine if there are significant differences between the low and medium similarity groups, and between the medium and high similarity groups.

The first step in applying Duncan's Test is to arrange the sample means in order of magnitude. The next step is to find the difference between adjacent sample means and record that difference, circled, between each pair of sample means. For the moment you may disregard the line drawn beneath the low and medium sample means. We shall explain its significance later. The sample means from the previous analysis have been arranged in this manner and they are presented below.

Low		*Medium*		*High*
13.2	(2.7)	15.9	(4.6)	20.5

Duncan's Multiple Range Test makes use of a "range of nonsignificance" symbolized by R_n. A range of nonsignificance is a range or gap that must be exceeded by the range of the k means being compared, if the

[3] Dunnett, C. W. "A multiple comparison procedure for comparing several treatments with a control." *Journal of the American Statistical Association*, 1955, **50**, 1096–1121.

[4] Scheffe, H. A. "A method for judging all contrasts in the analysis of variance." *Biometrika*, 1953, **40**, 87–104.

[5] Edwards, A. L. *Experimental Design in Psychological Research* 3rd ed. New York: Holt, Rinehart and Winston, 1968.

variation among those k means is significant. We have only three means in this problem, and we know from the analysis of variance that the three vary significantly. We shall apply Duncan's test to the two adjacent pairs of means 13.2–15.9, and 15.9–20.5. Since we are dealing with pairs, we have $k = 2$. The range of nonsignificance is determined by multiplying $S_{\bar{X}}$ by C_k, a coefficient obtained from Table D of the Appendix. Thus, the range of nonsignificance for comparisons of two means will be given by

$$R_{n_2} = C_2 \cdot S_{\bar{X}}$$

The value of $S_{\bar{X}}$ is obtained from the error term which is the within groups estimate of variance, calculated earlier when we conducted the analysis of variance for these data. We find $S_{\bar{X}}$ by using the relationship $S_{\bar{X}} = S/\sqrt{n} = \sqrt{S_w^2/n}$. Since we have already determined that $S_w^2 =$ 18.18, and that $n = 10$, we have $S_{\bar{X}} = \sqrt{18.18/10} = 1.35$.

The value of C_2 is obtained from Table D. We shall use the section of the table which gives the values of C_k appropriate for the 5% level of significance. The columns of Table D are headed from $k = 2$ to $k = 19$ for the number of groups to be compared. We are comparing groups of two, so we read down the column headed $k = 2$. The correct row of the table will be opposite the degrees of freedom for the error term, the within groups estimate of variance, from which we determined $S_{\bar{X}}$. In this problem we have 27 df, but since no row has that particular value we shall choose the row for the more conservative df = 24. The value of C_2 will then be at the intersection of the $k = 2$ column and the 24 df row. We have $C_2 = 2.92$. We can now determine R_{n_2}, the "range of nonsignificance," at the 5% level for the two adjacent pairs of means. This is given by

$$R_{n_2} = C_2 \cdot S_{\bar{X}}$$

$$R_{n_2} = 2.92 \cdot 1.35$$

$$R_{n_2} = 3.94$$

When R_{n_2} has been calculated, we underline the pairs of adjacent means whose *difference* fails to exceed that value. The difference between these underlined means will *not be significant* at the 5% level. The difference between a mean which is not underlined, and *all* means which are underlined *is* significant at the 5% level.

In our example, the mean trials to learn lists of medium and low intralist similarity do not differ significantly. These means are inside the range of nonsignificance for pairs. On the other hand, the difference between means for the medium and high similarity lists is larger than the range of nonsignificance, so this pair is significantly different. Since we have arranged the means in order, the difference between the high and low similarity list must also be greater than R_{n_2} and, therefore, this difference will be significant too.

In summary, means which are under the same line do not differ significantly among themselves. Means which are not under the same line do differ significantly from all other means not under the same line. Duncan's Test is fairly simple when only three groups are involved. It becomes somewhat more complicated as the number of groups are increased. We shall illustrate this by conducting an analysis of variance and following it with Duncan's Test for an experiment involving five experimental groups.

Suppose we investigate the effects of five different diets on the activity level of rats. We might randomly assign 35 animals so that seven are in each of five diet groups, *A* through *E*. One month later we give our subjects access to an activity wheel. The mean revolutions per hour for each animal over a 24-hour period have been recorded below.

A	*B*	*C*	*D*	*E*
7	14	9	22	24
9	15	8	21	27
13	19	11	23	30
17	19	9	27	15
23	24	16	14	18
14	22	15	19	22
15	21	16	26	17
$\Sigma X_A = 98$	$\Sigma X_B = 134$	$\Sigma X_C = 84$	$\Sigma X_D = 152$	$\Sigma X_E = 153$
$\overline{X}_A = 14.0$	$\overline{X}_B = 19.1$	$\overline{X}_C = 12.0$	$\overline{X}_D = 21.7$	$\overline{X}_E = 21.8$

When samples have equal n's, as they have in this example, Formulas 11.4, 11.5, and 11.6 can be simplified somewhat.

$$SS_t = \sum_N X^2 - \frac{\left(\sum_N X\right)^2}{N} = 12209 - \frac{(621)^2}{35} \qquad \text{Formula 11.4}$$

$$= 1190.7$$

$$SS_b = \frac{\sum_k \left(\sum_n X\right)^2}{n} - \frac{\left(\sum_N X\right)^2}{N} \qquad \text{Formula 11.5}$$

$$= \frac{(98)^2}{7} + \frac{(134)^2}{7} + \frac{(84)^2}{7} + \frac{(152)^2}{7}$$

$$+ \frac{(153)^2}{7} - \frac{(621)^2}{35} = 571.5$$

$$SS_w = \sum_N X^2 - \frac{\sum_k \left(\sum_n X\right)^2}{n} \qquad \text{FORMULA 11.6}$$

$$= 12209 - \left[\frac{(98)^2}{7} + \frac{(134)^2}{7} + \frac{(84)^2}{7} + \frac{(152)^2}{7} + \frac{(153)^2}{7}\right] = 619.2$$

Source of Variation	Sum of Squares	df	Estimate of Variance (Mean Square)	F
Between	571.5	4	142.9	6.94**
Within	619.2	30	20.6	
Total	1190.7	34		

$F = 6.94$, df $= 4/30$, $p < .01$.

For 4 and 30 df, Table F shows that $F = 4.02$ is required for significance at the 1% level. The obtained F exceeds this value, so we can reject the null hypothesis at the 1% level of significance and conclude that these five samples of activity behavior did not come from the same population. The design of the experiment would probably lead us to conclude that variations among the several diets were responsible for differences in activity. Unfortunately, from the analysis of variance alone, we cannot tell which diet effects differ significantly from which others. This question can be answered by using Duncan's Multiple Range Test.

Duncan's test is conducted by first arranging the means in order of magnitude, with the difference between each pair of means circled as illustrated below.

C		*A*		*B*		*D*		*E*
12.0	(2.0)	14.0	(5.1)	19.1	(2.6)	21.7	(0.1)	21.8

In the previous example of Duncan's test we had only three sample means. The F test demonstrated significant differences among this group of three, and Duncan's test was applied to pairs, the next smallest grouping unit. Duncan's test demonstrated that a significant difference existed between trials to learn a high-similarity list and lists of medium or low similarity, but there was no significant difference between trials to learn lists of low and medium similarity.

The F in our present problem is also significant; we know that these five means differ significantly among themselves. We shall now determine if there is significant variation within the two groups of four adjacently ordered means formed by groups *CABD*, and by groups *ABDE*. If either set of four shows *less* divergence than R_{n_4}, the range of nonsignificance for groups of four, we shall conclude that the variability of means within that group of four is random. Further testing of differences within that group of four is *not* conducted and the group is underlined. We shall determine R_{n_4} by

$$R_{n_4} = C_4 \cdot S_{\bar{X}}$$

where $S_{\bar{X}} = \sqrt{S_w^2/n} = \sqrt{20.6/7} = 1.71$, and C_4 for the 5% level is obtained from the third column of Table D, in the row designated for 30 df. These were the df available for the within variance estimate (S_w^2) used in obtaining $S_{\bar{X}}$. We find that $C_4 = 3.13$.

$$R_{n_4} = C_4 \cdot S_{\bar{X}}$$

$$R_{n_4} = 3.13 \times 1.71$$

$$R_{n_4} = 5.35$$

This range of nonsignificance, 5.35, is exceeded by the range of means in both of the blocks of four adjacent means. Block *CABD* has a range of 9.7 and block *ABDE* has a range of 7.8. We must therefore conclude that there are significant differences among the means of groups *CABD* and among the means of *ABDE*. *Since neither group of four falls within the nonsignificant range, neither group is underlined.*

The next step is to test the three groups of three adjacent means; these are *CAB*, *ABD*, and *BDE*. We find the range of nonsignificance for three means much as we did for four.

$$R_{n_3} = C_3 \cdot S_{\bar{X}}$$

We have already determined $S_{\bar{X}}$ for our previous calculations. The correct value of C_3 will be found at the intersection of the $k = 3$ column and the 30 df row of Table D. We have:

$$R_{n_3} = 3.04 \times 1.71$$

$$R_{n_3} = 5.20$$

Of the three triplets to be tested, the differences within two exceed $R_{n_3} = 5.20$. We find $CAB = 7.1$ and $ABD = 7.7$. However, the group consisting of *BDE* has a range of only 2.7. We conclude that significant variation exists among the means *CAB*, and among the means *ABD*, *but not among the means BDE*. Since there is no significant difference among means *BDE*, we have underlined this group. *No further testing is to be conducted entirely within this group, just as we would not apply Duncan's test if the original F had not been significant.*

We shall now test adjacent pairs of means. Since we can test none of the pairs falling entirely within the underlined *BDE* group, we need test only the adjacent pairs *CA* and *AB*. We may determine the range of nonsignificance for pairs just as we have previously done for triplets and groups of four. The value of $S_{\bar{X}}$ has already been calculated, and the value for C_2 is drawn from the 30 df row and the $k = 2$ column of Table D.

$$R_{n_2} = C_2 \cdot S_{\bar{X}}$$

$$R_{n_2} = 2.89 \times 1.71$$

$$R_{n_2} = 4.94$$

The first pair of means, *CA*, does not exceed R_{n_2} but the second pair, *AB*, does. The difference between the first pair is 2.0, and since this is less than 4.94 we have underlined this nonsignificant pair. The *AB* pair difference is 5.1, which is larger than the range of nonsignificance, so this pair is not underlined.

Now we can summarize the results of our analysis. It is rather complicated, so please follow closely and then reread it. First, and most obviously, we have isolated two groups of means consisting of the nonsignificantly different pair *CA*, and the nonsignificantly different triplet, *BDE*. Remember, there are *nonsignificant* differences among the means spanned by a common underline. Second, there are significant differences between members of *any* pair not spanned by a common underline, among any *set* of triplet means, or *any* group of four means not spanned by a common underline. For example, if *AB* is larger than the minimum range required for nonsignificance, then any pairing of nonadjacent means which includes the *AB* difference, such as *CD*, or *AD*, or *CE*, etc., will also exceed the range of nonsignificance for pairs. Notice, however, that no set of pairs within the nonsignificant triplet *BDE* exceeds the nonsignificant range for pairs. On the other hand, any triplet which spans the nonunderlined gap will exceed the nonsignificant range of triplets. Since both sets of four (*CABD* and *ABDE*) must span this gap, both sets of four differ significantly among themselves. You may now understand the reason for the underlining process. It provides a very convenient summary of the results.

C		*A*		*B*		*D*		*E*
12.0	(2.0)	14.0	(5.1)	19.1	(2.6)	21.7	(0.1)	21.8

We now find that the diets tested have really produced two distinct activity groupings. The alert researcher, having snooped about in his data, will try to find elements of diet or perhaps other aspects of treatment common to groups AB but not occurring in BDE, or vice versa. It is reasonable to suspect that some experimental treatment or procedure is producing the difference between these sets of homogeneous subgroups.

The illustrative problems using Duncan's test have all been carried out with $\alpha = .05$. Recent evidence suggests that the tabled values of C may not be sufficiently conservative for the 5% level; consequently, when using Duncan's test, results that just achieve the 5% level, but fall short of significance at the 1% level, should be viewed as suggestive of further research rather than as adequate evidence for the rejection of the null hypothesis.

In both of the simple analyses of variance, which are analyses involving only one set of experimental groups, we have followed the F test with Duncan's multiple range test. This was done to illustrate Duncan's test. The procedure of comparing a multiple set of means by analysis of variance does not necessarily require any additional testing even if F is significant. However, knowing only that a group of k means vary among themselves in an improbable fashion is not a very satisfying conclusion for an experiment.

ASSUMPTIONS FOR THE USE OF ANALYSIS OF VARIANCE

The analysis of variance is based on several assumptions. One of these assumptions is that the samples have been drawn from normally distributed populations. However, as with the t test, the analysis of variance is relatively unaffected by modest departures from the assumptions on which it is based. When the data are severely skewed, the investigator may resort to a transformation of the scale in order to normalize it, or he may use distribution-free tests of significance.

We have already mentioned in our discussion of the assumptions for the t test, that samples whose means are being compared should have come from populations with the same variance. The same situation exists in analysis of variance. The within variance estimate is calculated by pooling the sums of squares from each sample and dividing the result by the pooled degrees of freedom. This can only be done if it is reasonable to assume that all samples have been drawn from populations with the same variance.

When sample variances are unequal it may also be an indication that the treatment effects are not additive. We mentioned that the numerator of an F ratio can be conceptualized as representing the variation due to error plus the variation due to the influence of the independent variable. The analysis of variance assumes that this relationship is an additive one.

Ordinarily it is, but there are situations in which the independent variable acts as a multiplier. For example, if we are investigating the effect of a given drug dosage on anxiety we may find that the effect is proportional to the original level of anxiety, that is, the effect is multiplicative. If we had randomly assigned subjects to drug and no drug conditions, we would be treating initial level of anxiety as an error factor. Under these circumstances, error and drug dosage would be multiplicative rather than additive, and the assumption of additivity would be untenable. Heterogeneity of variance, regardless of its cause, can be determined by appropriate tests of significance that we shall discuss in the next section.

In spite of these precautions, it has been found that both the t test and analysis of variance can resist modest violations of the rigorous mathematical assumptions on which they are based and still produce percentages of t's and F's very close to the tabled α levels. In general, if the investigator suspects heterogeneous variances he should be sure his subgroups are approximately equal ($n_1 = n_2 = \cdots = n_k$) and as large as possible ($n > 20$). He might also increase the level of significance he requires from 1% to .5%, or .1%, before rejecting the null hypothesis. The alternative to this conservative approach is a transformation of scale, or the use of distribution-free tests of significance discussed in Chapter 14.

F TEST FOR SAMPLE VARIANCES

It is possible to test the significance of the difference between two or more sample variances just as we can test the difference between two or more sample means. If there are only two variances, we can simply divide the smaller into the larger to form an F ratio. This F ratio has $n_1 - 1$ and $n_2 - 1$ df, where n_1 and n_2 are the number of measures in each of the samples. A moment's thought however should convince you that we shall have to modify our use of the F table when we come to interpret this F ratio. We shall explain why this is so, but first let us review the nature of the F distribution.

You should recall that the F distribution is formed by dividing one randomly obtained sample variance by another. This means that we shall find some F ratios larger than 1.00 and some smaller than 1.00. The F table records only the upper half of the F distribution, the portion in which $F \geqq 1.00$. The lower section of the distribution, in which F is less than 1.00, is not ordinarily tabled. The reason for tabling only half of the distribution is that F has its most frequent use in analysis of variance where any value of F less than 1.00 indicates the absence of significance, regardless of the actual value of F. On the other hand, if F is greater than

1.00, we still cannot determine the level of significance unless we have tabled the values of F required for specific degrees of freedom.

When, as is ordinarily the situation, we have a table which includes only the portion of the F distribution with values larger than 1.00, we simply divide the smaller estimate of variance into the larger so that our obtained F's will also be larger than 1.00. Of course, when we do this we shall find twice as many significant F ratios as the table indicates we should. This means that if we are testing the hypothesis that $\sigma_1 = \sigma_2$, which is a two-tailed or two-sided test, and we arbitrarily divide the larger estimate of variance by the smaller, we must *double* the probability figure given by the F table for that F ratio.

Consider the following example. Suppose our null hypothesis states that two variance estimates differ only as a result of sampling error.

Sample$_1$			*Sample*$_2$		
8	14	9	12	20	10
16	15	17	25	19	26
15	11	14	9	10	9
14	12	19	15	12	15
13	15	12	12	9	10

$$S_1^2 = \frac{\Sigma(X - \overline{X}_1)^2}{n_1 - 1} = \frac{\Sigma X^2 - (\Sigma X)^2/n}{n_1 - 1} = \frac{2892 - (204)^2/15}{14} = 8.4$$

$$S_2^2 = \frac{\Sigma(X - \overline{X}_2)^2}{n_2 - 1} = \frac{\Sigma X^2 - (\Sigma X)^2/n}{n_2 - 1} = \frac{3487 - (213)^2/15}{14} = 33.0$$

$$F = \frac{S_2^2}{S_1^2} = 3.93 \qquad F = 3.93,\ \text{df} = 14/14,\ p < .02$$

This value of F would reach significance at the 1% level in an analysis of variance problem, but when we are comparing two sample variances in a nondirectional test, it is only significant at the 2% level.

Now consider what happens if we have a situation requiring a one-sided or a one-tailed test. Suppose we have an experimental procedure which we think will increase the variability of errors made by a sample of rats as they learn a maze. This experimental hypothesis predicts that σ_1^2, the population variance of the experimental group, will be larger than σ_2^2, the population variance of the control group. The null hypothesis, which the experimenter expects to reject, will claim that if S_1^2, is *observed* to be greater than S_2^2, it is simply the result of sampling error from two populations in which σ_1^2 is really equal to or less than σ_2^2. If we have predicted $\sigma_1^2 > \sigma_2^2$, and if we have two samples, each with 25 observations in which $S_1^2 = 300$ and $S_2^2 = 100$, we can conduct a one-tailed F test.

This will yield $F = 3.00$ with df $= 24/24$. The F ratio is significant at the 1% level in the F table, and this is the level of significance we should report. We can reject the null hypothesis and conclude that the population variance of the experimental group is larger than that of the control group.

While the F ratio in analysis of variance is based on only one tail of the F distribution, analysis of variance itself is not a one-sided or one-tailed test. The F ratio is a function of squared differences between sample means and a grand mean. The sign of the differences will disappear with the squaring operation. This is exactly the same situation we had with χ^2, where χ^2 was determined by the squared differences between observed and expected frequencies. As with the χ^2 distribution, when one uses the analysis of variance he uses only one tail of the F distribution to make a two-tailed test.

The F test can be used as a test of significance when comparing two estimates of population variance, whether we are comparing S_1^2 and S_2^2 from randomly obtained samples, or S_b^2 and S_w^2 from the analysis of variance. Even when we wish to compare more than two sample variances, the F ratio is still usable. If an investigator were to conduct an analysis of variance involving five groups, he might wish to test the within estimates of variance for heterogeneity before he proceeded with the analysis.

When the samples have equal numbers of cases we can use a rather simple statistical test called Hartley's Maximum F ratio. To conduct the test we find the maximum F ratio for our data. This will be the ratio obtained by dividing the smallest estimate of variance into the largest. If we know k, the number of samples, and the df in each sample, we can obtain the critical values for F_{max} at the 5% and 1% level from Table H of the Appendix.

Suppose we have five samples each based on 25 cases and calculate the following estimates of population variance:

Sample	S^2
1	600
2	924
3	891
4	372
5	586

We find F_{max} to be $924/372 = 2.48$. However, Table H shows that for $k = 5$, and df $= 20$, we need an $F_{max} = 3.54$ in order to reject the hypothesis that all estimates have come from the same population. We cannot reject the hypothesis of sampling error at the 5% level.

Another and extensively used, but more complicated, test for the significance of differences among estimates of population variance is called

Bartlett's test for homogeneity of variance. Bartlett's test yields a χ^2 which, if significant, requires us to reject the hypothesis that the samples all yield estimates of the same parameter. If Hartley's or Bartlett's test is significant the investigator should use transformations to reduce the heterogeneity or resort to nonparametric statistics.

THE RELATIONSHIP BETWEEN t AND F

Before proceeding to the more complex analysis of variance problems which are taken up in the next chapter, we shall devote a brief paragraph to the relationship between t and F. You may have surmised that the analysis of variance can also be applied to test the significance of the difference between two sample means. This is quite true, but the usual procedure is to use the t test in such situations. Actually, when only two samples are involved, $F = t^2$. When $k = 2$, and $n_1 = n_2$, we have

$$F = \frac{n\displaystyle\sum_k \frac{(\bar{X} - \bar{\bar{X}})^2}{k-1}}{\displaystyle\sum_k\sum_n \frac{(X - \bar{X})^2}{k(n-1)}} = \frac{n(\bar{X}_1 - \bar{\bar{X}})^2 + (\bar{X}_2 - \bar{\bar{X}})^2}{\dfrac{\Sigma(X - \bar{X})^2 + \Sigma(X - \bar{X})^2}{2(n-1)}} = \frac{\dfrac{n}{2}(\bar{X}_1 - \bar{X}_2)^2}{\dfrac{S_1{}^2 + S_2{}^2}{2}}$$

dividing by $n/2$, we have

$$\frac{(\bar{X}_1 - \bar{X}_2)^2}{\dfrac{S^2}{n} + \dfrac{S^2}{n}}$$

but t has been given as

$$\frac{\bar{X}_1 - \bar{X}_2}{\sqrt{\dfrac{S^2}{n} + \dfrac{S^2}{n}}}$$

therefore

$$t^2 = \frac{(\bar{X}_1 - \bar{X}_2)^2}{\dfrac{S^2}{n} + \dfrac{S^2}{n}}$$

which we have just shown is equal to F.

Even without algebra, an inspection of the formula for t shows its numerator $(\bar{X}_1 - \bar{X}_2)$ depends on variation between sample means, and its denominator, $\sqrt{S^2/n + S^2/n}$, depends on S^2, a variance estimate from within samples.

SUMMARY

While the t test is a satisfactory technique for determining if a significant difference exists between two sample means, it is not appropriate where more than two means are to be compared. Under these circumstances we make use of a technique called analysis of variance. In analysis of variance we obtain two estimates of the population variance, one from the variation among sample means about the grand mean of all samples S_b^2, and one from the variation of the measurements within each sample about their respective sample means S_w^2. Since the former can be thought of as representing *both* error and the influence of the independent variable, while the latter should represent only sampling error, ratios of S_b^2 to S_w^2, called F ratios, which are larger than 1.00, should be evidence of non chance fluctuation among sample means.

Comparisons of sample means following the discovery of a significant F ratio can take a variety of forms. This chapter discusses Duncan's Multiple Range Test, which yields ranges of *nonsignificance*, R_{n_1}. This range is a function of $S_{\bar{X}}$ calculated from S_w^2, and a coefficient which depends upon the df and number of samples being compared. If the range of k means *exceeds* the tabled value of R_{n_k}, then that group of k means shows significant variation. The F ratio can also be used to test the significance of the difference between two randomly obtained samples, but if more than two samples are involved one must resort to a test such as Hartley's $F_{\max}$ statistic.

EXERCISES

1.

Sample 1	*Sample* 2
26	34
24	19
19	11
12	9
11	11
27	21
26	23
29	19
9	11
12	8

(The computing formula for $S^2 = \dfrac{\Sigma(X - \bar{X})^2}{n - 1} = \dfrac{\Sigma X^2 - \dfrac{(\Sigma X)^2}{n}}{n - 1}$)

(a) Are there significant differences between the variances of Samples 1 and 2 above? F = ______, df = ______, p______.

(b) Conduct an analysis of variance on the samples above. Determine F = ______, df = ______, p______.

2.

Samples

A	B	C	D
14	11	19	18
15	12	18	22
10	9	20	24
10	6	22	27
15	9	23	26
19	12	27	25

Conduct a simple analysis of variance and, if appropriate, Duncan's Multiple Range Test on the data above. F = ______, df = ______, p______.

3. Describe how it is possible to determine the significance of the difference among several means by determining the significance of the difference between two variances.

4.

Samples

A	B	C	D	E
7	10	8	9	14
9	14	11	11	18
6	16	12	12	21
5	14	11	8	21
4	12	6	11	15
8	7	5	7	10
9	8	12	10	18

Conduct an analysis of variance on the data above. F = ______, df = ______, p______.

5.

Samples

A	B	C	
3	9	12	$n_A = 9$
2	1	11	$n_B = 7$
3	5	8	
0	7	4	$n_C = 6$
1	6	7	
5	5	9	
9	8		
6			
4			

Conduct an analysis of variance on the preceding data. $F =$ ______, df = ______, p______.

6. T—F. The total variance is equal to the within variance plus the between variance.
7. T—F. Duncan's Multiple Range Test should not be applied to a problem unless F is significant.
8. T—F. The terms "mean square" and "variance estimate" are synonymous.
9. T—F. The between groups df + the within groups df $= N - 1$, where N is the total number of observations.
10. T—F. The between, within, and total estimates of variance are each independent of the other.

CHAPTER 12

Analysis of Variance II

In the preceding chapter we illustrated the use of a one-way, or simple, analysis of variance with a study investigating the effect of diet on activity. If an investigator were to undertake such a study, he would vary some aspect of diet in a systematic fashion, and then take great care to see that there were no other sources of systematic variation among the groups.

THE FACTORIAL DESIGN

Until the use of analysis of variance in behavioral research, which began in the late 1930s and 1940s, investigators could examine the effects of only one independent variable at a time. Now, with the use of analysis of variance, we can conduct experiments which investigate the effects of each of several independent variables, and the joint effect of these variables acting together. Such investigations are called factorial experiments. With a factorial experiment we can investigate the effect of different amounts of reward, different degrees of task difficulty, and the effect of both variables operating jointly, on the number of trials required to learn a task.

The fundamental rule for a good experiment still applies. Somewhere in the study we must hold amount of reward constant and vary only task difficulty; at another point, we must hold task difficulty constant and vary only amount of reward. We will illustrate the use of analysis of variance in a factorial study, but with highly simplified data so that your attention will be focused on the statistics rather than the arithmetic.

We shall be investigating the effects of two different amounts of reward, and two different levels of difficulty on the trials to learn a task. We shall divide our subjects into four groups (cells) of three subjects each. This is called a 2 × 2 (two by two) factorial design, since there are two levels of each of two factors (independent variables), and there is a total of four cells. The four cells and the individual scores in each cell are recorded below.

1 *Low Reward* *Low Difficulty*	2 *High Reward* *Low Difficulty*	3 *Low Reward* *High Difficulty*	4 *High Reward* *High Difficulty*
5	1	7	9
6	2	8	10
7	3	9	11
$\overline{X}_1 = 6$	$\overline{X}_2 = 2$	$\overline{X}_3 = 8$	$\overline{X}_4 = 10$

THE MAIN EFFECTS

If we compare cells 1 and 2 with cells 3 and 4, we shall be comparing groups of six subjects which differ systematically only on task difficulty. If we compare cells 1 and 3 with cells 2 and 4, we shall be comparing groups of six subjects which differ systematically only on amount of reward. If the cells are arranged in a 2 × 2 table, we shall see these comparisons more clearly.

Difficulty		Reward *L*	Reward *H*	
	L	Cell 1 5 6 7 $\overline{X}_1 = 6$	Cell 2 1 2 3 $\overline{X}_2 = 2$	$\overline{X}_{DL} = 4$
	H	Cell 3 7 8 9 $\overline{X}_3 = 8$	Cell 4 9 10 11 $\overline{X}_4 = 10$	$\overline{X}_{DH} = 9$
		$\overline{X}_{RL} = 7$	$\overline{X}_{RH} = 6$	

Notice that the upper row of cells can be pooled to form a low difficulty group with a mean ($\overline{X}_{DL} = 4$), and the lower row of cells can be pooled to form a high difficulty group with a mean ($\overline{X}_{DH} = 9$). The *only* systematic difference between rows is due to the difference in task difficulty.

The effect of this difference, coupled with sampling error, is seen in the difference between the row means.

Similarly, the left column of cells may be pooled to give a low reward group whose mean is at the base of that column ($\overline{X}_{RL} = 7$). The high reward groups consists of the two cells in the right column whose mean is at the base of that column ($\overline{X}_{RH} = 6$). The only systematic difference between columns is due to the difference in reward magnitude. This difference, and sampling error, is reflected in the difference between column means.

The difference between the row means, for task difficulty, and the difference between the column means, for reward magnitude, are each called main effects. Thus, in this factorial experiment, one factor, task difficulty, supplies one main effect; the other factor, reward magnitude, supplies the other main effect.

THE INTERACTION

There is still another effect due to the joint action of task difficulty and reward magnitude. This is called an interaction effect. When interaction is present it means that the effect produced by Variable A depends upon the level, or magnitude, of Variable B. If Variable A does not produce the same effect at all levels of Variable B, we have an interaction of A and B. Such an interaction exists in this problem.

Notice the difference between the effects of low and high reward on trials to learn a task of low difficulty. This effect is seen in the difference between the means of cells 1 and 2. High reward *reduces* the mean trials required from $\overline{X}_1 = 6$ to $\overline{X}_2 = 2$. On the other hand, the effect of low and high reward on trials to learn a task of *high* difficulty is quite different. This effect is given by the difference between the means of cells 3 and 4. Here, high reward increases the mean trials required from $\overline{X}_3 = 8$ to $\overline{X}_4 = 10$. Increasing reward magnitude for an easy task apparently improves performance, but increasing it for a difficult task disrupts performance. That is to say, the effect of reward magnitude depends upon the level of task difficulty. Put in still another way, the effects of differences in reward magnitude are not the same at all levels of task difficulty. The interaction effect in this example is illustrated by the graph in Figure 12.1.

We have seen that there is an interaction effect in this problem. We can determine its magnitude by obtaining the difference between the means of pooled cells 1 and 4 and polled cells 2 and 3. These are the diagonal cells of the table. If difficulty exerts a constant effect regardless

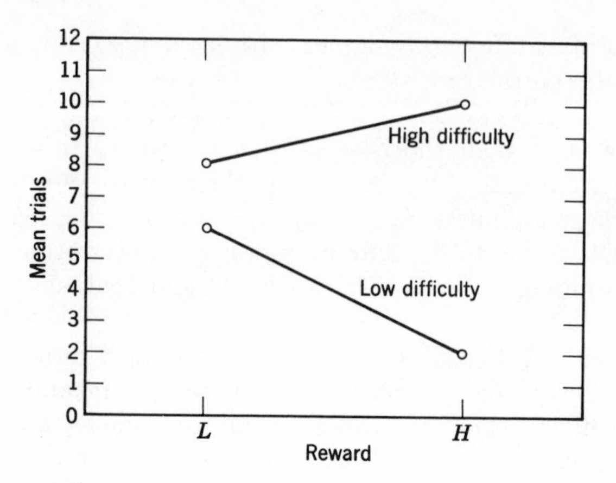

FIGURE 12.1 Graph of an interaction between magnitude of reward and task difficulty.

of reward magnitude, then the mean of pooled cells 1 and 4 will equal the mean of pooled cells 2 and 3. Since $\overline{X}_{1,4} = 8$, and $\overline{X}_{2,3} = 5$, interaction apparently exists in this problem.

INDEPENDENCE OF SOURCES OF VARIATION

The two main effects and the interaction each constitute independent sources of variation in this study, and together they account for all of the variation *between* the four cell means. There is, however, an additional source of variation in these data. That is the variation *within* cells that is produced by individual measurements varying about the means of the cells in which they fall.

We shall illustrate the independence of these sources of variation by removing each in turn, and showing that the remaining variation is unaffected. We shall first remove the variation due to differences between rows, or levels of difficulty. This may be done by finding the grand mean of all cells, which is 6.5, and then subtracting the row means, $\overline{X}_{DL} = 4$ and $\overline{X}_{DH} = 9$, from this value. The deviations are -2.5 and $+2.5$. We then add 2.5 to every score in the low difficulty row, and subtract 2.5 from every score in the high difficulty row. This will equalize the row means and yield the following data.

Difficulty	Reward L	Reward H	
L	[1] 7.5, 8.5, 9.5 $\bar{X} = 8.5$	[2] 3.5, 4.5, 5.5 $\bar{X} = 4.5$	$\bar{X}_{DL} = 6.5$
H	[3] 4.5, 5.5, 6.5 $\bar{X} = 5.5$	[4] 6.5, 7.5, 8.5 $\bar{X} = 7.5$	$\bar{X}_{DH} = 6.5$
	$\bar{X}_{RL} = 7$	$\bar{X}_{RH} = 6$	

Notice that we have removed the variation due to difficulty level, but the variation due to differences in reward magnitude remains ($\bar{X}_{RL} = 7$, $\bar{X}_{RH} = 6$), and so does the variation due to interaction ($\bar{X}_{1,4} = 8$, $\bar{X}_{2,3} = 5$). The variation within cells is also the same; each cell still consists of three consecutive numbers.

We shall next remove the variation due to differences in rewards. This is done in the same way, each number in the right column is increased by .5, and each number in the left column is decreased by .5. The result appears below.

Difficulty	Reward L	Reward H	
L	[1] 7, 8, 9 $\bar{X} = 8$	[2] 4, 5, 6 $\bar{X} = 5$	$\bar{X}_{DL} = 6.5$
H	[3] 4, 5, 6 $\bar{X} = 5$	[4] 7, 8, 9 $\bar{X} = 8$	$\bar{X}_{DH} = 6.5$
	$\bar{X}_{RL} = 6.5$	$\bar{X}_{RH} = 6.5$	

We have removed the variation due to each of the main effects, but the variation due to the interaction remains, as does the variation within cells. The interaction effect ($\bar{X}_{1,4} = 8$, $\bar{X}_{2,3} = 5$) will now be removed by adding 1.5 to the scores in cells 2 and 3, and subtracting 1.5 from those in cells 1 and 4. The result appears on the next page.

		Reward *L*	Reward *H*	
Difficulty	*L*	[1] 5.5, 6.5, 7.5 $\overline{X} = 6.5$	[2] 5.5, 6.5, 7.5 $\overline{X} = 6.5$	$\overline{X}_{DL} = 6.5$
	H	[3] 5.5, 6.5, 7.5 $\overline{X} = 6.5$	[4] 5.5, 6.5, 7.5 $\overline{X} = 6.5$	$\overline{X}_{DH} = 6.5$
		$\overline{X}_{RL} = 6.5$	$\overline{X}_{RH} = 6.5$	

All cell means are now equal. We have removed the variation between difficulty levels, between reward magnitudes, and the variation due to the interaction of these main effects. The only variation left is that within each cell and it should be apparent that this variation is the same as it was before we began removing the other sources of variation.

PARTITIONING THE SUMS OF SQUARES

We can test the significance of each main effect and their interaction by analysis of variance. Each source of variation yields a sum of squares and, with its associated degrees of freedom, provides an estimate of population variance. F ratios can be formed from each of these sources by dividing the appropriate variance estimate, or mean square by the within cells estimate of variance. Since the within cells estimate presumably is not affected by any of the experimental conditions, it can be used as the error term. It forms the denominator for *each* of the several F ratios. There will be a between task difficulties F ratio, a between reward magnitudes F ratio, and an F ratio for interaction. Each is formed by dividing the appropriate estimates of variance by the estimate obtained from the variation *within* samples. We are using exactly the same logic as was used in establishing F as a test of significance in the one-way analysis of the previous chapter.

The procedures for calculating component sums of squares for a factorial design are not different in principle from those we followed in the previous chapter. We can still partition the total sum of squares (SS_t) into two major components, a sum of squares between the k cells (SS_b) and a sum of squares within the k cells (SS_w). The important difference between the factorial design and the simple, or one-way, analysis of the preceding chapter is that in the factorial design we can *subdivide* SS_b into a sum of squares between task difficulties or rows (SS_R), a sum of squares between reward magnitudes or columns (SS_C), and a sum of squares for interaction (SS_I).

Using the data from the study above we shall first find the total sum of squares. This is obtained by subtracting the grand mean ($\bar{\bar{X}} = 6.5$) from each score, squaring, and summing the result over all N measures.

$$SS_t = \sum_N (X - \bar{\bar{X}})^2 = (5 - 6.5)^2 + (6 - 6.5)^2 + \cdots + (10 - 6.5)^2 + (11 - 6.5)^2 = 113$$

The sum of squares within cells is also obtained as it was in the last chapter. We square the deviation of each score from its cell mean and sum the result over all n_k measures in each cell, and then sum again for all k cells.

$$SS_w = \sum_k \sum_{n_k} (X - \bar{X}_k)^2 = (5 - 6)^2 + (6 - 6)^2 + \cdots + (10 - 10)^2 + (11 - 10)^2 = 8$$

The sum of squares between the four cells can be found by subtraction, since $SS_b = SS_t - SS_w$, or it can be calculated directly just as we did in the last chapter. There we squared the deviation of each cell mean from the grand mean, summed the result, and multiplied it by n_k where n_k was the number of measures in each cell.

$$SS_b = n_k \sum_k (\bar{X}_k - \bar{\bar{X}})^2 = 3[(6 - 6.5)^2 + \cdots + (10 - 6.5)^2] = 105$$

The sum of squares between cells is composed of sums of squares for each main effect represented by the column and row means, and the sum of squares for interaction. The sum of squares between reward magnitudes or columns is calculated just as we would any other between sum of squares. We find the deviation of $\bar{X}_{RH}$ and $\bar{X}_{RL}$ for the grand mean $\bar{\bar{X}}$, square these deviations, sum them, and then multiply by the n_C number of measures used in the calculation of each reward magnitude, or column, mean.

$$SS_C = n_C \sum_C (\bar{X}_C - \bar{\bar{X}})^2 = 6[(7 - 6.5)^2 + (6 - 6.5)^2] = 3$$

The sum of squares between difficulty levels, or rows, is obtained similarly.

$$SS_R = n_R \sum_R (\bar{X}_R - \bar{\bar{X}})^2 = 6[(4 - 6.5)^2 + (9 - 6.5)^2] = 75$$

We said earlier that $SS_b = SS_R + SS_C + SS_I$. Therefore, the sum of squares for interaction can be found by subtracting the two main effects' sums of squares from the sum of squares between cells. Thus

$$SS_I = SS_b - (SS_R + SS_C) = 105 - (75 + 3) = 27$$

Since the interaction of columns and rows may be represented as the residual variation among cells after the row and column effects have been accounted for, we can provide a formula for the interaction of two main effects each having any number of levels. The difference between a cell mean and the grand mean can be given by $\bar{X}_k - \bar{\bar{X}}$; subtracting from this expression the column effects $(\bar{X}_C - \bar{\bar{X}})$, and the row effects $(\bar{X}_R - \bar{\bar{X}})$, yields the expression

$$\bar{X}_k - \bar{\bar{X}} - [(\bar{X}_C - \bar{\bar{X}}) + (\bar{X}_R - \bar{\bar{X}})]$$

which can be simplified to

$$\bar{X}_k - \bar{X}_R - \bar{X}_C + \bar{\bar{X}}$$

This deviation of a cell mean from the grand mean is in excess of, and independent of, the row and column effects which also contribute to the value of that cell mean. When this deviation is squared, summed over all cells, and multiplied by n_k, the number of cases on which each cell mean is based, it becomes the interaction sum of squares.

$$n_k \sum_k (\bar{X}_k - \bar{X}_R - \bar{X}_C + \bar{\bar{X}})^2$$

In our problem, columns stand for reward, and rows stand for difficulty, so the interaction becomes

$$\left.\begin{aligned} 3(6 - 4 - 7 + 6.5)^2 &= 3(1.5)^2 = 6.75 \\ 3(8 - 9 - 7 + 6.5)^2 &= 3(-1.5)^2 = 6.75 \\ 3(2 - 4 - 6 + 6.5)^2 &= 3(-1.5)^2 = 6.75 \\ 3(10 - 9 - 6 + 6.5)^2 &= 3(1.5)^2 = 6.75 \end{aligned}\right\} = 27.00$$

This is the same value for the interaction sum of squares that we obtained earlier by the subtraction formula $SS_I = SS_b - (SS_R + SS_C)$.

This alternative method for calculating an interaction sum of squares may clarify the concept and provide a check on computations even though it may seem more complicated.

When we are dealing with a design consisting of just two levels of each of two factors, such as we have in this study, the interaction can be found in a third way which requires less arithmetic. This method makes use of the fact that when interaction is zero, the means of the pooled diagonal cells in a 2×2 table must be equal. The departure of these means from equality provides a measure of interaction. We shall determine the interaction for the present problem once again, by using the means of the two pooled diagonals. The grand mean is subtracted from the means of the pooled diagonal cells. The result is squared, summed, and multiplied by the number of cases on which the pooled diagonal means are based. Thus

$$SS_I = n_D \sum_D (\overline{X}_D - \overline{\overline{X}})^2 = 6[(5 - 6.5)^2 + (8 - 6.5)^2] = 27$$

The sources of the component sums of squares may now be a bit clearer. The total sum of squares is composed of the sum of squares between all cells and the sum of squares within all cells. The sum of squares between cells can be subdivided further into a sum of squares between columns (reward magnitude in our problem), between rows (difficulty level in our problem), and the interaction of rows and columns.

$$SS_{\text{total}} = SS_{\text{between cells}} + SS_{\text{within cells}}$$

$$SS_{\text{between cells}} = SS_{\text{between rows}} + SS_{\text{between columns}} + SS_{\text{interaction}}$$

PARTITIONING THE DEGREES OF FREEDOM

Each of these sums of squares is based upon a certain number of degrees of freedom. These may be determined just as they were in the last chapter. The total df $= kn - 1$ (or $N - 1$), where we have k cells of n subjects each. The total df may be subdivided into df between cells $(k - 1)$ and df within cells $k(n - 1)$. These are exactly the same formulas we used in the previous chapter. However, the between cells df of $k - 1$ can be further subdivided into a between columns df of $C - 1$, between rows df of $R - 1$, and an interaction df of $(R - 1)(C - 1)$. For our problem $k - 1 = 3$, giving us a total of 3 df between the four cells. We have two columns (two reward conditions) having 1 df, two rows (two levels of difficulty) having 1 df, and 1 df for interaction.

A RESUME OF THE ANALYSIS

We may now construct the complete analysis of variance table.

Source		*SS*		*df*	*MS*	*F*
Between cells		105		3	35	
Between rows (difficulty)	75		1		75	75**
Between columns (reward)	3		1		3	3
Interaction	27		1		27	27**
Within cells (error)		8		8	1	
Total (between cells + within cells)		113				

In this analysis of variance problem there are three F ratios, an F ratio between rows (difficulty), between columns (reward) and for interaction. Each F is based upon a variance estimate from within cells (error), divided into a variance estimate based on a main effect or interaction. We evaluate these F's just as we did in the previous chapter.

The F between rows has 1/8 df, and the value required for significance at the 1% level from Table F is 11.26. Our obtained F of 75 is, therefore, highly significant and indicates that the difference in means produced by problem difficulty across the range of rewards is quite unlikely to occur by sampling error alone.

The between columns F of 3 has the same number of degrees of freedom, but is clearly less than the tabled value required for significance at the 5% level. We have no evidence that magnitude of reward produces systematic effects when applied across these levels of problem difficulty.

The final F, for interaction, is again based on 1 and 8 df and is significant beyond the 1% level. We, therefore, conclude that there is a joint effect of difficulty and reward operating together.

Notice how these data have been interpreted. We cannot interpret the nonsignificant F for rewards as an indication that magnitude of reward has no effect on trials to learn. This would be much too broad a conclusion, particularly in the light of a significant interaction between reward and difficulty. Rather, we must emphasize that magnitude of reward produces no significant effect *when averaged over the two problem difficulty levels being investigated.* Of course, as with any other piece of research, our results should not be generalized beyond the kinds of subjects, the methods of procedure, and both qualitative and quantitative aspects of the independent and dependent variables used in our investigation. For example, the nature of the significant interaction between amount of reward and problem difficulty may be based on children solving concept problems for toys as a reward. We might find entirely different effects with college students learning Spanish vocabulary for quiz grades as a reward.

COMPUTATIONAL FORMULAS AND EXAMPLES

Once again we have used what we have chosen to call conceptual, rather than computational, formulas for obtaining the sums of squares. Less cumbersome formulas for calculating the various sums of squares for actual data are given below.

$$SS_{\text{total}} = \sum_{N} X^2 - \frac{\left(\sum_{N} X\right)^2}{N}$$

FORMULA 12.1
SS total
Where there are a total of *N* measures

$$SS_{\text{within cells}} = \sum_{N} X^2 - \sum_{k} \frac{\left(\sum_{n_k} X\right)^2}{n_k}$$

FORMULA 12.2
SS within cells
Where there are *k* cells with *n* measures in each cell

$$SS_{\text{between columns}} = \sum_{C} \frac{\left(\sum_{n_C} X\right)^2}{n_C} - \frac{\left(\sum_{N} X\right)^2}{N}$$

FORMULA 12.3
SS between columns
Where there are *C* columns with n_C measures in each column

$$SS_{\text{between rows}} = \sum_{R} \frac{\left(\sum_{n_R} X\right)^2}{n_R} - \frac{\left(\sum_{N} X\right)^2}{N}$$

FORMULA 12.4
SS between rows
Where there are *R* rows and n_R measures in each row

We need not give computational formulas for the between cells and the interaction sums of squares since both of these can be obtained from the formulas above by subtraction. We have already shown that

$$SS_{\text{total}} = SS_{\text{within cells}} + SS_{\text{between cells}}$$

Since we have formulas for SS_{total} and $SS_{\text{within cells}}$, we find

$$SS_{\text{between cells}} = SS_{\text{total}} - SS_{\text{within cells}}$$

We also know that

$$SS_{\text{between cells}} = SS_{\text{between rows}} + SS_{\text{between columns}} + SS_{\text{interaction}}$$

therefore,

$$SS_{\text{interaction}} = SS_{\text{total}} - (SS_{\text{within cells}} + SS_{\text{between columns}} + SS_{\text{between rows}})$$

We are rarely interested in an overall between cells test of significance in factorial designs, but the interaction sum of squares is quite necessary.

We shall use these formulas in another factorial design analysis of variance. This time our hypothetical experiment will involve three levels of each independent variable giving us a 3 × 3, instead of a 2 × 2, design. The procedures and logic are exactly the same as they were for the 2 × 2 design, and would continue to be the same for any number of levels for either variable. It would be perfectly feasible to have a 6 × 2 design or a 4 × 7 design. These numbers simply designate the number of levels on each independent variable.

There is an advantage to using more than two levels of each independent variable, because we can then determine the approximate shape of the function relating each independent variable to the dependent variable. If only two levels of the independent variable are used, the only function that can result is a straight line. This could be quite misleading. If three or more levels of the independent variables are used, we will be able to determine the shape of the function more accurately.

Suppose we wish to study the effect of length of intertrial interval and difficulty of material on trials to learn lists of nonsense syllables. We shall use intertrial intervals of 1 minute, 3 minutes, and 9 minutes, and vary the difficulty level of the material by varying the meaningfulness of the syllables. We shall use syllables with 0%, 30% and 70% association values (association value is a measure of meaningfulness). With a 3 × 3 design there are 9 cells to which 45 subjects are randomly assigned, with the restriction that there shall be 5 subjects per cell. The data could take the following form. (You can probably see why multilevel factorial designs are uncommon. If this were a 6 × 6 design, we would have to fill 36 cells. This would require four times as many subjects as a 3 × 3 design, and would probably yield little additional information.)

0%	0%	0%	30%	30%	30%	70%	70%	70%
1 min.	3 min.	9 min.	1 min.	3 min.	9 min.	1 min.	3 min.	9 min.
30	23	12	20	12	13	10	14	14
20	10	15	26	14	14	22	19	8
19	15	9	23	16	20	16	9	13
29	25	6	17	19	15	10	10	13
36	20	15	16	24	11	8	8	9

The nine cells can be ordered into a 3 × 3 table. We shall assign difficulty level to columns and intertrial interval to rows.

Intertrial Interval	Difficulty (Association Value) 0% (High)	30% (Medium)	70% (Low)	
9 min.	12, 15, 9, 6, 15; $\Sigma X = 57$; $\overline{X} = 11.4$	13, 14, 20, 15, 11; $\Sigma X = 73$; $\overline{X} = 14.6$	14, 8, 13, 13, 9; $\Sigma X = 57$; $\overline{X} = 11.4$	$\Sigma_{R_{9\,\text{min.}}} = 187$; $\overline{X}_{R_{9\,\text{min.}}} = 12.5$
3 min.	23, 20, 15, 25, 20; $\Sigma X = 103$; $\overline{X} = 20.6$	12, 14, 16, 19, 24; $\Sigma X = 85$; $\overline{X} = 17.0$	14, 19, 9, 10, 8; $\Sigma X = 60$; $\overline{X} = 12.0$	$\Sigma_{R_{3\,\text{min.}}} = 248$; $\overline{X}_{R_{3\,\text{min.}}} = 16.5$
1 min.	30, 20, 19, 29, 36; $\Sigma X = 134$; $\overline{X} = 26.8$	20, 26, 23, 17, 16; $\Sigma X = 102$; $\overline{X} = 20.4$	10, 22, 16, 10, 8; $\Sigma X = 66$; $\overline{X} = 13.2$	$\Sigma_{R_{1\,\text{min.}}} = 302$; $\overline{X}_{R_{1\,\text{min.}}} = 20.1$
	$\Sigma_{C_{0\%}} = 294$; $\overline{X}_{C_{0\%}} = 19.6$	$\Sigma_{C_{30\%}} = 260$; $\overline{X}_{C_{30\%}} = 17.3$	$\Sigma_{C_{70\%}} = 183$; $\overline{X}_{C_{70\%}} = 12.2$	

$$SS_t = \sum_N X^2 - \frac{\left(\sum_N X\right)^2}{N} = (12)^2 + (15)^2 + \cdots + (10)^2 + (8)^2 - \frac{(737)^2}{45} = 1896.6$$

$$SS_w = \sum_N X^2 - \sum_k \frac{\left(\sum_{n_k} X\right)^2}{n_k}$$

$$= (12)^2 + (15)^2 + \cdots + (10)^2 + (8)^2 - \left[\frac{(57)^2}{5} + \frac{(103)^2}{5} + \cdots + \frac{(60)^2}{5} + \frac{(66)^2}{5}\right] = 771.6$$

$$SS_C = \sum_C \frac{\left(\sum_{n_C} X\right)^2}{n_C} - \frac{\left(\sum_N X\right)^2}{N}$$

$$= \left[\frac{(294)^2}{15} + \frac{(260)^2}{15} + \frac{(183)^2}{15}\right] - \frac{(737)^2}{45} = 431.3$$

$$SS_R = \sum_R \frac{\left(\sum_{n_R} X\right)^2}{n_R} - \frac{\left(\sum_N X\right)^2}{N}$$

$$= \left[\frac{(187)^2}{15} + \frac{(248)^2}{15} + \frac{(302)^2}{15}\right] - \frac{(737)^2}{45} = 441.4$$

$$SS_I = SS_t - [SS_w + SS_C + SS_R]$$

$$= 1896.6 - [771.6 + 431.3 + 441.4] = 252.3$$

After obtaining the sums of squares for each source of variation, we can obtain the associated df by the following formulas.

Between columns	= number of columns − 1, or $C - 1$	= 2
Between rows	= number of rows − 1, or $R - 1$	= 2
Interaction	= $(C - 1)(R - 1)$	= 4
Within cells	= (number of cells) × (number of measures in each cell − 1), or $k(n - 1)$	= 36
Total	= total measures − 1, or $N - 1$, or $kn - 1$	= 44

Once we know the component sums of squares and the degrees of freedom available for each we can construct the analysis of variance table below.

Source	*SS*	*df*	*MS*	*F*
Between columns (assoc. level)	431.3	2	215.6	10.1**
Between rows (intertrial int.)	441.4	2	220.7	10.3**
Interaction	252.3	4	63.1	2.9*
Within cells	771.6	36	21.4	
Total	1896.6	44		

The between columns F shows a significant effect for difficulty over the three intertrial interval levels tested, and the between rows F shows a significant intertrial interval effect over the difficulty levels tested.

The interaction F is significant at the 5% level. Figure 12.2 illustrates this interaction. We have graphed the mean trials to criterion for the 0%, 30%, and 70% groups at each of the intertrial intervals used in the study. (We could also illustrate this interaction by graphing mean trials for the 1, 3, and 9 minute groups at each of the difficulty levels.) Notice, there is a progressive decline in mean trials to criterion for each association level group as we increase the intertrial interval from 1 to 9 minutes. This decline produces the significant F between intertrial intervals. However, the decline does not occur at a constant rate for all groups. It is substantially more pronounced for the 0% association level group, moderate for the 30% association group, and slight to nonexistent for the 70% as-

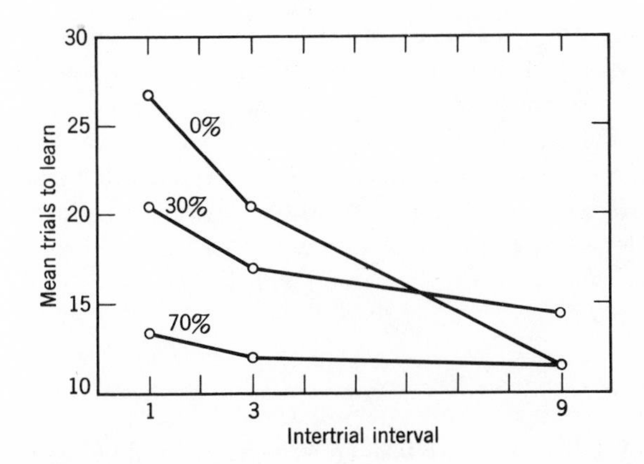

FIGURE 12.2 Interaction of intertrial interval and difficulty.

sociation level group. A longer intertrial interval seems to have a greater facilitative effect on performance with low association value syllables than with syllables of high association value. The differences among the association value levels are not constant across the three intertrial intervals. This is the effect that produces the significant F for interaction.

Consider the advantages of the factorial design over the simple or one-way analyses described in Chapter 11. It is true that we could have obtained the information about significant differences between intertrial intervals by using a one-way, or simple, analysis of variance, and we could have conducted another one-way analysis to investigate the effects of association level on trials to learn. But if we had conducted these as two separate studies, we would have required 90 subjects instead of 45, and we would have gained no information on the interaction of the two variables. The use of a factorial design does require a more elaborate statistical treatment; but it can reveal the existence of interactions which separate studies cannot demonstrate; and it is a much more efficient design, because it requires far fewer subjects than would be required for several separate one-way analyses.

Another point can be made concerning the interaction effects. Notice what would happen if someone had conducted a study investigating the effects of intertrial interval on trials to learn a group of nonsense syllables and decided to use 0% association level stimuli. From Figure 12.2 it appears that he would probably obtain significant differences. A student decides to check the results, but wishes to use easier lists so that his subjects will remain friendly after the experiment is over. The student uses 70% association value syllables and, according to Figure 12.2, would probably find no significant differences between the intertrial interval groups.

The considerable degree of interaction present between many independent variables in behavioral research makes it imperative that when an experiment is replicated, we use *exactly* the same levels of the independent variables as those used by the original investigator. Had there been no interaction between association level and intertrial interval, we could have demonstrated the effect of the latter by using only the most convenient levels of the former. Since an interaction does exist between the variables, we must use the same levels of the factors as were used in the original study.

THE 2^n DESIGN

At this point our factorial designs have been confined to just two independent variables or factors. Initially, we discussed a design involving two levels of each of two factors, then we discussed two factor designs with several levels of each factor. Now we shall discuss designs using several factors with each factor at two levels. We shall be investigating the effects of two levels of each of three independent variables on a dependent variable. The study is thus a $2 \times 2 \times 2$ design, or a 2^3 design. It is also possible to have a $4 \times 3 \times 2 \times 3$ design where we investigate several levels of each of four independent variables. Variations on this theme are limited only by the experimenter's ambition, the size of his research grant, and the capacity of his computer.

In the study to be described we shall investigate the effect of viewing two kinds of films, and participating in two kinds of discussion groups on the willingness of students classified as "adjusted" or "maladjusted" to give up smoking. We shall first give a paper and pencil test of "psychological adjustment" to a large group of male university students. On the basis of the test scores, we shall select the upper (well adjusted) and the lower (poorly adjusted) quarters of the distribution, and from each quarter we shall randomly select 24 students. The 24 S's from each end of the distribution will then be assigned randomly to the four treatments so that six S's will fall in each cell.

We shall have students from each adjustment group randomly assigned to one of two films, a "scare" film that vividly illustrates the terminal phases of lung cancer, heart disease, and emphysema; and an "educational" film that shows how smoking affects the oxygen carrying capacity of blood and reviews the Surgeon General's report on the statistical evidence linking smoking to lung cancer and other diseases.

Students from each adjustment group will also be assigned randomly to one of two 20-minute discussion groups; in one discussion group the leader will keep the discussion focused on the reasons why people smoke and the reasons why it is difficult to stop. In the other discussion group the leader will focus attention on the reasons why people should stop, and on the advantages of stopping. At the end of one week each student will be asked to fill out a questionnaire designed to measure changes in his smoking habits. Scores above 20 indicate increased smoking, scores below 20 indicate decreased smoking. The data, like that presented in previous illustrations is fictional, although the general design is not uncommon in the literature.

We shall have the following main effects: A_1—high adjustment, A_2—low adjustment; B_1—"scare" film, B_2—"educational" film; C_1—"drawback" discussion, C_2—"why smoke" discussion. Within the A_1 group, 24 subjects are randomly assigned to B_1–C_1, B_2–C_1, B_1–C_2, and B_2–C_2 cells. Similar random assignment is used for the 24 subjects in the A_2 group. Notice that we *cannot randomly assign* subjects to the A_1 and A_2 groups.

This variable is a characteristic of the subject and cannot be assigned to him. Our 48 measures distribute themselves as follows.

A_1				A_2			
B_1		B_2		B_1		B_2	
C_1	C_2	C_1	C_2	C_1	C_2	C_1	C_2
17	9	15	17	26	20	16	19
18	11	13	15	28	18	14	13
15	15	11	17	16	17	13	15
21	17	17	15	23	21	15	15
14	12	19	12	21	22	17	18
17	14	15	20	18	16	15	16
$\Sigma X = 102$	$\Sigma X = 78$	$\Sigma X = 90$	$\Sigma X = 96$	$\Sigma X = 132$	$\Sigma X = 114$	$\Sigma X = 90$	$\Sigma X = 96$
$\bar{X} = 17$	$\bar{X} = 13$	$\bar{X} = 15$	$\bar{X} = 16$	$\bar{X} = 22$	$\bar{X} = 19$	$\bar{X} = 15$	$\bar{X} = 16$

While this is a more complicated problem than the previous one, we shall partition the sums of squares in much the same way. We have SS_{total}, which is the sum of $SS_{\text{between cells}}$ and $SS_{\text{within cells}}$. The $SS_{\text{between cells}}$ is in turn composed of the sum of $SS_{\text{between } A\text{'s}}$ (adjustment groups), $SS_{\text{between } B\text{'s}}$ (film type), and $SS_{\text{between } C\text{'s}}$ (lecture type); as well as the interactions between each pair of factors $SS_{A \times B}$, $SS_{A \times C}$, $SS_{B \times C}$, and the triple interaction term $SS_{A \times B \times C}$. We shall discuss the meaning of each term as well as describe the procedures for calculating it.

It is possible to represent a $2 \times 2 \times 2$ design by a three dimensional solid, but the branching "tree" above is somewhat more efficient. Each column represents a cell, and by combining sums from appropriate cells we can arrive at the sums for each level of the independent variables. These sums are listed on the left. The calculation of sums of squares proceeds as it did in the previous problem until we arrive at the interaction sums of squares. Please follow the calculations carefully so that you know how each one follows from the formulas.

$$\begin{aligned}
\Sigma X &= 798 \\
\Sigma_{A_1} X &= 366 \\
\Sigma_{A_2} X &= 432 \\
\Sigma_{B_1} X &= 426 \\
\Sigma_{B_2} X &= 372 \\
\Sigma_{C_1} X &= 414 \\
\Sigma_{C_2} X &= 384
\end{aligned}$$

$$SS_{\text{total}} = \sum_N X^2 - \frac{\left(\sum_N X\right)^2}{N} = (17)^2 + (18)^2 + \cdots$$

$$+ (18)^2 + (16)^2 - \frac{(798)^2}{48} = 639.3$$

$$SS_{\text{within cells}} = \sum_N X^2 - \sum_k \frac{\left(\sum_{n_k} X\right)^2}{n_k}$$

$$= [(17)^2 + (18)^2 + \cdots + (18)^2 + (16)^2]$$

$$- \left[\frac{(102)^2}{6} + \frac{(78)^2}{6} + \cdots + \frac{(90)^2}{6} + \frac{(96)^2}{6}\right] = 316.0$$

$$SS_{\text{between } A\text{'s}} = \sum_{A\text{'s}} \frac{\left(\sum_{n_A} X\right)^2}{n_A} - \frac{\left(\sum_N X\right)^2}{N} = \left[\frac{(366)^2}{24} + \frac{(432)^2}{24}\right] - \frac{(798)^2}{48} = 90.8$$

$$SS_{\text{between } B\text{'s}} = \sum_{B\text{'s}} \frac{\left(\sum_{n_B} X\right)^2}{n_B} - \frac{\left(\sum_N X\right)^2}{N} = \left[\frac{(426)^2}{24} + \frac{(372)^2}{24}\right] - \frac{(798)^2}{48} = 60.8$$

$$SS_{\text{between } C\text{'s}} = \sum_{C\text{'s}} \frac{\left(\sum_{n_C} X\right)^2}{n_C} - \frac{\left(\sum_N X\right)^2}{N} = \left[\frac{(414)^2}{24} + \frac{(384)^2}{24}\right] - \frac{(798)^2}{48} = 18.8$$

The sums of squares for each main effect, and the sum of squares for error, have now been calculated. We can calculate the sum of squares for interaction $A \times B$ by setting up a 2×2 table of the ΣX values for A_1B_1. A_1B_2, A_2B_1 and A_2B_2 calculated across both C_1 and C_2 levels.

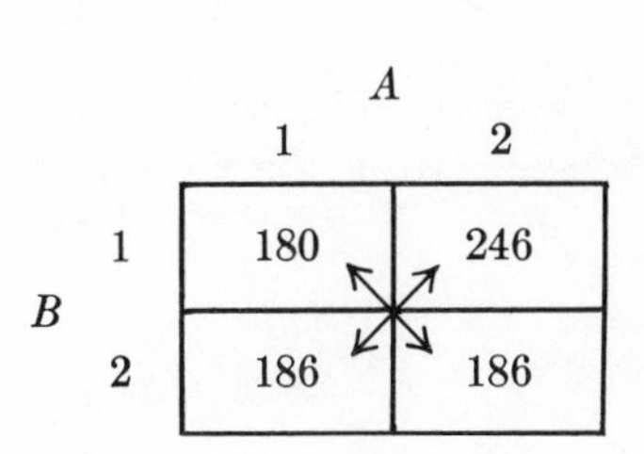

By summing the diagonals we arrive at

$$\sum_{n_{A_1B_2+A_2B_1}} X = 432,$$

$$\sum_{n_{A_1B_1+A_2B_2}} X = 366$$

These sums may be used in computational formulas similar to those used for obtaining the interaction sums of squares in the preceding problem.

$$SS_{A\times B} = \sum_{A\times B} \frac{\left(\sum_{n_{A\times B}} X\right)^2}{n_{A\times B}} - \frac{\left(\sum_{N} X\right)^2}{N} = \left[\frac{(432)^2}{24} + \frac{(366)^2}{24}\right] - \frac{(798)^2}{48} = 90.8$$

Similarly, a 2 × 2 table for A_1C_1, A_2C_1, A_1C_2 and A_2C_2 summed across both B levels yields

		A 1	A 2
C	1	192	222
	2	174	210

$$SS_{A\times C} = \sum_{A\times C} \frac{\left(\sum_{n_{A\times C}} X\right)^2}{n_{A\times C}} - \frac{\left(\sum_{N} X\right)^2}{N} = \left[\frac{(402)^2}{24} + \frac{(396)^2}{24}\right] - \frac{(798)^2}{48} = 0.8$$

A 2 × 2 table for $B \times C$ summed across A levels yields

		B 1	B 2
C	1	234	180
	2	192	192

$$SS_{B\times C} = \sum_{B\times C} \frac{\left(\sum_{n_{B\times C}} X\right)^2}{n_{B\times C}} - \frac{\left(\sum_{N} X\right)^2}{N} = \left[\frac{(426)^2}{24} + \frac{(372)^2}{24}\right] - \frac{(798)^2}{48} = 60.8$$

The interaction of $A \times B \times C$ is most easily obtained by subtraction. (However, a more direct method for calculating the triplet interaction will be given shortly.)

$$SS_{A\times B\times C} = SS_t - (SS_{\text{within}} + SS_{A\text{'s}} + SS_{B\text{'s}} + SS_{C\text{'s}} + SS_{A\times B} + SS_{A\times C} + SS_{B\times C})$$

$$SS_{A\times B\times C} = .5$$

We are now in possession of all sums of squares. The derivation of the df's follow the logic previously given.

$SS_{A's} = 90.8,$ df = 1 $(A\text{'s} - 1)$
$SS_{B's} = 60.8,$ df = 1 $(B\text{'s} - 1)$
$SS_{C's} = 18.8,$ df = 1 $(C\text{'s} - 1)$
$SS_{A\times B} = 90.8,$ df = 1 $(A\text{'s} - 1)(B\text{'s} - 1)$
$SS_{A\times C} = .8,$ df = 1 $(A\text{'s} - 1)(C\text{'s} - 1)$
$SS_{B\times C} = 60.8,$ df = 1 $(B\text{'s} - 1)(C\text{'s} - 1)$
$SS_{A\times B\times C} = .5,$ df = 1 $(A\text{'s} - 1)(B\text{'s} - 1)(C\text{'s} - 1)$
$SS_{\text{within cells}} = 316.0,$ df = 40 $k(n - 1)$, for k cells with n observations in each cell
$SS_{\text{total}} = 639.3$ df = 47 $N - 1$, for N observations

The analysis of variance table for this problem has been constructed below.

Source	*SS*	*df*	*MS*	*F*
Between A's	90.8	1	90.8	11.49**
Between B's	60.8	1	60.8	7.69**
Between C's	18.8	1	18.8	2.37
$A \times B$	90.8	1	90.8	11.49**
$A \times C$	.8	1	.8	<1.00
$B \times C$	60.8	1	60.8	7.69**
$A \times B \times C$	.5	1	.5	<1.00
Within cells (error)	316.0	40	7.9	
Total	639.3	47		

The difference between A's, good adjustment versus poor adjustment, is highly significant. The interpretation of this significant difference is somewhat different from the interpretation that may be given other significant differences in this problem. The reason is that subjects were not randomly assigned *to* the A categories, but were classified in this way on the basis of attributes they already possessed. Good adjustment and poor adjustment are, therefore, not experimental treatments in the same sense that type of discussion group and type of film are experimental treatments.

Even though the F ratio of 11.49 indicates that this is a highly significant difference, we did not randomly assign our subjects to these two "treatment" groups and, therefore, we cannot assume that the difference in their tendencies to stop smoking is due entirely, or even in part, to their having been classified into good versus poor adjustment groups on a psychological test. It is quite reasonable to assume that some *other* systematic difference, such as early childhood experience, current success in college, or a host of related factors, could lead to *both* level of test score and the difference in response to our questionnaire on smoking.

The significant difference between B treatments is another matter. Here the random assignment of subjects to conditions B_1 or B_2 has been carried out, and presumably there are no other systematic differences between these groups. We can therefore conclude that the type of film presentation has a significant effect on the dependent variable across the adjustment levels and the discussion group levels being investigated.

The failure of the F ratio for C treatments to reach significance means that the different types of discussion groups do not produce differences in tendency to stop smoking when applied across the adjustment levels and types of films we have used. Keep in mind, however, that this F ratio is substantially larger than 1.00, and is, in fact, "significant" between the 25% and the 10% levels. There would probably be a substantial risk of a type II error if we concluded that the difference in discussion groups produced no effect across the other treatment levels. We might better reserve judgment about the effectiveness of C.

The $A \times B$ interaction is significant at the 1% level. This interaction is best illustrated in a 2×2 table similar to that from which the $A \times B$ sum of squares was calculated.

		A: 1	A: 2
B	1	$\bar{X} = 15.0$	$\bar{X} = 20.5$
	2	$\bar{X} = 15.5$	$\bar{X} = 15.5$

The $A \times B$ interaction results from the fact that the difference between the B_1 and B_2 films is not the same at both levels of adjustment. Of course, it is equally accurate to say that the difference between the A_1–A_2 adjustment groups is not the same for both types of film. Remember that these are exactly equivalent statements. The crux of interaction is that we have a difference between differences.

Practically speaking, in terms of this particular study, the interaction results from the fact that smoking by poorly adjusted students seems to be more readily modified by the "educational" film than by the "scare" film. On the other hand, smoking by well adjusted students seems to be modified to the same extent by both types of films. With the better adjusted student there is apparently no difference between the effects of the two films. It is this difference between differences that produces the significant $A \times B$ interaction.

The $A \times C$ interaction is diagrammed in the 2×2 table below. This interaction is very slight and is not significant.

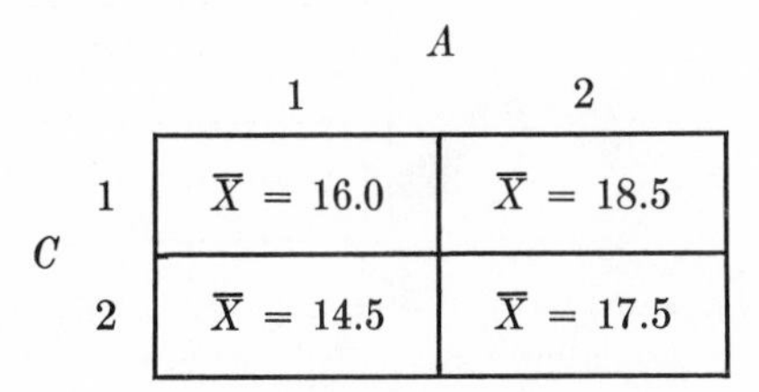

		A 1	2
C	1	$\bar{X} = 16.0$	$\bar{X} = 18.5$
	2	$\bar{X} = 14.5$	$\bar{X} = 17.5$

The effect of the C factor is essentially the same for both levels of A. Notice that the C_1–C_2 difference for A_1 is about the same as it is for A_2. In this instance there is no appreciable difference between the differences, so the interaction is not significant.

The $B \times C$ interaction is significant at the 1% level.

		B 1	2
C	1	$\bar{X} = 19.5$	$\bar{X} = 15.0$
	2	$\bar{X} = 16.0$	$\bar{X} = 16.0$

Apparently the more rational film (B_2) is about equally effective with either discussion group, C_1 or C_2. The more emotional film (B_1) is not equally effective in both discussion groups. The film with emotional appeal (B_1) loses effectiveness if it is paired with the discussion group which also emphasized reasons for giving up cigarettes (C_1). Again, the significant interaction reflects this difference between differences.

Before we discuss the meaning of an $A \times B \times C$ interaction, notice once more the $B \times C$ interaction illustrated in the 2×2 table above. This table shows differences between B_1 and B_2 means, at each level of C, across pooled levels of A. Similarly, the $A \times B$ interaction was based on sums calculated across pooled levels of C, and the $A \times C$ interaction was derived from pooled levels of B. If the $A \times B \times C$ interaction is significant, it will mean that the $B \times C$ interaction is different at *each* level of A; the $A \times B$ interaction different at *each* level of C, and the $A \times C$ interaction different at *each* level of B. Each of these statements will be true if the $A \times B \times C$ interaction is significant. Each statement emphasizes a different aspect of the $A \times B \times C$ interaction. If the interaction is significant, each of them will be true. In our problem $A \times B \times C$ is not significant, so we should expect to find the $A \times B$ interaction about the same at each level of C. The two 2×2 tables on the next page show the $A \times B$ interactions to be about the same at C_1 and C_2.

C_1

		A: 1	A: 2
B	1	$\overline{X} = 17$	$\overline{X} = 22$
	2	$\overline{X} = 15$	$\overline{X} = 15$

C_2

		A: 1	A: 2
B	1	$\overline{X} = 13$	$\overline{X} = 19$
	2	$\overline{X} = 16$	$\overline{X} = 16$

This conception of a triplet interaction provides us with another way to calculate its sum of squares. If we calculate an $A \times B$ sum of squares for each level of C separately, and then subtract the $A \times B$ sum of squares with C collapsed, which has already been obtained, we shall have the $A \times B \times C$ sum of squares. Thus

C_1

		A: 1	A: 2
B	1	$\Sigma_{A_1B_1C_1}X$ 102	$\Sigma_{A_2B_1C_1}X$ 132
	2	$\Sigma_{A_1B_2C_1}X$ 90	$\Sigma_{A_2B_2C_1}X$ 90

C_2

		A: 1	A: 2
B	1	$\Sigma_{A_1B_1C_2}X$ 78	$\Sigma_{A_2B_1C_2}X$ 114
	2	$\Sigma_{A_1B_2C_2}X$ 96	$\Sigma_{A_2B_2C_2}X$ 96

$$SS_{A \times B(C_1)} = \frac{(222)^2}{12} + \frac{(192)^2}{12} - \frac{(414)^2}{24} \qquad SS_{A \times B(C_2)} = \frac{(210)^2}{12} + \frac{(174)^2}{12} - \frac{(384)^2}{24}$$

$$SS_{A \times B(C_1)} = 37.5 \qquad SS_{A \times B(C_2)} = 54.0$$

And, since $SS_{A \times B} = 90.8$, we have $SS_{A \times B \times C} = (37.5 + 54.0) - 90.8 = .7$, which (within limits of rounding) is what we calculated earlier when we subtracted the sum of squares associated with all other sources of variation from the total sum of squares.

It should be clear that we can also find the sum of squares attributable to triplet interaction by calculating $SS_{A \times C}$ at each level of B, and then subtracting $SS_{A \times C}$ with B collapsed, or finding $SS_{C \times B}$ for each level of A and then subtracting the $SS_{C \times B}$ with A collapsed. Any of these procedures will yield $SS_{A \times B \times C}$.

EQUAL NUMBERS OF SUBJECTS IN THE SUBGROUPS

In this chapter we have dealt with research designs in which each cell contained the same number of subjects. We should always plan the research so this will be the case, but occasionally, even with careful planning, we may have unequal subgroups. This may happen for a variety of reasons, some of which can introduce an uncorrectable bias. For example, suppose a study involved learning nonsense syllables of different difficulty levels in order to test later retention. We might find that relatively few subjects from the higher difficulty groups are willing to return for retention tests. The subjects who do return might be those who had the least trouble and embarrassment initially; they may show better retention than the subjects who "forgot" their appointments. We cannot correct this bias by simply testing more subjects. On the other hand, if a recorder fails to operate properly on some trial, it is probably safe to assume that the failure was not correlated with some level of an independent variable and the subject can be replaced.

While unequal n's in themselves pose no particular problem for the calculations in a one-way analysis of variance, we are forced to make some rather complex corrections when unequal n's appear in the subgroups of factorial designs. More advanced texts treat this topic at some length, but the best plan is to make *every effort* to keep the cells the same size in the first place.

INDEPENDENT OBSERVATIONS

We must now call your attention to another feature of the research designs in this and the preceding chapter. Notice that we had *n independent* observations in each cell. This was accomplished for the most part by randomly assigning a *different* subject to each cell and making only one observation per subject. The procedures we discussed in this and the previous chapter *absolutely* require that the observations be independent.

In the next chapter we shall discuss a very common form of analysis of variance which is designed for use with repeated measures on the same subjects, or with different individuals matched into cells on the basis of their performance on some variable correlated with the dependent variable. For example, one of our main effects could consist of trials in a learning study. We may be interested in the significance of *changes* in performance for a group of subjects over a series of 10 trials. Each trial would then constitute a level of the trials' effect. In this situation a subject's performance on trial 10 is *not independent* of his performance on trial 9. We cannot apply the procedures outlined in this chapter to repeated measures problems. The appropriate procedures will be discussed in Chapter 13.

MODELS IN THE ANALYSIS OF VARIANCE

Another feature of these factorial designs has been the use of predetermined levels of the factors, the independent variables, under investigation. For example, we deliberately selected levels of each factor which were of particular interest. We chose to investigate 1, 3, and 9 minute intertrial intervals, and 0%, 30%, and 70% association values of nonsense syllables. When the "levels" are deliberately chosen, we are using a "fixed effects" or "fixed constants" model of analysis of variance. If we had selected the levels at random from populations of intertrial intervals and association values, we would have been using a "random" model of analysis of variance.

It is also possible and quite common in factorial designs to have the levels of one factor chosen randomly while the levels of the other factor are fixed. For example, we might assign subjects to two different kinds of sensitivity training (fixed effects), and then randomly select 5 judges from among a group of clinical psychologists (random) to rate the subjects' personalities. When the levels of one factor are fixed, and levels of the other factor are random, we have what is called a "mixed effects" model. Examples of fixed effects and mixed effects models are fairly common in behavioral literature, but the random model is relatively rare.

The conclusions drawn from a fixed effects model apply *only* to the specific levels tested. The random model permits conclusions to be generalized to the population of possible levels from which the tested levels were randomly drawn. In the one-way analysis of variance, the F ratio is calculated the same way whether the model is fixed or random. (A mixed model is not possible in a one-way analysis.) However, when a factorial design is used there are differences between the models regarding which sum of squares is appropriate for testing the significance of the various effects. In the fixed effects model, which we have illustrated, the within cells variance is the appropriate error term for all main effects and all interactions.

In a factorial design using the mixed effects model, the within cells variance is used to test both the interaction term and the main effect term in which levels were derived from random sampling. However, the main effect term with the fixed levels is tested by using the interaction term as the estimate of error.

In a random effects model, the interaction term is tested using the within cells term for error, but *both* main effects are then tested by using the interaction term as the estimate of error.

The theory underlying the choice of an error term for the various models of analysis of variance is beyond the scope of this book, even though we shall dip into some rather complex "mixed" designs in the next chapter. For additional information on models and an excellent treatment of more complex analysis of variance designs, the interested student should consult Winer.[1]

Exercises

1. Conduct a 2 × 2 factorial analysis of variance for the data below.

		A: 1	A: 2
B	1	8 9 7 6	12 17 18 19
	2	12 14 16 15	5 7 9 12

2. Conduct a 2 × 2 factorial analysis of variance for the data below.

A_1B_1	A_2B_1	A_1B_2	A_2B_2
17	42	15	29
27	29	17	36
15	31	19	29
19	35	12	47
12	36	23	34
$\Sigma X = 90$	$\Sigma X = 173$	$\Sigma X = 86$	$\Sigma X = 175$

$\Sigma X^2 = 15726$ $\qquad$ $\Sigma X = 524$

3. Conduct a 2 × 3 factorial analysis of variance for the data below.

A_1B_1	A_2B_1	A_3B_1	A_1B_2	A_2B_2	A_3B_2
14	15	26	15	16	14
16	19	29	19	15	12
20	23	33	22	19	9
15	27	37	21	18	10
17	21	31	24	12	17
15	24	28	17	24	20
$\Sigma X = 97$	$\Sigma X = 129$	$\Sigma X = 184$	$\Sigma X = 118$	$\Sigma X = 104$	$\Sigma X = 82$

Determine the significance of the main effects and interpret the interaction.

[1] Winer, J. B. *Statistical Principles in Experimental Design.* New York: McGraw-Hill, 1962.

4. Conduct a $2 \times 2 \times 2$ analysis of variance for the data given below.

$A_1B_1C_1$	$A_2B_1C_1$	$A_1B_2C_1$	$A_2B_2C_1$	$A_1B_1C_2$	$A_2B_1C_2$	$A_1B_2C_2$	$A_2B_2C_2$
14	15	11	9	9	15	20	17
6	11	9	16	8	17	12	18
7	9	19	16	7	13	14	9
9	7	13	14	14	8	10	21
$\Sigma X = 36$	$\Sigma X = 42$	$\Sigma X = 52$	$\Sigma X = 55$	$\Sigma X = 38$	$\Sigma X = 53$	$\Sigma X = 56$	$\Sigma X = 65$

5. Give examples of two variables between which you would expect to find interaction effects.
6. What are the advantages and disadvantages of multilevel factorial designs?
7. If, in Exercise 4, you found $SS_{A \times B \times C}$ by subtracting all other sums of squares from the total, check your calculations by using an alternate method.
8. Find an example of a factorial analysis of variance in the literature of your field. Make sure that the data involves independent measures; the subjects should not be matched, nor should repeated measures be taken, as would be done if "trials" were a main effect. You will find that the repeated measures design is very widely used, and we will discuss it in the next chapter.

CHAPTER 13

Analysis of Variance III

You may recall that in Chapter 10 we discussed two kinds of t tests. One of these was designed to be used with independent groups, while the other was to be used with correlated observations such as might be obtained in a test-retest, or matched pairs, investigation. We mentioned that the matched pairs t test was normally a more powerful test of significance than the independent groups t test, because the positive correlation between the paired measurements tended to decrease the variability of the difference scores on which the standard error of the mean difference was based. But we also pointed out that the matched pairs procedure cost $n - 1$ degrees of freedom, since there were $n_1 + n_2 - 2$ df available for evaluating an independent groups t, and only half that number $(n - 1)$ for evaluating a matched pairs t. (If the section on the independent groups and matched pairs t tests is not familiar from the brief discussion in the preceding paragraph, it would be advisable to return to Chapter 10 for a review before continuing with this chapter.)

ONE-WAY REPEATED MEASURES AND RANDOMIZED BLOCKS ANALYSIS OF VARIANCE

The rationale for a matched pairs t test can be extended to comparisons of more than two group means by the use of a procedure called repeated measures analysis of variance. Just as the one-way analysis of variance discussed in Chapter 11 is used to compare the means of k independent groups, a repeated measures one-way analysis of variance is used to compare the means resulting from k tests of the same, or matched, subjects.

Suppose we conduct a learning experiment and wish to determine if there is a change in our subjects' performance as a function of trials. If we use the independent groups technique we will randomly divide 40 subjects among five groups so as to have eight subjects in each group, and then record the Trial 1 performance of one group of subjects, the Trial 2 performance of another, the Trial 3 performance of a third and so on for the five trials. This procedure would provide us with the *independent* measurements which would be required if we were to use the independent groups analysis of variance.

However, there is a much easier way to conduct such a study. We can use the *same* eight subjects for all five trials and subject the resulting data to a repeated measures analysis of variance. The fact that the same eight subjects are used on every trial means that some correlation will probably exist between the subjects' scores from one trial to the next. Just as the presence of correlation makes the matched pairs t test more powerful than the independent groups t test, it also makes the repeated measures analysis of variance more powerful than the analysis for independent groups.

To illustrate this, the data for a repeated measures analysis of variance are given in Table 13.1. The five subjects, A through E, have their performance measured on each of three trials. Notice the correlation between the subjects' performances over trials.

Before we show the actual calculations, consider for a moment the logic by which the sums of squares might be partitioned. Following the procedures we used for the one-way analysis with independent groups, we shall have $SS_{\text{total}} = SS_{\text{between trials}} + SS_{\text{within trials}}$. If we were conducting an independent groups analysis we would use $SS_{\text{within trials}}$ to form the error term, against which we would evaluate the significance of the between trials effect. With a repeated measures analysis, however, a *portion* of the variation within trials is the result of having used the *same subjects*

TABLE 13.1

	Trials			
Subjects	1	2	3	$\bar{X}$ Subjects
A	12	10	8	10
B	15	11	7	11
C	9	9	6	8
D	6	5	4	5
E	8	5	5	6
$\bar{X}$ trials	10	8	6	$\bar{\bar{X}} = 8$

for all trials. This had produced the correlation between the sets of trial scores. Since we have several measurements on each subject, we can calculate a between subjects sum of squares and *subtract* this source of systematic variation from $SS_{\text{within trials}}$. The result is an error term which is the interaction of trials × subjects.

We have partitioned the $SS_{\text{within trials}}$ into an $SS_{\text{between subjects}}$ and an $SS_{\text{trials} \times \text{subjects}}$ (error). By using repeated measures on the same subjects we are able to extract a between subjects sum of squares that is really a measure of the consistency of individual differences in response to the task; in short, it is a measure of correlation. This correlation, as in the matched pairs t test, permits a reduction in the error term, and a more powerful test of significance results.

We begin the calculations for a repeated measures analysis of variance by finding the total sum of squares. This will be the sum of squared deviations of each measurement taken from the grand mean. The conceptual formulas, calculations, and the associated degrees of freedom for the data in Table 13.1 appear below.

$$SS_{\text{total}} = \sum_{N} (X - \bar{\bar{X}})^2$$

$$= (12 - 8)^2 + (15 - 8)^2 + \cdots + (5 - 8)^2 = 132$$

$$\text{df} = N - 1 \text{ (where } N \text{ is the number of } \textit{measures}) = 14$$

The sum of squares *between* trials can be found by squaring the deviation of each trial mean from the grand mean, summing over the trials (t), and multiplying the result by the number of subjects (n_s) on which each trial mean is based.

$$SS_{\text{b. trials}} = n_s \sum_{t} (\bar{X}_t - \bar{\bar{X}})^2 = 5[(10 - 8)^2 + (8 - 8)^2 + (6 - 8)^2] = 40$$

df $= t - 1 = 2$, where t is the number of trials.

The sum of squares *within* trials can be found by squaring the deviation of each score from its trial mean, summing over the scores within trials (n_s), and then summing the result for all trials (t).

$$SS_{\text{w. trials}} = \sum_{t} \sum_{n_s} (X - \bar{X}_t)^2$$

$$= (12 - 10)^2 + (15 - 10)^2 + \cdots + (4 - 6)^2 + (5 - 6)^2 = 92$$

$$\text{df} = t(n_s - 1) = 12$$

The partitioning of the sums of squares up to this point has proceeded *exactly* as it did in the independent groups analysis of variance described in

Chapter 11. We have calculated SS_{total}, which is the sum of $SS_{\text{between trials}}$ and $SS_{\text{within trials}}$. If we continue with an independent groups analysis, the within sum of squares divided by $t(n_s - 1) = 12$ df will provide the error term. This will be the error term because it provides an estimate of variance uninfluenced by the effect of trials. In the independent groups analysis, the variation within columns is assumed to be entirely random. It is precisely on this point that the repeated measures analysis differs from the independent groups analysis.

In the repeated measures analysis there is a source of *systematic* variation within trials, and it can be extracted as a sum of squares. This source of systematic variation is the result of using the *same* subjects across all trials. It is called between subjects (S's) variation. This systematic variation results from the fact that the same subjects, measured over a series of trials, differ consistently from one another. We expect to find systematic differences between subjects on a learning task, or any other reliably measured task. In fact, this is a definition of reliable measurement.

We can calculate the between subjects sum of squares by finding the squared difference between each subject's mean (over trials) and the grand mean, summing the result over (n_s) subjects, and then multiplying by t the number of trials on which the subject means are based. The mean for each subject over the three trials appears at the right of Table 13.1.

$$SS_{\text{between subjects}} = t \sum_{n_s} (\bar{X}_s - \bar{\bar{X}})^2$$

$$= 3[(10 - 8)^2 + (11 - 8)^2 + \cdots + (6 - 8)^2] = 78$$

$$\text{df} = (\text{subjects} - 1) = 4$$

The $SS_{\text{between subjects}}$ is 78. If this is subtracted from the $SS_{\text{within trials}}$ of 92 the residual error sum of squares will be reduced to 14. This is a sum of squares associated with the interaction of subjects and trials. Interaction of subjects and trials occurs when the differences among subjects are inconsistent from one trial to the next. Notice that the score for subject A declines from 12 to 10 between Trial 1 and Trial 2, the score for subject B declines from 15 to 11, and the score for subject C remains at 9 for both trials. The interaction of subjects and trials is produced by the inequality of these differences between trials. The interaction sum of squares can be found most easily by subtraction.

$$SS_{\text{trials x subjects}} = SS_{\text{total}} - (SS_{\text{between } Ss} + SS_{\text{between trials}}) = 14$$

$$\text{df} = (\text{trials} - 1)(\text{subjects} - 1) = 8 \text{ df}$$

There is no way to test the significance of this interaction, because the "cells" in this study contain only one measurement each and, therefore, we do not have a within cells sum of squares. In fact, we simply assume that this interaction is composed entirely of error and then use it, divided by the appropriate df, as an error term to test the between trials effect. The repeated measures analysis of variance is actually an example of a "mixed" model that we discussed briefly in the preceding chapter. In this illustration the fixed variable is trials, the random variable is subjects, but there is only one measure per cell. When the mixed model is used, you will recall that the fixed effect should be tested against the interaction term.

We shall set up the analysis of variance (see Table 13.2), and then comment more fully on "interaction" as an error term. Notice that the rules have not changed for establishing the df associated with the various sums of squares in a repeated measures analysis. Three trials produce 2 df between trials; five subjects have 4 df between subjects; the interaction of these (subjects) rows and (trials) columns has $(R - 1)(C - 1)$, or 8 df, and the total df $= N - 1$, which is 14.

The F for the between trials main effect is 11.43, and is significant at the 1% level for 2/8 df. This is the effect that would concern us in a research study. Notice, however, that the between S's "effect" is also significant. This means that there is a significant difference among subjects in the way they respond to the total task, a finding which reflects the reliability of the task.

We have *assumed* that the subjects $\times$ trials interaction is really just random error. What would be the meaning of a *significant* trials $\times$ subjects interaction if one could be demonstrated? Such a finding would mean that the trials do not exert the same effect on all subjects. Perhaps high ability subjects show a rate of improvement over trials which is consistently different from the rate of improvement over trials shown by low ability subjects. Unfortunately, we have no way to tell if this interaction

Table 13.2 Analysis of Variance

Source		*SS*		df	*MS*	*F*
Between trials (Columns)		40		2	20.00	11.43**
Within trials		92		12		
Between S's (Rows)	78		4		19.50	11.14**
Trials $\times$ S's (error)	14		8		1.75	
Total		132		14		

is significant, because we have no estimate of variance to use as an error term. In Chapter 11 we tested the interaction terms with the estimate of variance from within cells. We can't do that in this situation, because we have only one observation in each cell. Since we can't test the significance of the trials × subjects interaction, we shall assume that it is not significant, that it is entirely the result of error.

Consider for a moment what happens if our assumption about the error term is correct, that it reflects only the effect of sampling error. If that is the case we will have Situation 1 shown below. The F ratio, shown as F_1, will have as a divisor an estimate of variance based entirely on error.

On the other hand, if our assumption about the interaction is *incorrect*, if it is composed of both true interaction and error, then we shall have Situation 2 shown below, and the resulting F ratio, F_2, will have as a divisor an estimate of variance based on error *and* the interaction.

Situation 1

$$F_1 = \frac{\text{Variance (error + between trials)}}{\text{Variance (error)}}$$

Situation 2

$$F_2 = \frac{\text{Variance (error + between trials)}}{\text{Variance (error + interaction)}}$$

Notice that if true interaction occurs it will reduce the power of our test, because it will inflate the error term; but it does not render the test invalid. A "significant" F ratio in either situation is still significant.

COMPARISON OF THE INDEPENDENT GROUPS AND REPEATED MEASURES ANALYSES

When the data in Table 13.1 are treated as they should be, by a repeated measures analysis of variance, we obtain the significant F shown in Table 13.2. Suppose the same data are treated by an independent groups analysis of variance. Would the between trials F still be significant? We shall conduct an independent groups analysis with the data exactly as they were in Table 13.1. This means that the correlation between trials, while still present in the data, is ignored in the analysis because the independent groups procedure uses the total within trials SS as the error term. Since the same numbers are in exactly the same order, the calculations will be exactly the same, except that we will not partition the within trials SS into the between S's and interaction components.

TABLE 13.3 Analysis of Variance

Source	*SS*	df	*MS*	*F*
Between trials	40	2	20.00	2.73
Within trials	92	12	7.67	
Total	132	14		

The summary table appears above. Compare it with Table 13.2 from the previous analysis. The between trials F is no longer significant, because the estimate of variance based on error has increased from 1.75 to 7.67. We have 4 additional df for this error term, but that does not compensate for the increased error sum of squares.

A comparison of Tables 13.2 and 13.3 should clarify what has happened. In the independent groups analysis reported in Table 13.3 we used the entire within trials *SS* as error, even though it included a portion of variance known to be associated with consistent individual differences across trials. (The between *S*'s variation is significant in Table 13.2.) We mentioned previously that when an error term contains a source of variance other than error we reduce the power of the test. Just such a situation has happened here as a result of pooling (or failing to partition) the significant individual differences sum of squares and the subjects × trials interaction sum of squares.

$$F_1 = \frac{\text{Variance (error + between trials)}}{\text{Variance (error)}}$$

$$F_2 = \frac{\text{Variance (error + between trials)}}{\text{Variance (error + between subjects)}}$$

By applying the independent groups analysis to a repeated measures experiment, as we have just done, we have the situation shown in F_2. This less powerful test fails to produce a significant F ratio for the between trials effect, although the repeated measures analysis had previously shown it to be significant.

The repeated measures analysis is a more powerful test than the independent groups analysis *if* we have repeated measures on the same or matched subjects and a reliable measuring device. But, if both tests are applied to truly independent data, the independent groups analysis will be somewhat more powerful. We shall examine this situation in the example below.

If subjects were randomly assigned to trials and we recorded only the first trial performance of the Trial 1 group, the second trial performance of the Trial 2 group, and the third trial performance of the Trial 3 group, we could no longer assign a source of systematic variation to the between rows sum of squares. This is because the between rows sum of squares would be measuring differences among means with each mean based on the performance of *several different subjects*. In the previous analysis the row means were each based on the performance of *the same subject*. Since individual differences no longer provide a source of systematic variation between rows, there is no way to isolate and remove an individual differences sum of squares. Individual differences between subjects still exist, but that source of variation cannot be separated from "error" because of the random assignment of subjects to trials. If we ignore this situation and subtract the between rows SS from the within trials SS we shall, of course, get an interaction of "rows" $\times$ "columns," but this will measure only randomly assigned individual differences and error. As a result, we will have a situation in which *either* the variance estimates from between rows, or the interaction of rows $\times$ columns could be used as an error term since *each* reflects only random variation. Of course, it would be more sensible to leave the within trials sum of squares intact and have available the additional df for evaluating the between trials F. This is exactly the method employed in the simple independent groups analysis of variance that would be appropriate for these data. (The student can demonstrate the approximate equality of the between rows, and interaction variance for himself, by randomizing the entries within trials from Table 13.1, and then finding the between rows and the rows $\times$ columns interaction estimates of variance based on the randomized data. The rows and interaction variances should differ only randomly. Conduct an F test between the variances as a check.)

COMPUTATIONAL FORMULAS AND EXAMPLES OF THE REPEATED MEASURES ANALYSIS

In Chapter 10 we pointed out that the matched pairs t test could be used when the same subjects were tested twice, as in a pre-post test design, or it could be used when different subjects were matched or paired on some variable known to be correlated with the dependent variable. We can also use the repeated measures analysis of variance in the same two ways, designs in which the same subjects are measured over a series of trials, or designs in which subjects are matched into triplets, or larger blocks. The procedures for the analysis are the same whether a row contains matched subjects or repeated measures on the same subjects.

Suppose we wish to investigate the effect of four different drug dosages on problem-solving ability. If we test a group of subjects on the same problem at four different dosage levels, we can expect that differences between scores on the problem-solving tests will be influenced by drug dosage and by having solved the same problem on the previous trial. We may, therefore, decide to match subjects on the basis of intelligence and then randomly assign subjects from each matched group of four to the four drug dosages. The 24 subjects would be given an intelligence test and then rank ordered on the basis of their scores. These data appear below.

A	145	*E*	129	*I*	119	*M*	110	*Q*	99	*U*	91
B	142	*F*	128	*J*	116	*N*	109	*R*	97	*V*	91
C	132	*G*	127	*K*	114	*O*	100	*S*	94	*W*	90
D	130	*H*	122	*L*	112	*P*	100	*T*	92	*X*	89

Subjects whose scores fall in the first column above can be randomly assigned to the four dosage or treatment levels below. Thus, subject *A* may be assigned level 1, subject *B* level 4, subject *C* level 2, and subject *D* level 3. This assignment has been made for the top block (row) of Table 13.4. The same procedure is then repeated for all six blocks. Such a technique gives repeated measures analysis of variance an alternate label, randomized blocks design. The different drug levels can now be administered to these four groups of subjects and their performance tested on the same problem with assurance that practice effects will not produce systematic differences among the groups as would surely have been true had we used repeated measurements. The analysis appears below.

Table 13.4

	Conditions (Dosage Levels) 1	2	3	4	$\sum_{n_B} X$
Block					
1	(Subject *A*) 19	(Subject *C*) 15	(Subject *D*) 11	(Subject *B*) 20	65
2	(Subject *F*) 25	(Subject *G*) 19	(Subject *H*) 16	(Subject *E*) 24	84
3	36	29	17	27	109
4	42	36	27	21	126
5	46	38	21	26	131
6	44	39	33	34	150
$\sum_{n_C} X =$	212	176	125	152	665

We may use Formula 12.1 for the total sum of squares.

$$SS_{\text{total}} = \sum_N X^2 - \frac{\left(\sum_N X\right)^2}{N} = (19)^2 + (25)^2 + (36)^2 + \cdots$$

$$+ (26)^2 + (34)^2 - \frac{(665)^2}{24} = 2263.0$$

$$\text{df} = N - 1 = 23$$

The sum of squares between conditions follows Formula 12.3.

$$SS_{\text{between conditions}} = \sum_C \frac{\left(\sum_{n_C} X\right)^2}{n_C} - \frac{\left(\sum_N X\right)^2}{N}$$

$$= \frac{(212)^2}{6} + \frac{(176)^2}{6} + \frac{(125)^2}{6} + \frac{(152)^2}{6} - \frac{(665)^2}{24} = 682.2$$

$$\text{df} = \text{Conditions} - 1 = 3$$

The between blocks sum of squares can be found from Formula 12.4. We have simply substituted blocks for rows.

$$SS_{\text{between blocks}} = \sum_B \frac{\left(\sum_{n_B} X\right)^2}{n_B} - \frac{\left(\sum_N X\right)^2}{N}$$

$$= \frac{(65)^2}{4} + \frac{(84)^2}{4} + \cdots + \frac{(131)^2}{4} + \frac{(150)^2}{4} - \frac{(665)^2}{24} = 1248.7$$

$$\text{df} = \text{Blocks} - 1 = 5$$

$$\underset{B \times C}{SS_{\text{interaction}}} = SS_{\text{total}} - (SS_{\text{conditions}} + SS_{\text{blocks}}) = 332.1$$

$$\text{df} = (\text{Conditions} - 1)(\text{Blocks} - 1) = 15$$

Source	SS	df	MS	F
Between conditions (drug levels)	682.2	3	227.4	10.3**
Between blocks	1248.7	5	249.7	
Interaction (error)	332.1	15	22.1	
Total	2263.0	23		

An $F = 10.3$ is significant at the 1% level for 3/15 df. We can conclude that the various dosage levels have significantly different effects on problem-solving ability.

REPEATED MEASURES IN A FACTORIAL DESIGN

We have illustrated the repeated measures, or randomized blocks analysis of variance in a one-way design. The logic of the analysis is seen more clearly in this simple form; but the repeated measures, or the randomized blocks, feature can also be combined with a factorial design. In the next example we shall illustrate the use of a randomized blocks design in a 2×2 factorial experiment.

Suppose we wish to investigate the effects of difficulty level and delay of recall after learning certain kinds of material. Our study might involve two levels of difficulty, A_1 and A_2 varied independently of two different intervals between learning and recall, B_1 and B_2. In such a study it would be unwise to have one subject serve in all four conditions, since having served in one would probably affect his performance in the next. So, we will use the randomized blocks procedure. Subjects will be matched in blocks of four from which they can be assigned randomly one to each of the conditions.

Since this is a study of retention we shall match the subjects on the basis of their performance on a prior retention task similar to that required of them in this study. Such a performance measure should be at least moderately correlated with the dependent variable in this research design. If the subjects are matched on some task which does not correlate with the dependent variable, we shall lose the advantage in power which the randomized blocks design has over the independent groups analysis. In fact, with the loss of df in the error term, an ineffective matching technique will make the randomized blocks design slightly less powerful than the independent groups analysis.

We shall assume that the data below resulted from testing 28 S's who were first matched into blocks and then randomly assigned to the experimental conditions. This would yield seven different blocks, or levels of ability. The procedure is quite similar to that which we described for assigning S's to blocks in the previous study.

Let us first look at the data in Table 13.5 as a randomized blocks design with seven blocks of matched S's and four conditions. We shall ignore for the moment that the four conditions also form a 2×2 factorial design.

The partitioning of the sums of squares can proceed according to the pattern we used for the one-way randomized blocks experiment we described earlier. The SS_{total} will be composed of an $SS_{\text{between conditions}}$, an $SS_{\text{between blocks}}$, and an $SS_{\text{blocks} \times \text{conditions}}$. The latter sums of squares will form the error term for *all* between conditions effects.

Now let us consider the 2×2 factorial design aspect of this study. It

TABLE 13.5

Block Number	Conditions				
	A_1		A_2		
	B_1	B_2	B_1	B_2	Σ Blocks
1	14	15	17	20	66
2	17	20	15	22	74
3	29	32	34	40	135
4	34	36	46	49	165
5	38	40	49	52	179
6	41	46	51	57	195
7	50	54	64	70	238
Σ Conditions =	223	243	276	310	1052

means that the $SS_{\text{between cooditions}}$ can be partitioned into an $SS_{\text{between } A\text{'s}}$, an $SS_{\text{between } B\text{'s}}$, and an $SS_{A \times B}$ interaction. Each of these effects will be tested against an error term formed from $SS_{\text{blocks} \times \text{conditions}}$ interaction divided by its associated df.

We shall proceed with the calculations of these component sums of squares and their associated degrees of freedom. The first step is the calculation of the total sum of squares.

$$SS_{\text{total}} = \sum_N X^2 - \frac{\left(\sum_N X\right)^2}{N} = (14)^2 + (17)^2 + \cdots$$

$$+ (57)^2 + (70)^2 - \frac{(1052)^2}{28} = 6780.9$$

The next step is to calculate the various "between" sums of squares. Notice that all of the formulas below follow the same form to obtain the betweens. We find ΣX over the n measures constituting each level of the variable. This sum is squared, and the result is divided by the n measures on which it was based. We then sum this value over all levels of the variable. The final step is to subtract the squared sum of all measures divided by n. The subscripts of the summation signs should make clear exactly what is being summed.

$$SS_{\text{between blocks (rows)}} = \sum_{B} \frac{\left(\sum_{n_B} X\right)^2}{n_B} - \frac{\left(\sum_{N} X\right)^2}{N}$$

$$= \frac{(66)^2}{4} + \frac{(74)^2}{4} + \cdots + \frac{(195)^2}{4} + \frac{(238)^2}{4} - \frac{(1052)^2}{4} = 5972.9$$

with df for blocks $= B - 1 = 6$.

$$SS_{\substack{\text{between conditions}\\ \text{(columns)}}} = \sum_{C} \frac{\left(\sum_{n_C} X\right)^2}{n_C} - \frac{\left(\sum_{N} X\right)^2}{N}$$

$$= \frac{(223)^2}{7} + \cdots + \frac{(310)^2}{7} - \frac{(1052)^2}{28} = 625.4$$

with df for conditions $= C - 1 = 3$.

$$SS_{\substack{\text{interaction}\\ (\text{B}\times\text{C})}} = \text{total} - (\text{blocks} + \text{conditions}) = 182.6$$

with df for blocks $\times$ conditions interaction $= (B - 1)(C - 1) = 18$.

We shall now subdivide the between conditions sum of squares into the two main effects and interaction which are of interest in this experiment.

$$SS_{\text{between } A\text{'s}} = \sum_{A\text{'s}} \frac{\left(\sum_{n_A} X\right)^2}{n_A} - \frac{\left(\sum_{N} X\right)^2}{N}$$

$$= \frac{(466)^2}{14} + \frac{(586)^2}{14} - \frac{(1052)^2}{28} = 514.3$$

$$\left(\sum_{n_{A_1}} X = 466, \sum_{n_{A_2}} X = 586\right) \qquad \text{with } (A - 1) = 1 \text{ df}$$

$$SS_{\text{between } B\text{'s}} = \sum_{B\text{'s}} \frac{\left(\sum_{n_B} X\right)^2}{n_B} - \frac{\left(\sum_{N} X\right)^2}{N}$$

$$= \frac{(499)^2}{14} + \frac{(553)^2}{14} - \frac{(1052)^2}{28} = 104.2$$

$$\left(\sum_{n_{B_1}} X = 499, \sum_{n_{B_2}} X = 553\right) \qquad \text{with } (B - 1) = 1 \text{ df}$$

The $A \times B$ interaction can be obtained as a sum of squares between diagonals or directly by subtraction. We shall obtain it both ways.

$$SS_{A\times B} = \sum_{A\times B} \frac{\left(\sum_{n_{A\times B}} X\right)^2}{n_{A\times B}} - \frac{\left(\sum_{N} X\right)^2}{N}$$

$$= \frac{(519)^2}{14} + \frac{(533)^2}{14} - \frac{(1052)^2}{28} = 7.1 \qquad \sum_{n_{A_1B_2+A_2B_1}} X = 519$$

$$\text{with } (A - 1)(B - 1) = 1 \text{ df} \qquad \sum_{n_{A_1B_1+A_2B_2}} X = 533$$

and by subtraction

$$SS_{A\times B} = SS_{\text{conditions}} - (SS_{A\text{'s}} + SS_{B\text{'s}}) = 6.9$$

These figures for the interaction sum of squares agree within limits of rounding. The summary analysis of variance table appears below.

Source	*SS*	df	*MS*	*F*
Between *A*'s	514.3	1	514.3	50.72**
Between *B*'s	104.2	1	104.2	10.28**
$A \times B$	6.9	1	6.9	<1.00
Total between conditions	625.4	3		
Between blocks	5972.9	6		
Blocks × conditions (error)	182.6	18	10.14	
Total	6780.9	27		

It would be possible to further partition the blocks × conditions interaction into three components: blocks × *A*'s, blocks × *B*'s, and blocks × $A \times B$. Each of these components is based on 6 df, and each could be used to test a different main effect, Blocks × *A*'s would be the error term for *A*'s, blocks × *B*'s for *B*'s, and blocks × $A \times B$ for the $A \times B$ interaction. The blocks × conditions term, then, is really a "pooled" interaction but, since there is no reason to suppose these separate interactions (blocks × *A*, blocks × *B*, and blocks × $A \times B$) represent other than random error, pooling is legitimate and, of course, it has the advantage of increasing the df on which the error term is based.

On the basis of this analysis, we may conclude that both the degree of difficulty and the delay of recall used in this study exert a significant effect on amount of recall, but that there is no evidence for interaction of these effects at the levels we have used in this study.

DESIGNS WITH RANDOM ASSIGNMENT AND REPEATED MEASURES

In the previous illustration of a randomized blocks factorial design, we tested all main effects with interaction terms, because all main effects were imposed on blocks of matched subjects. We can also have factorial designs in which some main effects, such as trials, are based on repeated measures of the same subjects, while other main effects are based on conditions to which subjects have been randomly assigned. These designs will have *two* error terms, one for each kind of main effect. We shall illustrate two such designs.

Suppose we wish to investigate the effect of trials and amount of incentive on performance. We might randomly assign five subjects to each of two incentive groups, and then give each of these five subjects three trials on the task of interest. Data which might result appear below.

TABLE 13.6

Subjects	Trials 1	2	3	ΣX Subject	
Incentive 1					
A	10	9	8	27	
B	16	14	12	42	$\Sigma X = 215$ Incentive$_1$
C	19	15	11	45	
D	20	17	13	50	
E	25	16	10	51	
	$\Sigma X = 90$	71	54		
Incentive 2					
F	17	14	11	42	
G	19	16	12	47	$\Sigma X = 313$ Incentive$_2$
H	28	22	16	66	
I	32	24	15	71	
J	37	29	21	87	
	$\Sigma X = 133$	105	75	$\sum_N X = 528$	
	ΣX Trials = 223	176	129		

This is a relatively complex analysis of variance when compared with those we have previously undertaken, but no new principles are involved. We shall simply combine the procedures for a repeated measures analysis and an independent groups analysis. Even so, you will probably have to study this problem quite carefully if you are to keep track of all the component sums of squares. Before we begin calculating these, it might be advisable to describe the major sources of variation which exist in this problem, and how these may be partitioned into their components. This really becomes a matter of bookkeeping. We shall go about it by first partitioning the degrees of freedom, then we shall go back over the same ground and calculate the various sums of squares.

With 10 subjects, each measured on three trials, we have a total of 30 measurements and $(N - 1) = 29$ df. This total variation can be partitioned into variation between columns (trials) with (trials $- 1) = 2$ df, variation between rows (S's) with (S's $- 1) = 9$ df, and the interaction of trials $\times$ S's with (trials $- 1$)(S's $- 1) = 18$ df. Notice that these component degrees of freedom sum to the total df for this problem.

The between rows (S's) df of 9 is subdivided just as if we were conducting a one-way analysis of variance based on two groups of randomly assigned subjects with one measure (the sum of the three trial scores) obtained on each S. We have a between incentives and a within incentives source of variation. The between incentives has (incentives $- 1) = 1$ df, and the within incentives, or between subjects in the same incentive groups, has [incentives groups $\cdot$ (S's$_{\text{within groups}} - 1$)] $= 8$ df. The between subjects within incentive groups provides the estimate of error for testing the between incentives effect. Notice that the between incentives df, and the between S's within incentive groups df, sum to the between S's df.

The between columns (trials) source of variation is not further subdivided, but remains as it is, with 2 df.

The trials $\times$ S's interaction with 18 df can be partitioned into trials $\times$ incentives interaction, with (trials $- 1$)(incentives $- 1) = 2$ df, and a trials $\times$ S's interaction *within each incentive group* with

$$[\text{incentives (trials} - 1)(S\text{'s} - 1)] = 16 \text{ df}$$

This last source of variation is based on the pooled interaction of trials $\times$ S's within each incentive group, and constitutes the error term for the repeated measures portion of the analysis, for example, the trials effect, and the trials $\times$ incentives interaction. Observe that the df for trials by incentives interaction, plus that for pooled trials $\times$ S's within incentive groups, sums to the total trials $\times$ S's df. The degrees of freedom for this problem are partitioned in Table 13.7. Please review this section carefully, until you understand why the arrangement of data in Table 13.6

TABLE 13.7 The Component Degrees of Freedom for the Data in Table 13.6

Source		df
Between S's		9
Between incentives	1	
Between S's in the same incentive group (error)	8	
Between trials		2
Trials × S's		18
Trials × incentives	2	
Trials × S's within incentive groups (error)	16	—
Total		29

produces the component df in Table 13.7. This section should be carefully reviewed before proceeding to the calculation of the component sums of squares.

We shall now compute the various sums of squares, but we shall henceforth dispense with all conceptual formulas and deal directly with computing formulas.

The main effects of this study will be contributed by "trials" and "incentive conditions." We shall also be interested in finding if the trials × incentives interaction is significant. This interaction would tell us if the trials produce different effects for the two incentive conditions.

The total sum of squares is found by summing the squared deviation of each score from the grand mean. The computational formula for this sum of squares yields

$$SS_{\text{total}} = \sum_{N} X^2 - \frac{\left(\sum_{N} X\right)^2}{N} = (10)^2 + (16)^2 + \cdots$$

$$+ (15)^2 + (21)^2 - \frac{(528)^2}{30} = 1461.2$$

This sum of squares has 29 df $(N - 1)$.

The $SS_{\text{between rows}}$ represents variation between subjects. Notice that the computational formula given below requires that we find ΣX over the n_s measurements on each subject. The n_s in this instance is 3, since it refers to the scores made by each subject on the 3 trials to which he was subjected.

$$SS_{\text{between } S\text{'s}} = \sum_s \frac{\left(\sum_{n_s} X\right)^2}{n_s} - \frac{\left(\sum_N X\right)^2}{N} = \frac{(27)^2}{3} + \frac{(42)^2}{3} + \cdots$$

$$+ \frac{(71)^2}{3} + \frac{(87)^2}{3} - \frac{(528)^2}{30} = 893.2$$

$$\text{df} = \text{subjects} - 1 = 9$$

The $SS_{\text{between columns}}$ represents variation between trials. These trial means are each based on the ten subjects in both incentive groups. The computational formula for this sum of squares, therefore, will require that we obtain ΣX for each trial across all $n_t = 10$ measurements obtained on each trial.

$$SS_{\text{between trials}} = \sum_t \frac{\left(\sum_{n_t} X\right)^2}{n_t} - \frac{\left(\sum_N X\right)^2}{N}$$

$$= \frac{(223)^2}{10} + \frac{(176)^2}{10} + \frac{(129)^2}{10} - \frac{(528)^2}{30} = 441.8$$

$$\text{df} = \text{trials} - 1 = 2$$

The trials × subjects interaction is obtained by subtraction.

$$SS_{\text{trials}\times\text{subjects}} = SS_{\text{total}} - (SS_{\text{columns}} + SS_{\text{rows}}) = 126.2$$

$$\text{df} = (\text{subjects} - 1)(\text{trials} - 1) = 18$$

Please refer again to Table 13.6 and note the partitioning of the sums of squares up to this point. We have found $SS_{\text{total}} = 1461.2$ which is composed of three parts. We have $SS_{\text{between } S\text{'s}} = 893.2$, an $SS_{\text{between trials}} = 441.8$, and by subtraction we can then obtain $SS_{\text{trials}\times\text{subjects}}$ interaction $= 126.2$. We can now proceed with the analysis by partitioning certain of these sums of squares into their components. Please keep the data of Table 13.6 before you as you follow the description in the text.

The $SS_{\text{between } S\text{'s}}$ can be partitioned into two sources of variation, one of these is a sum of squares between incentive groups and the other is a sum of squares within incentive groups, or between S's in the same incentive group. The first of these sums of squares can be found directly, and the second may then be obtained by subtraction. Notice that the ΣX for each incentive group is based on all 15 of the measurements within that incentive condition.

$$SS_{\text{between incentives}} = \sum_{\text{incentive}} \frac{\left(\sum_{n_{\text{incentive}}} X\right)^2}{n_{\text{incentive}}} - \frac{\left(\sum_N X\right)^2}{N}$$

$$= \frac{(215)^2}{15} + \frac{(313)^2}{15} - \frac{(528)^2}{30} = 320.1$$

$$\text{df} = \text{incentives} - 1 = 1$$

Since $SS_{\text{between } S\text{'s}} = SS_{\text{between incentives}} + SS_{\text{between } S\text{'s within incentive groups}}$, the $SS_{\text{between } S\text{'s within incentive groups}}$ can now be obtained by subtraction.

$$SS_{\text{between subjects within incentive groups}} = SS_{\text{between } S\text{'s}} - SS_{\text{between incentives}}$$
$$= 893.2 - 320.1 = 573.1$$

$$\text{df} = \text{incentive groups (subjects within groups} - 1) = 8$$

This completes the partitioning of the $SS_{\text{between } S\text{'s}}$, and our second major component, the $SS_{\text{between trials}}$, is not to be partitioned further, so we shall turn our attention to the $SS_{\text{trials}\times\text{subjects interaction}}$. This term represents two separable sources of variation. One of these is trials $\times$ incentive interaction, and the other is the trials $\times$ subjects interaction within each incentive group. In order to calculate the trials $\times$ incentives interaction it will be necessary to classify the measurements obtained under each incentive condition and on each trial as constituting a cell k. We can then obtain an $SS_{\text{between cells}}$ by the formula below.

$$SS_{\text{between cells}} = \sum_{k_{\text{cells}}} \frac{\left(\sum_{n_k} X\right)^2}{n_k} - \frac{\left(\sum_N X\right)^2}{N} = \frac{(90)^2}{5} + \frac{(71)^2}{5} + \cdots$$

$$+ \frac{(105)^2}{5} + \frac{(75)^2}{5} - \frac{(528)^2}{30} = 786.4$$

This $SS_{\text{between cells}}$ can be analyzed into three components, an $SS_{\text{between incentives}}$ and an $SS_{\text{between trials}}$, both of which we have already calculated, and an $SS_{\text{trials}\times\text{incentives}}$ interaction. Since $SS_{\text{between cells}} = SS_{\text{between incentives}} + SS_{\text{between trials}} + SS_{\text{trials}\times\text{incentives}}$, and since we have previously calculated

$$SS_{\text{between incentives}} = 320.1$$

and

$$SS_{\text{between trials}} = 441.8$$

we can obtain

$$SS_{\text{trials}\times\text{incentives}} = \text{between cells} - (\text{between incentives} + \text{between trials})$$
$$= 786.4 - (320.1 + 441.8) = 24.5$$
$$\text{df} = (\text{trials} - 1)(\text{incentives} - 1) = 2$$

The degrees of freedom for this interaction is found in the usual way.

This interaction can also be found by direct calculation, and this should probably be done as a check on the accuracy of the arithmetic. Table 13.8 consists of the k cell means, C column means, and R row means from the sums in Table 13.6. The formula for this interaction was developed previously

$$n_k \sum_k (\bar{X}_k - \bar{X}_R - \bar{X}_C + \bar{\bar{X}})^2$$

For this problem we have

$$5\Sigma(18.0 - 22.3 - 14.3 + 17.6)^2 + \cdots + (15.0 - 12.9 - 20.9 + 17.6)^2 = 24.5$$

This additional check on the $SS_{\text{trials}\times\text{incentives}}$ interaction confirms the sum of squares we previously obtained by subtraction.

The $SS_{\text{trials}\times\text{incentives}}$ of 24.5 is one portion of the gross $SS_{\text{trials}\times\text{subjects}}$ of 126.2 that we obtained earlier. The other portion is the $SS_{\text{trials}\times\text{subjects}}$ within each incentive group. This is a pooled SS and can now be obtained by subtraction. We have

$$SS_{\text{trials}\times\text{subjects within incentive groups (pooled)}} = SS_{\text{trials}\times\text{subjects}} - SS_{\text{trials}\times\text{incentives}}$$
$$SS_{\text{trials}\times\text{subjects within incentive groups (pooled)}} = 126.2 - 24.5 = 101.7$$
$$\text{df} = \text{incentive groups (trials} - 1)(\text{subjects} - 1) = 16$$

TABLE 13.8

		Trials 1	Trials 2	Trials 3	$\bar{X}_R$
Incentive	1	18.0	14.2	10.8	14.3
	2	26.6	21.0	15.0	20.9
	$\bar{X}_C =$	22.3	17.6	12.9	$17.6 = \bar{\bar{X}}$

TABLE 13.9

Source	*SS*		df		*MS*	*F*
Between incentives	320.1		1		320.1	4.47
Between *S*'s within incent. groups (error)	573.1		8		71.6	
Between *S*'s		893.2		9		
Between trials		441.8		2	220.9	34.52**
Trials × incentives	24.5		2		12.2	1.91
Pooled trials × subjects (error)	101.7		16		6.4	
Trials × subjects		126.2		18		
Total		1461.2		29		

The total sum of squares and df have now been partitioned, and the complete analysis of variance is summarized in Table 13.9. We shall discuss this table in some detail.

The principle difference between this analysis and previous ones lies in the use of two error terms. These are required because some of the main effects are between independent groups of randomly assigned subjects, and other main effects and interactions are between treatments based on the same subjects.

The between incentives main effect uses the between subjects within incentive groups as its error term, because subjects are randomly assigned to these two groups. If the design were "collapsed," that is, if all other aspects of the study were ignored except subjects and incentive groups, we would have only the two incentive conditions with five subjects randomly assigned to each. This design, using the sums across trials as "score" for each subject, appears below.

Incentive 1		*Incentive 2*	
A	27	*F*	42
B	42	*G*	47
C	45	*H*	66
D	50	*I*	71
E	51	*J*	87

Considered alone, the between incentives main effect would be tested by an *F* ratio based on the between incentives estimate of variance divided by a within incentives estimate of variance as the error term.

This "within incentives" is the same as the "between subjects in the same incentives group" error term we have used to test the between incentives effect in Table 13.9.

The between trials effect is tested by an error term composed of the pooled trials × subjects interaction for both incentive groups. We have five subjects measured across three trials in each of two incentive groups. The pooled interaction of trials × subjects permits us to test the effect of trials over both incentive groups, just as the interaction of conditions × subjects provided the error term for the between conditions effect in the one-way repeated measures analyses previously discussed.

Testing the trials × incentives interaction with the pooled trials × subjects interaction may be troublesome to understand because we are testing the significance of one interaction term with another interaction term. The independence of the trials × incentives interaction from the pooled trials × subjects interaction, which we are using for error, can be understood more clearly by reference to Table 13.8 from which the trials × incentives interaction was calculated. The trials × S's interaction *within* each of the incentive conditions cannot affect the entries in this table because the entries consist of sums for the five subjects at each incentive level on each trial. Individual subject's performance is thus lost. The pooled trials × subjects interaction is not influenced by incentive differences, because these interactions occur entirely within the incentive groups.

We find from the analysis of variance table above that the between incentives $F = 4.47$, and is just short of significance at the 5% level for 1/8 df. The between trials effect yields $F = 34.52$, which is significant at the 1% level for 2/16 df. The interaction of trials × incentives is not close to significance for 2/16 df.

The analysis we have just completed is quite similar to many of the more commonly used analyses of variance studies in the behavioral literature. The repeated measures feature is found quite often, though it sometimes appears as a true randomized blocks rather than a repeated measures design. We shall illustrate one more factorial design using repeated measures; this one just a bit more complex than the last because an additional dimension has been added; again, no new principles are involved. Please follow the analysis carefully, but don't expect to understand it until you have gone through it several times.

In the previous analysis we have two treatment groups to which subjects were randomly assigned and then tested over a series of trials. The next study has four groups in a factorial design to which subjects are randomly assigned and then tested over a series of trials. The basic plan and rationale of the two designs are quite similar, only the statistical bookkeeping of the various sums of squares has made this study more complex than the previous one.

The following analysis could be used to investigate the effect of a chemical assumed to increase learning ability, and the effect of magnitude of incentive on the performance of a group of animals over a series of trials. Such a study has three independent variables, presence, or absence of the chemical (A_1 and A_2), low and high incentive (B_1 and B_2) and trials (1–4). Each of these constitutes a main effect and the combinations constitute interactions which could influence the number of errors per trial.

The data from such a study might look like that pictured in Table 13.10. We shall assume that 20 subjects have been randomly assigned to the four groups and that their performance has been measured over a series of four trials. We shall continue to show the computations of sums of squares, but we will no longer repeat formulas unless some new technique is required. The initial steps in this analysis follow the familiar pattern. We begin by obtaining a total sum of squares and then determine its components between trials, between subjects, and trials $\times$ subjects interaction.

$$SS_{\text{total}} = (14)^2 + (16)^2 + \cdots + (14)^2 + (12)^2 - \frac{(1527)^2}{80} = 5336.4$$

$$\text{df} = N - 1 = 79$$

$$SS_{\text{between trials}} = \frac{(491)^2}{20} + \frac{(422)^2}{20} + \frac{(351)^2}{20} + \frac{(263)^2}{20} - \frac{(1527)^2}{80} = 1430.1$$

$$\text{df} = \text{trials} - 1 = 3$$

$$SS_{\text{between } S\text{'s}} = \frac{(42)^2}{4} + \frac{(40)^2}{4} + \cdots + \frac{(99)^2}{4} + \frac{(85)^2}{4} - \frac{(1527)^2}{80} = 3511.6$$

$$\text{df} = S\text{'s} - 1 = 19$$

$$SS_{\text{trials}\times S\text{'s}} = SS_{\text{total}} - (SS_{\text{between } S\text{'s}} + SS_{\text{between trials}}) = 394.7$$

$$\text{df} = (\text{trials} - 1)(S\text{'s} - 1) = 57$$

The between trials sum of squares cannot be analyzed further, so we shall turn our attention to the between S's variation. The between S's sum of squares is partitioned in the same way as the total sum of squares for a 2×2 factorial design with independent groups. We shall have two major partitions, a between conditions sum of squares and a within conditions sum of squares, which we shall refer to more

TABLE 13.10

		S's		Trials 1	2	3	4	Σ
A_1	B_1	(1)		14	10	10	8	42
		(2)		16	9	10	5	40
		(3)		18	16	12	7	53
		(4)		17	14	11	8	50
		(5)		22	19	17	10	68
			Σ	87	68	60	38	253
	B_2	(6)		18	19	17	15	69
		(7)		25	24	20	16	85
		(8)		37	35	32	24	128
		(9)		21	18	15	14	68
		(10)		29	29	28	26	112
			Σ	130	125	112	95	462
A_2	B_1	(11)		17	15	10	7	49
		(12)		24	12	9	9	54
		(13)		28	20	14	12	74
		(14)		30	22	16	12	80
		(15)		22	16	9	4	51
			Σ	121	85	58	44	308
	B_2	(16)		42	34	30	22	128
		(17)		29	28	26	22	105
		(18)		26	25	20	16	87
		(19)		30	30	25	14	99
		(20)		26	27	20	12	85
			Σ	153	144	121	86	504
			ΣΣ	491	422	351	263	1527

specifically as $SS_{\text{between subjects within conditions}}$. We can in turn partition the $SS_{\text{between conditions}}$ into the major effects of interest, $SS_{\text{between } A\text{'s}}$, $SS_{\text{between } B\text{'s}}$, and $SS_{A \times B \text{ interaction}}$. Each of these main effects and their interaction will be tested against an estimate of error variance based on $SS_{\text{between subjects within conditions}}$.

The partitioning of the between S's sum of squares proceeds as follows.

$$SS_{\text{between } S\text{'s}} = 3511.6, \text{ df} = 19 \text{ (previously calculated)}$$

$$SS_{\text{between conditions}} = \frac{(253)^2}{20} + \frac{(462)^2}{20} + \frac{(308)^2}{20} + \frac{(504)^2}{20} - \frac{(1527)^2}{80}$$

$$= 2170.0$$

$$\text{df} = \text{conditions} - 1 = 3$$

$$SS_{\text{between } S\text{'s within conditions}} = SS_{\text{between } S\text{'s}} - SS_{\text{between conditions}} = 1341.6$$

$$\text{df} = \text{conditions (subjects in conditions} - 1) = 16$$

The between subjects within conditions term can be obtained by subtraction because once the between conditions variation is removed from the between subjects variation, only the between subjects within conditions variation can remain. This term can also be calculated directly for each group, but the procedure will be left as an exercise for the student and explained in the answer section.

The between conditions sum of squares can be partitioned into three components; between A's, between B's, and the $A \times B$ interaction.

$$SS_{\text{between } A\text{'s}} = \frac{(715)^2}{40} + \frac{(812)^2}{40} - \frac{(1527)^2}{80} = 117.6$$

$$\text{df} = A\text{'s} - 1 = 1$$

$$SS_{\text{between } B\text{'s}} = \frac{(561)^2}{40} + \frac{(966)^2}{40} - \frac{(1527)^2}{80} = 2050.3$$

$$\text{df} = B\text{'s} - 1 = 1$$

$$SS_{A \times B} = SS_{\text{conditions}} - (SS_{A\text{'s}} + SS_{B\text{'s}}) = 2.1$$

$$\text{df} = (A\text{'s} - 1)(B\text{'s} - 1) = 1$$

(The $A \times B$ interaction can also be calculated directly from the diagonals.)

$$\frac{(757)^2}{40} + \frac{(770)^2}{40} - \frac{(1527)^2}{80} = 2.1$$

This completes the partitioning of the between S's sum of squares.

In Table 13.11 we have shown how this same partitioning affects the degrees of freedom for the components of the between S's variation.

We originally partitioned our data into a between trials component which we did not partition further, a between S's component whose par-

TABLE 13.11

Source			df
Between S's			19
Between conditions		3	
Between A's	1		
Between B's	1		
$A \times B$	1		
Between S's within conditions (error)		16	

titioning we have just explained, and a trials $\times$ subjects interaction component to which we shall now turn our attention. The interaction of trials and subjects can be partitioned into a trials $\times$ conditions and a trials $\times$ subjects within conditions effect. The latter is an error term that results from pooling the interaction of trials $\times$ subjects within each condition. The trials $\times$ conditions effect can also be partitioned further into trials $\times$ A, trials $\times$ B, and trials $\times$ $A \times B$ interactions.

As a preliminary to the partitioning of the trials $\times$ S's sum of squares into its several components, we shall chart the partitioning of the trials $\times$ S's degrees of freedom in Table 13.12. You may then be in a position to follow the analysis of the sums of squares more easily.

The first step in partitioning the $SS_{\text{trials}\times\text{subjects}}$ is to obtain the $SS_{\text{trials}\times\text{conditions}}$ interaction. This is done in the same roundabout way we accomplished it in the last problem. There we obtained a $SS_{\text{between cells}}$, where a cell consisted of the n_k measurements obtained under a given condition and on a given trial. Then, since the $SS_{\text{between cells}}$ can be partitioned into $SS_{\text{between conditions}} + SS_{\text{between trials}} + SS_{\text{conditions}\times\text{trials}}$, we ob-

TABLE 13.12

Source			df
Trials $\times$ subjects			57
Trials $\times$ conditions		9	
Trials $\times$ A	3		
Trials $\times$ B	3		
Trials $\times$ $A \times B$	3		
Trials $\times$ subjects within conditions		48	

tained the needed $SS_{\text{conditions}\times\text{trials}}$ interaction by subtraction. We shall follow the same procedure here. The $SS_{\text{between cells}}$ is given by the following formula.

$$SS_{\text{between cells}} = \sum_k \frac{\left(\sum_{n_k} X\right)^2}{n_k} - \frac{\left(\sum_N X\right)^2}{N} + \frac{(87)^2}{5} + \frac{(130)^2}{5} + \cdots$$

$$+ \frac{(44)^2}{5} + \frac{(86)^2}{5} - \frac{(1527)^2}{80} = 3789.2$$

When $SS_{\text{between cells}}$ has been obtained we have all the components required to calculate $SS_{\text{trials}\times\text{conditions}}$ because

$$SS_{\text{trials}\times\text{conditions}} = SS_{\text{between cells}} - (SS_{\text{between trials}} + SS_{\text{between conditions}})$$

We have just calculated $SS_{\text{between cells}} = 3789.2$, and we previously calculated $SS_{\text{between trials}} = 1430.1$, and $SS_{\text{between conditions}} = 2170.0$. We can therefore find

$$SS_{\text{trials}\times\text{conditions}} = 3789.2 - (1430.1 + 2170.0) = 189.1$$

$$\text{df} = (\text{trials} - 1)(\text{conditions} - 1) = 9$$

Once the trials × conditions interaction has been obtained, it is possible to find the trials × subjects within conditions interaction by subtraction. Please refer to Table 13.12 and note that

$$SS_{\text{trials}\times\text{subjects}} = SS_{\text{trials}\times\text{conditions}} + SS_{\text{trials}\times\text{subjects within conditions}}$$

so that

$$SS_{\text{trials}\times\text{subjects within conditions}} = SS_{\text{trials}\times\text{subjects}} - SS_{\text{trials}\times\text{conditions}}$$

And, since the values on the right side of the equality are known, we have

$$SS_{\text{trials}\times\text{subjects within conditions}} = 394.7 - 189.1 = 205.6$$

$$\text{df} = \text{conditions (subjects} - 1)(\text{trials} - 1) = 48$$

This is the error term for testing the significance of all effects involving repeated measures on the same subjects. It is a pooled interaction of trials × S's within each condition.

Return once again to Table 13.12 and note that the trials × conditions interaction can be partitioned into trials × A's, trials × B's, and trials × $A \times B$ interactions.

When we obtained the conditions × trials interaction, it was accomplished by subtracting between conditions and between trials from between cells, when each cell was defined as a cluster of 5 individual

Table 13.13

	T_1	T_2	T_3	T_4	Σ
A_1	217	193	172	133	715
A_2	274	229	179	130	812
Σ	491	422	351	263	1527

measures under one condition and on one trial. A similar technique will be used here, but we shall collapse the B condition so it will include both levels of B in each "cell." Cells in this analysis will consist of the sum of all measures in each $A \times T$ condition, regardless of the level of B. These sums are given in Table 13.13.

Redefining cells as they appear in this table, the

$$SS_{\text{between cells }(A \cdot T)} = \frac{(217)^2}{10} + \frac{(274)^2}{10} + \cdots + \frac{(133)^2}{10} + \frac{(130)^2}{10} - \frac{(1527)^2}{80} = 1660.3$$

Since $SS_{\text{between cells}} = SS_{\text{between }A\text{'s}} + SS_{\text{between trials}} + SS_{\text{trials}\times A}$, we have $SS_{\text{trials}\times A\text{'s}} = SS_{\text{between cells}} - (SS_{\text{between }A\text{'s}} + SS_{\text{between trials}})$. We have already obtained $SS_{\text{between }A\text{'s}}$ and $SS_{\text{between trials}}$, so we have $SS_{\text{trials}\times A\text{'s}} = 1660.3 - (117.6 + 1430.1) = 112.6$.

The $SS_{\text{trials}\times B\text{'s}}$ interaction can be obtained in a similar way. We shall collapse the A conditions to give the sums in Table 13.14. Defining a

Table 13.14

	T_1	T_2	T_3	T_4	Σ
B_1	208	153	118	82	561
B_2	283	269	233	181	966
Σ	491	422	351	263	1527

cell as it appears in this table we have

$$SS_{\text{between cells } (B\cdot T)} = \frac{(208)^2}{10} + \frac{(283)^2}{10} + \cdots + \frac{(82)^2}{10} + \frac{(181)^2}{10} - \frac{(1527)^2}{80} = 3535.5$$

This $SS_{\text{between cells}}$ is composed of the already obtained $SS_{\text{between trials}}$, $SS_{\text{between } B\text{'s}}$, and the to be obtained $SS_{\text{trials}\times B\text{'s}}$. Therefore, since

$$SS_{\text{between cells}} = SS_{\text{between } B\text{'s}} + SS_{\text{between trials}} + SS_{\text{trial}\times B\text{'s}}$$

we have

$$SS_{\text{trials}\times B\text{'s}} = SS_{\text{between cells}} - (SS_{\text{between } B\text{'s}} + SS_{\text{between trials}})$$

We have already obtained $SS_{\text{between } B\text{'s}}$ and $SS_{\text{between trials}}$, so we have

$$SS_{\text{trials}\times B\text{'s}} = 3535.5 - (2050.3 + 1430.1) = 55.1$$

$$\text{df} = (B\text{'s} - 1)(\text{trials} - 1) = 3$$

The final interaction with trials is the $SS_{\text{trials}\times A\text{'s}\times B\text{'s}}$ which can be obtained by subtraction from the relationship $SS_{\text{trials}\times\text{conditions}} = SS_{\text{trials}\times A\text{'s}} + SS_{\text{trials}\times B\text{'s}} + SS_{\text{trials}\times A\text{'s}\times B\text{'s}}$. We have

$$SS_{\text{trials}\times A\text{'s}\times B\text{'s}} = SS_{\text{trials}\times\text{condition}} - (SS_{\text{trials}\times A\text{'s}} + SS_{\text{trials}\times B\text{'s}})$$

$$= 189.1 - (112.6 + 55.1) = 21.4$$

$$\text{df} = (\text{trials} - 1)(A\text{'s} - 1)(B\text{'s} - 1) = 3.$$

This completes the partitioning of the trials × subjects interaction, so we can construct the summary analysis of variance in Table 13.15.

This completes the statistical bookkeeping. The results of the analysis show that the chemical is without direct effect when measured across incentive magnitudes and trials, but it does interact with trials to a significant extent. The nature of this interaction can be examined in the table used to calculate that interaction term. Incentive magnitude does exert a considerable effect and also interacts with trials; the nature of this interaction can also be seen in the table from which the interaction

TABLE 13.15

Source	SS			df	MS	F
Between A's	117.6			1	117.6	1.40
Between B's	2050.3			1	2050.3	24.47**
$A \times B$	2.1			1	2.1	<1.00
Between conditions		2170.0				
Between S's within conditions (error)		1341.6		16	83.8	
Total between subjects			3511.6			
Total between trials			1430.1	3	476.7	110.86**
A's × trials	112.6			3	37.5	8.72**
B's × trials	55.1			3	18.4	4.28**
A's × B's × trials	21.4			3	7.1	1.65
Conditions × trials		189.1				
Subjects × trials within conditions (error)		205.6		48	4.3	
Total subjects × trials			394.7			
Grand total			5336.4	79		

was obtained. The $A \times B$ interaction, and the $A \times B \times$ trials interactions are not significant.

THE NATURE OF COMPLEX DESIGNS IN ANALYSIS OF VARIANCE

Investigators in the behavioral sciences use analysis of variance more frequently than any other inferential statistic. By now, you should have some appreciation of the kinds of information it can yield and the complexity of some of the designs. We have only provided an introduction to the technique in these chapters, even though the text devotes more space to this research tool than it does to any other.

There are a variety of other research designs which lend themselves to analysis of variance techniques but they are beyond the scope of this text. The student is advised to consult Winer[1] or Lindquist[2] for a thorough coverage of more elaborate designs and more elegant analyses.

A word of caution should be injected. Analysis of variance is a powerful tool for answering questions which simply could not be answered at all before the technique was developed. However, it is easy to begin to add variables of no real interest to a research project just because the techniques are now available to measure their effects and their interactions with other variables. One occasionally sees very elaborate analyses of variance which produce conceptual monstrosities, such as five-way interactions. The student should be careful not to substitute elaborate statistical analyses for thoughtful research questions.

Exercises

1. T—F. The interaction of trials with subjects is the same as the interaction of subjects with trials.
2. T—F. The repeated measures and randomized blocks analyses of variance are conducted in the same way.
3. Conduct a repeated measures analysis of variance on the following data.

	Trials			
S's	1	2	3	Σ
1	10	9	7	26
2	8	6	3	17
3	7	5	4	16
4	5	6	3	14
5	11	9	8	28
6	15	13	10	38
Σ	56	48	35	139

4. In the study whose data are presented below, we have matched 16 *S*'s into blocks of two and, then, randomly assigned one member of each block to level 1 of Variable A, and the other member of the block to level 2 of Variable A. We then randomly assign 4 blocks to each of two levels of Variable B. Conduct the appropriate analysis of variance to determine the significance of the effects produced by different levels of A, levels of B, and interaction of A and B.

[1] Winer, J. B. *Statistical Principles in Experimental Design.* New York: McGraw-Hill, 1962.

[2] Lindquist, E. F. *Design and Analysis of Experiments in Psychology and Education.* Boston: Houghton Mifflin, 1953.

			A			
		Blocks	1	2	Σ	
B	1	*A*	10	6	16	Σ = 45
		B	8	4	12	
		C	7	3	10	
		D	5	2	7	
		Σ	30	15		
	2	*E*	12	13	25	Σ = 118
		F	13	16	29	
		G	17	18	35	
		H	14	15	29	
		Σ	56	62		
		Σ Σ	86	77		163

5. Conduct an analysis of variance on the randomized blocks data below.

Block	A_1: B_1	A_1: B_2	A_2: B_1	A_2: B_2	Σ
A	8	5	12	9	34
B	9	6	13	10	38
C	10	9	16	11	46
D	14	12	16	16	58
E	12	10	15	12	49
Σ	53	42	72	58	225

$\Sigma A_1 = 95$ $\Sigma A_2 = 130$

$\Sigma B_1 = 125$ $\Sigma B_2 = 100$

$\Sigma A_1B_1, A_2B_2 = 111$ $\Sigma A_1B_2, A_2B_1 = 114$

6. In the following block, each cell is the sum of 10 measures. Find the $A \times B$ interaction sum of squares.

		T: 1	*T*: 2	Σ
A_1	B_1	160	180	340
	B_2	170	190	360
A_2	B_1	200	230	430
	B_2	150	180	330
	Σ	680	780	1460

7. Extrapolating from the procedure for direct calculation of a two-way interaction, develop a method for calculating the *pooled* $A \times B$, $A \times T$, $B \times T$, $A \times B \times T$ sum of squares for the data in Exercise 6. Check the accuracy of your results by obtaining each interaction sum of squares separately.
8. Find two published studies from your field which make use of an analysis of variance technique covered in this chapter. Describe the results of the study in your own words.

CHAPTER 14

Nonparametric Tests of Significance

In spite of the robust nature of the t and F tests, there are situations where their use is inappropriate. If distributions are clearly skewed, if variances to be pooled for error are radically different, and particularly if these tendencies exist in small samples ($n < 10$), we should make use of distribution-free, or nonparametric statistical tests.

Most of these tests, as their name implies, make no assumptions about the shape of distributions from which the comparison samples have been drawn. The tests are also easy to understand and simple to conduct. One of their disadvantages is that they tend to be somewhat less powerful than parametric tests of the same data. However, all nonparametric tests are not equally powerful, so it is important that the research worker use the most powerful test consistent with the nature of the data he has collected. Another, perhaps more important disadvantage of the nonparametric tests is that they cannot be used to reveal the subtle interaction effects for which the analysis of variance is so well suited.

THE SIGN TEST

Suppose we wish to know which of two laboratory procedures is most effective for teaching principles of experimental design. We have paired 40 students on a measure of academic ability and assigned one member from each pair to a flexible laboratory course that permits students to modify the laboratory exercises if they can give a good reason for doing so. The remaining students are assigned to a nonflexible program; they are required to carry out the laboratory exercises exactly as they appear in their manuals. At the end of the term all students are to submit an original research design on an assigned topic. A faculty judge compares the designs submitted by the students in each pair and awards a "+" to the student with the superior design. The resulting data appear below.

	Pairs																			
	A	*B*	*C*	*D*	*E*	*F*	*G*	*H*	*I*	*J*	*K*	*L*	*M*	*N*	*O*	*P*	*Q*	*R*	*S*	*T*
Nonflexible	+			+					+					+			+			
Flexible		+	+		+	+	+	+		+	+	+	+		+	+		+	+	+

If the flexible and nonflexible treatments do not differentially affect the student's ability to construct a good research design, then we should find the numbers of plus signs for the two groups differing only as a result of sampling error. The hypothesis may be tested by χ^2, as it was developed in Chapter 9 for correlated observations. (Please review pp. 159–161 and the answer to Exercise 12 on page 169.) For the data above we have

	Flexible	Nonflexible	
Observed	15	5	20
Expected	10	10	20

$$\chi^2 = \sum \frac{(|o - e| - .5)^2}{e} = \frac{(4.5)^2}{10} + \frac{(4.5)^2}{10} = 4.05$$

$$\chi^2 = 4.05, \text{ df} = 1, p < .05.$$

We would probably conclude from these data that the more flexible approach to the laboratory exercises facilitates performance when the student is subsequently required to initiate his own research design.

Considerable power is lost when the sign test is used because information about the magnitude of the differences between pairs is not used; the sign of a pair is, of course, based only on the direction of the difference between the members of that pair. There are situations in which all we can say about a comparison is "more than" or "less than," "higher" or "lower." In these situations the sign test is the *only* test available. If, however, we can measure the magnitude of the difference between pairs, another nonparametric test is available which is more powerful than the sign test. This is the Wilcoxon matched pairs signed ranks test.

THE WILCOXON MATCHED PAIRS SIGNED RANKS TEST

This test, like the sign test, is based upon paired observations where one member from each pair falls in a different treatment or category group. The hypothesis to be tested is that the samples come from stochastically different populations. This means that the majority of one population is higher (or lower) than the majority of the other population. If a one-tailed test were appropriate, the hypothesis might state that population A was stochastically higher than population B. An example problem using the Wilcoxon test appears below.

We might study the effect of heat on the hoarding behavior of two groups of rats divided into matched pairs by weight. Hoarding could be measured by the number of pellets of food an animal stores in its nest during a five-hour period. One group might be tested after 24 hours at 55°F., and the other group tested after 24 hours at 80°F. Example data appear in Table 14.1.

The Wilcoxon test makes use of the statistic T. The statistic is calculated by finding the ranks of the *absolute* values (disregarding sign) of the differences between pairs, with the smallest difference given a rank of 1, and then summing those ranks based on the differences with the *less* frequent sign. When tied ranks occur, the average of the tied ranks is given to all members of the tie; when the *original* measures yield a tied pair, that pair is excluded from the analysis.

The value of T for our example study is 5.5. This T is evaluated in Table W of the Appendix; Table W gives values required for a significant T depending upon n, the number of pairs, and α, the level of significance required. The table is designed for a two-tailed test. If a directional

Table 14.1

	Pellets Hoarded				
Pair	60°F	80°F	Difference D	Rank of $\|D\|$	T
1	3	12	−9	4	4
2	12	10	2	1.5	
3	19	15	4	3	
4	18	4	14	8	
5	12	2	10	5.5	
6	16	18	−2	1.5	1.5
7	27	14	13	7	
8	29	19	10	5.5	
9	26	5	21	9	
					$T = 5.5$

hypothesis has been used, the α levels of the table should be halved. From Table W, the value of 5.5 is significant at the 5% level for a two-tailed test. Unlike previous tests of significance, the value of T must be equal to or *smaller* than the tabled value if it is to reach the specified level of significance.

For samples larger than $N = 25$, the statistic T is normally distributed about $M_T = \frac{N(N+1)}{4}$, with $\sigma_T = \frac{N(N+1)(N+2)}{24}$. A z_T can thus be calculated and interpreted by reference to a table of the normal curve (Table N in the Appendix). Since the Wilcoxon test is more powerful than the sign test, it should always be the preferred nonparametric test for matched pairs when the scale of measurement permits its use.

The following three nonparametric tests are designed to be used with independent groups.

THE MEDIAN TEST

Even though the median test was discussed in some detail in Chapter 9, we shall give another illustration of its use here. Unlike the sign test, the median test makes use of a χ^2 based on uncorrelated or independent observations.

We may wish to investigate the effect of two types of speedreading courses on the comprehension of relatively difficult material. We have randomly assigned 22 students to a course of one type, and 14 to a course of a different variety. Their comprehension scores following the course are given below:

Type I				*Type II*		
18	20	29	32	12	16	14
17	21	27	34	13	17	34
16	24	29	34	13	17	16
22	25	29	35	14	18	
14	26	30		15	22	
19	26	32		26		

When the two sets of scores are combined into one array, the median of the *combined* groups will be 21.5. If these two samples have come from populations with the same median, we should expect approximately the same proportions in each group to fall above (or below) the median of the combined group. Dichotomizing the two groups into the classes "above 21.5" and "below 21.5," we have

	Type I	Type II	
Above 21.5	11 *a* 15	7 *b* 3	18
Below 21.5	11 *c* 7	7 *d* 11	18
	22	14	36

The data in this 2 × 2 table may be subjected to a χ^2 test. The expected frequency for cell "a" will be $(18 \times 22)/36 = 11$, the remaining frequencies can then be determined by subtraction from the appropriate border totals.

$$\chi^2 = \sum \frac{(|o - e| - .5)^2}{e} = \frac{(3.5)^2}{11} + \frac{(3.5)^2}{11} + \frac{(3.5)^2}{7} + \frac{(3.5)^2}{7} = 5.73$$

$$\chi^2 = 5.73, \text{ df} = 1, p < .02$$

We can conclude that the two types of instruction yield significantly different comprehension scores.

As we pointed out in Chapter 9, the median test lacks power because it does not take into account the magnitude of the difference between a measure and the median of the combined groups. The next test does take into account at least the relative magnitudes of the measures.

THE MANN-WHITNEY U TEST

As with the Wilcoxon test, the Mann-Whitney U test also involves ranks, but it can be used with independent groups, and the groups need *not* have equal numbers of cases. To apply the test, the measures in both groups $(n_1 + n_2)$ are pooled and, then, all are ranked with the smallest measure in the pool given the rank of 1. Tied measures are given the average rank for the tie. When all $n_1 + n_2$ measures have been ranked, we obtain the sum of ranks (R_1) based only on the n_1 measures, and then calculate U_1 where

$$U_1 = n_1 n_2 + \frac{n_1(n_1 + 1)}{2} - R_1.$$

There are two possible U's, the other U, which we shall designate as U_2, is based on the remaining ranks summed for the n_2 measures. Once

U_1 has been obtained (regardless of which variable has been arbitrarily designated U_1) we can obtain $U_2 = n_1n_2 - U_1$. It is the *smaller* U for which a significance level can be determined from Table U. The obtained U must be equal to or *less* than the tabled value in order to be significant at the indicated level for a two-tailed test. This table is entered with n_1 and n_2. Levels of significance can be determined from Table U as long as neither n_1 *nor* n_2 exceed 20. When n's are larger than 20, a normal deviate may be obtained where $M_U = \frac{n_1n_2}{2}$ and $\sigma_U = \sqrt{\frac{n_1n_2(n_1 + n_2 + 1)}{12}}$ so that

$$z_U = \frac{U - \frac{n_1n_2}{2}}{\sqrt{\frac{n_1n_2(n_1 + n_2 + 1)}{12}}}$$

This value of z_U may be interpreted by reference to a table of the normal curve.

We shall conduct a Mann-Whitney U test for the data presented below:

Group I	R_1	*Group II*	R_2
16	7	14	6
42	14	7	4
19	9	1	1.5
18	8	1	1.5
12	5	3	3
26	10	28	11
29	12		$R_2 = 27$
33	13		$n_2 = 6$
	$R_1 = 78$		
	$n_1 = 8$		

To determine the value of U.

$$U_1 = n_1n_2 + \frac{n_1(n_1 + 1)}{2} - R_1$$

$$U_1 = 8 \times 6 + \frac{8(8 + 1)}{2} - 78$$

$U_1 = 6$.

To determine if U_1 is the smaller.

$U_2 = n_1n_2 - U_1$

$U_2 = 48 - 6$

$U_2 = 42$, therefore,

$U_1 = 6$ is the smaller.

Table U shows that for n's of 8 and 6, a value of $U = 6$ is significant at the 5% level for a two-tailed test.

THE KRUSKAL-WALLIS ONE-WAY ANALYSIS OF VARIANCE BY RANKS

When we have data from several *independent* groups it is possible to conduct a *one-way analysis of variance* by ranks. This is the Kruskal-Wallis H test. As in the Mann-Whitney test, the data are pooled for ranking with the smallest measure ranked 1.

Group I	*Ranks I*	*Group II*	*Ranks II*	*Group III*	*Ranks III*
14	8	29	14	18	11
16	9.5	31	15	10	7
27	13	47	17	4	3
6	5.5	6	5.5	1	1
19	12	41	16	3	2
16	9.5	5	4		
	$\Sigma R_{\text{I}} = 57.5$		$\Sigma R_{\text{II}} = 71.5$		$\Sigma R_{\text{III}} = 24$

When the summed ranks have been obtained for each group, the formula below is used to obtain the statistic H.

$$H = \frac{12}{N(N+1)} \cdot \sum_k \frac{(\Sigma R)^2}{n_k} - 3(N+1)$$

For our data this becomes

$$H = \frac{12}{17(17+1)} \cdot \left[\frac{(57.5)^2}{6} + \frac{(71.5)^2}{6} + \frac{(24.0)^2}{5}\right] - 3(17+1)$$

$$H = 5.57$$

When all n_k's exceed 5, we can evaluate H as a χ^2 with $k - 1$ df. The value of H in our example is just short of significance at the 5% level for 2 df.

RELATIVE POWER OF THE NONPARAMETRIC TESTS

Establishing the power of a statistical test is a complex matter because it can depend upon the size of the difference to be detected, and the size and other characteristics of the samples. In general, however, the nonparametric tests are less powerful than their parametric equivalents.

Siegel[1] indicates that, of the "rank" tests we have discussed in this chapter, the Wilcoxon, Mann-Whitney, and Kruskall-Wallis, are about 95% as powerful as the equivalent parametric tests, t, and the analysis of variance. The sign and median tests are equivalent in power to these "rank" tests for very small samples, but their relative power declines to about 66% as the sample size increases. This means that a "just significant difference," obtained by using an appropriate t test and 100 observations, would require about 105 observations for the rank based tests, and 150 observations for the sign and median tests.

[1] Siegel, S. *Nonparametric Statistics for the Behavioral Sciences*. New York: McGraw-Hill, 1956.

Since the ranks based nonparametrics are so powerful, there should be little hesitancy about using them in place of t tests, particularly in the sensitive situation where samples are small and the distributions of t and F may be most easily distorted by skew and unequal sample variances.

The nonparametric techniques are also quite important for many kinds of observations that, by their nature, make measurement on a continuous scale impossible. Yet, in spite of their increasing use in research, nonparametrics have not been developed which can extract and test the significance of complex interaction effects. Analysis of variance remains the single, most important statistical tool in behavioral research. For additional information about nonparametric statistics, the student is referred to Siegel's *Nonparametric Statistics*,[2] a very readable book and one easily followed by the novice.

Exercises

1. Students, matched on intelligence, are assigned to one of two sections. The first section actively participates in a group dynamics program for three hours a week while the second section spends the same amount of time hearing lectures on the theory of group dynamics. At the end of the program the students are asked to comment on the most effective course of action a leader might take in a particular structured situation. Clinicians compare the remarks of each pair of students, deciding which member of the pair has suggested the most effective course of action. The data appear below.

Pair	*Lecture*	*Participation*	*Pair*	*Lecture*	*Participation*
A	Superior		*I*		Superior
B		Superior	*J*	Superior	
C		Superior	*K*		Superior
D		Superior	*L*	Superior	
E	Superior		*M*		Superior
F		Superior	*N*		Superior
G		Superior	*O*		Superior
H		Superior	*P*		Superior

Conduct an appropriate statistical test to determine if the procedures have produced significant differences.

2. In the previous problem one member of each pair was judged "superior." Assume we have developed a scale which permits us to measure the adequacy of the respondent's answer about the effective course of action. The following data are the result of the application of this scale.

[2] *Ibid.*

Pair	*I*	*II*	*Pair*	*I*	*II*	*Pair*	*I*	*II*
A	1	7	*F*	22	11	*K*	41	12
B	6	4	*G*	34	19	*L*	17	20
C	9	4	*H*	27	12	*M*	49	14
D	18	10	*I*	16	14	*N*	36	16
E	19	21	*J*	19	24	*O*	28	10
						P	24	4

Conduct a Wilcoxon matched pairs signed ranks test on these data.

3.

I	*II*
16	2
19	9
27	18
24	14
23	16
12	19
21	21
14	8
16	7
27	4
35	16
42	11
71	14

Conduct a median test on these data. We assume that subjects have been randomly assigned to the two groups. If they had been paired on some variable, the Wilcoxon test would have been appropriate.

4. Conduct a Mann-Whitney U test on the data for Exercise 3.

5. Conduct a median test on the data below.

I			*II*	
26	29	48	19	22
29	34	51	12	21
34	42	60	36	24
41	55		29	23
36	64		27	18
24	45		16	12

6.

I	*II*
16	18
12	9
14	6
13	3
9	7
8	5
24	18
27	26
16	
21	
30	

Conduct a Mann-Whitney U test on these data.

7. Conduct the Kruskall-Wallis one-way analysis of variance by ranks on the following raw data.

I	*II*	*III*	*IV*
47	40	17	1
41	50	14	8
29	21	16	4
36	26	19	9
35	34	26	7
26	31	32	20
19			
15			

8. Conduct the Kruskall-Wallis one-way analysis of variance by ranks on the following raw data.

I	*II*	*III*
127	16	140
16	14	184
19	3	176
24	3	64
100	36	161
47	42	98
81		121
29		

9. Find research reported in the literature which makes use of one of the nonparametric statistics discussed in this chapter.

Answers

CHAPTER 3

1. The modal interval is 35–39.

2. If we select the midpoint of the interval 53–55 as the constant (C) to subtract from each of the other intervals, and the interval size $i = 3$ as the dividing constant, we shall obtain

$$M = i\frac{\Sigma fX''}{N} + C$$

$$= 3 \times \tfrac{309}{172} + 54$$

$$= 59.39$$

Notice that we could let C equal any interval midpoint. We have selected this particular interval because it is close to the center of the distribution and should yield small values of X'' which must then be multiplied by f to yield $\Sigma fX''$. If you have chosen to subtract some other interval midpoint you should arrive at the same answer.

3. If we subtract $C = 48$ from each score we shall have the following.

$$M = \frac{\Sigma fX'}{N} + C$$

$$= \tfrac{23}{128} + 48$$

$$= 48.18$$

Note that any score value may be subtracted from the others; scores near the center of the distribution will yield easiest computations for $\Sigma fX'$.

4. $\Sigma fX''$ will be negative if the value for C is chosen from the upper end of the distribution. Any value of C larger than the mean of the true scores will yield a negative value for $\Sigma fX''$.

5. M'' will be zero if $C = M$.

6. The median = 55.5 (lower limit of the interval containing the point between the 86th and 87th cases) + $\tfrac{13}{17}$ (proportionate distance into that interval to include 73 + 13 cases) $\times$ 3 (the class interval on the measurement scale) = 57.79. The 25th centile ($.25N = 43$) = $49.5 + \tfrac{8}{13} \times 3 = 51.35$. The 75th centile ($.75N = 129$) = $64.5 + \tfrac{11}{18} \times 3 = 66.33$.

7. (a) Only in a symmetrical distribution.
   (b) The mean is more readily affected by extremely high or extremely low scores than the median. If a distribution is negatively skewed the few extremely low scores will have lowered the mean more than the median, as the skew increases so will the difference between these two measures of central tendency. Notice the relationship of the mean and median for distribution 2.3.

8. $\Sigma(X^2 - 2MX + M^2)$ — Sum over all terms.

   $\Sigma X^2 - 2M\Sigma X + NM^2$ — Substitute $\frac{\Sigma X}{N}$ for M.

   $\Sigma X^2 - \frac{2\Sigma X \Sigma X}{N} + \frac{N\Sigma X \Sigma X}{N^2}$ — Clear third term by dividing numerator and denominator by N, and collect terms.

   $\Sigma X^2 - \frac{(\Sigma X)^2}{N}$

9. The median will fall midway in the gap between the fourth and fifth highest running times. We have given 8 trials but two running times are known only to exceed 4 minutes. We cannot calculate a mean but we can arrive at a median since we use only the values of the two measures which bracket it.

10. *Either* could be misleading, depending upon the shape of the particular distribution. The most accurate depiction of the wages of employees would be a simple frequency distribution including *both* mean and median.

CHAPTER 4

1. Table 3.2, the determination of Q.

$$\text{75th centile} = 39.5 + \frac{10.5}{12} \times 5 = 43.88.$$

$$\text{25th centile} = 29.5 + \frac{2.5}{8} \times 5 = 31.06.$$

Therefore $$Q = \frac{Q_3 - Q_1}{2} = \frac{43.88 - 31.06}{2} = 6.41.$$

The determination of σ^2 and σ.

$$\sigma^2 = i^2\left[\frac{\Sigma fX''^2}{N} - \left(\frac{\Sigma fX''}{N}\right)^2\right]$$

$$\sigma^2 = 25[\tfrac{150}{58} - (\tfrac{6}{58})^2] = 64.375$$

$$\sigma = \sqrt{64.375} = 8.02$$

If you have begun coding from a different interval your values for $\Sigma fX''^2$ and $\Sigma fX''$ will not be the same as the ones above, but your *answer* should agree with the one given here regardless of the coding interval you used.

Table 2.3, the determination of Q.

From the previous set of questions at the end of Chapter 3 we have: 75th centile = 66.33, 25th centile = 51.35.

Therefore $$Q = \frac{Q^3 - Q_1}{2} = \frac{66.33 - 51.35}{2} = 7.49.$$

The determination of σ^2 and σ.

Coding from the midpoints of the interval 53–55

$$\sigma^2 = i^2 \left[\frac{\Sigma fX''^2}{N} - \left(\frac{\Sigma fX''}{N} \right)^2 \right] = 9 \left[\tfrac{3047}{172} - \left(\tfrac{309}{172}\right)^2 \right] = 130.39$$

$$\sigma = \sqrt{130.39} = 11.42$$

2. (a) $M = 59.39, \sigma = 11.42, X = 51; z = \dfrac{51 - 59.39}{11.42} = -.73$

(b) centile equivalent from Table N for $z = -.73$ is the 23rd centile.

(c) $X = 51$, centile calculated *directly* from Table 2.3 is the 24th centile.

(d) The difference in centile values is due to the skew of distribution 2.3. Note that in spite of this skew the centiles are quite close to the values expected on the basis of a precisely normal distribution.

3. $$\frac{\sum \dfrac{X - M}{\sigma}}{N} = 0$$ Sum over terms in numerator.

$$\frac{\Sigma X - NM}{\sigma N} = 0$$ Simplifying.

$$\frac{\Sigma X}{\sigma N} - \frac{NM}{\sigma N} = 0$$ Since $M = \dfrac{\Sigma X}{N}$

$$\frac{\Sigma X}{\sigma N} - \frac{\Sigma X}{\sigma N} = 0$$

5. Yes

6. $M = 64, \sigma = 8$ $X_a = 51, X_b = 66, X_c = 70, X_d = 80$

$$z_a = \frac{51 - 64}{8} = -1.62$$

$$z_b = \frac{66 - 64}{8} = .25$$

$$z_c = \frac{70 - 64}{8} = .75$$

$$z_d = \frac{80 - 64}{8} = 2.00$$

7. Centiles

$z_a =$	-1.62	5th centile
$z_b =$	$.25$	60th centile
$z_c =$	$.75$	77th centile
$z_d =$	2.00	98th centile

8. $z = 1.29$
9. False

CHAPTER 5

1. $\Sigma X^2 = 986 \qquad \Sigma X = 106$

$\Sigma XY = 1358$

$\Sigma Y^2 = 1922 \qquad \Sigma Y = 148$

$$r = \frac{1358 - \dfrac{106 \times 148}{14}}{\sqrt{\left[986 - \dfrac{(106)^2}{14}\right]\left[1922 - \dfrac{(148)^2}{14}\right]}} = .93$$

2.

Person	*Raw Y*	*Rank Y*	*Raw X*	*Rank X*	*D*	D^2
N	20	1	16	1	0	0
D	16	2.5	10	3.5	1	1
K	16	2.5	9	5	2.5	6.25
E	14	4.5	12	2	2.5	6.25
M	14	4.5	10	3.5	1	1
C	12	6.5	8	6.5	0	0
J	12	6.5	8	6.5	0	0
B	10	8	7	8.5	.5	.25
L	9	9	7	8.5	.5	.25
G	8	10	4	11	1	1
F	6	11.5	6	10	1.5	2.25
A	6	11.5	3	13	1.5	2.25
I	3	13	3	13	0	0
H	2	14	3	13	1	1

$$\rho = 1 - \frac{6\Sigma D^2}{N(N^2 - 1)} = 1 - \frac{129}{14(195)} = .95$$

3. Jones' z score on the college boards

$$z_{\text{CEEB}} = \frac{\text{CEEB} - M_{\text{CEEB}}}{\sigma_{\text{CEEB}}} = \frac{600 - 500}{100} = 1.00$$

Regression equation is

$$z \text{ score on grade point average} = rz_{\text{CEEB}} = .50 \times 1.00 = .50$$

$z_{\text{GPA}} = .50$ is equivalent to what raw grade point average?

$$z_{\text{GPA}} = \frac{\text{GPA} - M_{\text{GPA}}}{\sigma_{\text{GPA}}} = .50 = \frac{\text{GPA} - 2.20}{.60}; \text{ solve for GPA}$$

$$\text{GPA} = .50 \times .60 + 2.20$$

$$\text{GPA} = 2.50$$

4. If a GPA = 3.00 is required, $z_{\text{GPA}} = 1.33$.

$$z_{\text{GPA}} = \frac{3.00 - 2.20}{.60} = \frac{.80}{.60} = 1.33$$

We wish to be able to predict a $\hat{z}_{\text{GPA}} = 1.33$. $\hat{z}_{\text{GPA}} = rz_{\text{CEEB}}$

$$\frac{\hat{z}_{\text{GPA}}}{r} = z_{\text{CEEB}} = \frac{1.33}{.50} = 2.66 = z_{\text{CEEB}}$$

A $z_{\text{CEEB}} = 2.66$ is equivalent to a college board raw score of

$$z_{\text{CEEB}} = \frac{\text{CEEB} - M_{\text{CEEB}}}{\sigma_{\text{CEEB}}} = 2.66 = \frac{\text{CEEB} - 500}{100}; \text{ solve for CEEB}$$

$$\text{CEEB} = 2.66 \times 100 + 500$$

$$\text{CEEB} = 766$$

Since only 1 freshman in about 200 will have such a high score, we will give very few scholarships. The requirements are unrealistically high.

5. The predicted GPA for a college board score of 500 will be:

$$\hat{z}_{\text{GPA}} = rz_{\text{CEEB}} = .50 \times .00$$

$$= .00 \text{ or, in raw score terms, a GPA} = 2.20$$

$$\sigma_{Y \cdot X} = \sigma_Y \sqrt{1 - r^2}$$

$$= .60 \sqrt{1 - .50^2}$$

$$= .52$$

The standard deviation of the errors of prediction is .52. The probability of a GPA in excess of 3.00 can then be determined by finding its z value where $z = (3.00 - 2.20)/.52 = 1.54$. From the normal curve table we find about 6% of the curve above $z = 1.54$. Therefore, about 6% of students with college board scores of 500 can be expected to earn GPA's in excess of 3.00.

6. $$\phi = \frac{15 \times 15 - 5 \times 5}{\sqrt{20 \times 20 \times 20 \times 20}} = .50$$

8. $\hat{z}_2 = 1.00$, $r = .80$ $\qquad \hat{z}_1 = rz_2$

$\hat{z}_2 = rz_1$ $\qquad \hat{z}_1 = .80 \times 1.00$

$\frac{\hat{z}_2}{r} = z_1$ $\qquad \hat{z}_1 = .80$

$$\frac{1.00}{.80} = 1.25 = z_1$$

Predicting $\hat{z}_2$ uses rz_1 as the regression equation, but predicting $\hat{z}_1$ uses rz_2 as the regression equation. The difference in the z_1 scores resulted from the difference between the two kinds of questions, they each depend on different regression equations.

9. −.87. Remember that the magnitude of the number, not the sign, determines the closeness of the association. Choice 4 cannot be correct since r cannot exceed 1.00 if it is correctly calculated.
10. Nothing, it will be exactly the same.

CHAPTER 7

1. $(\frac{1}{2})^4 = \frac{1}{16}$. The events all have a probability of $\frac{1}{2}$. It is true that two heads and two tails are more probable than four heads, but only because there are six different ways for two heads and two tails to occur.
2. Heads on successive coins are by no means mutually exclusive events. As in the example in the text, the correct probability is .75. $p(A \text{ or } B) = p(A) + p(B) - p(A \text{ and } B) = .5 + .5 - .25 = .75$.
3. False. The sun will rise tomorrow or it will not. Only these two events are possible, but they are not equally likely.
5. $\frac{10!}{(10-4)!} = 10 \times 9 \times 8 \times 7 = 5040$. The probability is $\frac{1}{5040}$, or about .0002.
6. $\frac{45}{1024}$. This is the probability of getting exactly 8 right and 2 wrong. It is the value of the third (or 9th) term of the expansion $(p + q)^{10}$.
7. .50. The median is *defined* as the point which separates the distribution into two equal halves.
8. This cannot be determined from the information given because the events are not independent. If being above the median on the second test were independent of one's standing on the first test, we would use the multiplicative rule for independent events, and the probability would be .50 ×

.50 = .25. However, we might find that if a student is above the median on the first test, the probability that he will be above the median on the second test too is about .80. In this more realistic situation we have $p(A \text{ and } B) = p(A) \cdot p(B/A) = .50 \cdot .80 = .40$. Of course, one does not pull such conditional probabilities out of the air. They would be determined by experience.

9. (c) is the only incorrect choice.
10. We must evaluate the first three terms in the expansion of $(p + q)^{12}$. These are $p^{12} = \frac{1}{4096}$, $12p^{11}q = \frac{12}{4096}$, and $66p^{10}q^2 = \frac{66}{4096}$. The sum of these terms is $\frac{79}{4096}$, or .019.
11. The events are probably not independent. If they were, the product rule for independent events that has been applied would be appropriate. However, to find the probability of the compound event, we need to know the conditional probability of purchasing a new car *given* an annual income in excess of $18,000. This conditional probability has not been given.

CHAPTER 8

1. 3.92%, 1.70%, 76.11%, 52.39%.
2. $M_f = np - 60 \times .40 = 24$

$$\sigma_f = \sqrt{npq} = \sqrt{60 \times .40 \times .60} = 3.79$$

We wish to find the z_f equivalent to 31 or more events, the lower limit of which is 30.5. Therefore, $z_f = \dfrac{30.5 - 24}{3.79} = \dfrac{6.50}{3.79} = 1.72.$

z_f required for a one-tailed test at the 5% level is 1.645, therefore, we can reject the hypothesis that the population proportion is .40 or less.

3. $M_p = p = .50$

$$\sigma_p = \sqrt{\frac{pq}{n}} = \sqrt{\frac{.50 \times .50}{64}} = .062$$

Correcting the obtained proportions for continuity, we first convert to a frequency; .375 × 64 = 24. Thus, the equation for z_p will use an obtained proportion of 24.5/64 = .383.

$z_p = \dfrac{.383 - .500}{.062} = \dfrac{-.117}{.062} = -1.89$ For a two-tailed test, we require $z_p = \pm 1.96$. The hypothesis cannot be rejected at the 5% level of significance.

4. (a) The null hypothesis may be stated as, "The proportion of small-flowered plants in the population from which this sample has been drawn is .25."

(b) Two-tailed

(c) $M_f = np = 100 \times .25 = 25$

$$\sigma_f = \sqrt{npq} = \sqrt{100 \times .25 \times .75} = 4.33$$

$$z_f = \frac{20.5 - 25}{4.33} = -1.04$$

Using a two-tailed test, a z of ± 2.58 or larger is required for significance. z_f for this sample is -1.04 and, therefore, does not provide us with sufficient evidence to reject the null hypothesis.

5. Failure to correct for continuity will overestimate the value of z. Therefore, we will reject more than 1% of null hypotheses when they are true and increase our Type I errors.
6. Should ESP be decided solely on the basis of correctly calling a *larger* than expected frequency? Or, should we allow the possibility of a "negative" ESP when S calls fewer cards correctly than we would predict by chance? Again, debate may be offered, but the two-tailed test is probably safer. The minimum required z is, therefore, ± 2.58.

$$M_f = np = 25 \times .20 = 5.0$$

$$\sigma_f = \sqrt{npq} = \sqrt{25 \times .20 \times .80} = 2.0$$

In this problem we want to determine the number of correct calls required to give us a $z = \pm 2.58$. We therefore already have a value of z and we solve for X in the equation.

$$z = \frac{X_f - M_f}{\sigma_f}$$

$$2.58 = \frac{X_f - 5.0}{2}$$

$X_f = 10.16$, but correcting for continuity adds .5 to the required deviation, so we would expect S to get either 11 correct out of 25, or 0 correct.

7. A one-tailed test is required

$$z \geqq 1.645 \text{ for } \alpha = .05$$

$$M_f = np = 200 \times .5 = 100$$

$$\sigma_f = \sqrt{npq} = \sqrt{200 \times .5 \times .5} = 7.07$$

$$z = \frac{118.5 - 100}{7.07} = 2.62$$

We can safely predict victory for our candidate.

CHAPTER 9

1.

	A	B	C	D	E	F	
Germinating	55 / 65	55 / 60	55 / 50	55 / 55	55 / 55	55 / 45	330
Not germinating	45 / 35	45 / 40	45 / 50	45 / 45	45 / 45	45 / 55	270
	100	100	100	100	100	100	600

$$\chi^2 = \tfrac{100}{55} + \tfrac{100}{45} + \tfrac{25}{55} + \tfrac{25}{45} + \tfrac{25}{55} + \tfrac{25}{45} + \tfrac{100}{55} + \tfrac{100}{45} = 10.10$$

Expected frequencies

$$\text{Germinating, all cells} = \frac{100 \times 330}{600} = 55$$

$$\text{Not germinating, all cells} = \frac{100 \times 270}{600} = 45$$

Since df > 1, we do not correct for continuity.

$$\chi^2 = 10.10, \text{ df} = 5, p > .05$$

The deviation among the samples in proportion of seed germinating is not sufficient to reject the hypothesis that the differences have occurred as a result of random sampling from a common population. Since the χ^2 is close to significance, we might wish to repeat the study with larger samples because the probability of a type II error is probably substantial.

2.

	A	F	
Germinating	65	45	110
Not germinating	35	55	90
	100	100	200

$$\text{The expected frequency for Germinating, all cells} = \frac{110 \times 100}{200} = 55$$

$$\chi^2 = 7.29, \text{ df} = 1, p < .01$$

Rejection of the null hypothesis forces us to conclude that the observed discrepancy between two proportions is unlikely to be the result of comparing two randomly obtained samples from a common population. Quite true, since we deliberately selected for comparison the two samples with the largest difference. We can hardly be surprised to find that difference is due to something other than "chance." The fact that these were "Brand *A*" and "Brand F" was quite incidental to the fact that they represented the two samples with the greatest discrepancy. If we suspect a true difference (population difference) between brands A and F, we can easily test the hypothesis by drawing two new samples. Then the comparison is made on the basis of brand names rather than on the basis of differences in proportion of seeds germinating for these particular samples.

4. Median = 34.0

	One year math	No math	
Above Median	9.9 / 14	8.1 / 4	18
Not above Median	11.1 / 7	8.9 / 13	20
	21	17	38

$$\chi^2 = \frac{(3.6)^2}{9.9} + \frac{(3.6)^2}{11.1} + \frac{(3.6)^2}{8.1} + \frac{(3.6)^2}{8.9} = 5.54$$

$$\chi^2 = 5.54, \text{ df} = 1, p < .02$$

We would reject the hypothesis that this difference between medians occurred as a result of random sampling. We would conclude that there is an association between achievement in statistics and prior college work in mathematics.

5.

	Yes	?	No	
M	33.3 / 30	46.7 / 40	120 / 130	200
W	16.7 / 20	23.3 / 30	60 / 50	100
	50	70	180	300

$$\chi^2 = \sum \frac{(o - e)^2}{e} = \frac{(3.3)^2}{33.3} + \frac{(3.3)^2}{16.7} + \cdots + \frac{(10)^2}{60} = 6.37$$

$\chi^2 = 6.37$, df $= 2$, $p < .05$

We could conclude that type of response is related to sex of respondent. This is the same as saying that the proportion of "yes," "?," and "no" answers vary between men and women more than might be expected (at the 5% level) due to sampling error. Notice that we are not testing the significance of the difference between the proportions of "yes," "?," and "no" answers *within* either group although this could be done too.

6. This problem contains the same kind of faulty assumption present in Exercise 2. The hypothesis to be tested is that the difference in proportions is unlikely

to be as great as that observed if we are sampling randomly from a common population. If we *deliberately select the two samples having the greatest difference* we certainly have a difference occasioned by other than random sampling. March and September enter the picture *after* we have observed the months with the greatest difference between proportions. New samples for March and September must be compared for conclusive results. Comparisons among observations cannot depend on the differences observed because the hypothesis that these differences are the result of random selection is obviously untenable.

7. We cannot calculate χ^2 until we know the sample size.
8. Two observations on each of 100 rats does not yield 200 independent observations. This is a repeated observation problem. We must know the number of rats turning right_1–right_2, right_1–left_2, left_1–right_2, left_1–left_2. Since we have the border totals for this problem we need the frequency in only one of these classes. The others can be obtained by subtraction.
9. This is a one-tailed test, so the probability value associated with the tabled value of χ^2 must be halved. The result is significant at the 2.5% level.
10. We do not know the number of independent observations. The 147 signals on 192 left hand turns were not made by 192 women. The independent units are men signalling, men not signalling, women signalling, and women not signalling.
11. (a) There are 2 df, but the χ^2 is corrected for continuity. This should not be done when df > 1.
    (b) Same problem as 2 and 6. The investigator has made a comparison between two cells solely on the basis of the larger discrepancy. The assumption of randomness has been violated.
12. Each subject, $A, B, \ldots, R$, can be given a (+) if he has answered more questions correctly from the first 5 than from the second 5. He receives a (−) if the situation is reversed, and a zero if he answers equal numbers of questions from the two groups. On the basis of random sampling from a population with the same median, the frequency of + and − entries should be the same. Significant departures from equality can be evaluated by χ^2. The data are analyzed below.

	$A+$	$B+$	$C+$	$D-$	$E°$	$F+$	$G°$	$H+$	$I+$
1st.	5	4	3	2	5	1	3	4	3
2nd.	4	2	1	3	5	0	3	2	1

	$J+$	$K+$	$L+$	$M+$	$N-$	$O°$	$P+$	$Q-$	$R+$
1st.	4	4	3	3	2	3	5	4	4
2nd.	2	2	1	1	3	3	4	5	3

	+	−	
o	12	3	15
e	7.5	7.5	15

$$\chi^2 = \frac{(4)^2}{7.5} + \frac{(4)^2}{7.5} = 4.27$$

$$\chi^2 = 4.27, \text{ df} = 1, p < .05$$

We would reject the hypothesis that this difference between (+) and (−) observations occurred as the result of random sampling from a common population at the 5% level. Notice that this procedure is quite similar to the χ^2 for correlated observations discussed earlier. Here, as there, we are dealing with the "changes" and determining if there are a disproportionate number of changes in one direction. This is also a "repeated observation" problem. The median test would not be appropriate, because the samples are not independent. This test is called the sign test, and we shall discuss it more fully in Chapter 14.

CHAPTER 10

1. $$\sum x_1^2 = \sum X_1^2 - \frac{(\Sigma X_1)^2}{n} = 3007 - \frac{(195)^2}{14} = 291$$

$$\sum x_2^2 = \sum X_2^2 - \frac{(\Sigma X_2)^2}{n} = 2038 - \frac{(162)^2}{14} = 163$$

$$S_{\bar{X}_1 - \bar{X}_2} = \sqrt{\frac{\Sigma x_1^2 + \Sigma x_2^2}{n_1 + n_2 - 2} \times \frac{n_1 + n_2}{n_1 n_2}} = \sqrt{2.48} = 1.57$$

$$t = \frac{\bar{X}_1 - \bar{X}_2}{S_{\bar{X}_1 - \bar{X}_2}} = \frac{13.93 - 11.57}{1.57} = 1.50$$

$t = 1.50$, df $= 26$, $p > .05$

2. $\Sigma x^2 = 15000$

$$S_{\bar{X}} = \sqrt{\frac{\Sigma x^2}{n(n-1)}} = \sqrt{\frac{15000}{(101)(100)}} = 1.22$$

$$t = \frac{\bar{X} - M}{S_{\bar{X}}} = \frac{-5}{1.22} = -4.10$$

$t = -4.10$, df $= 100$, $p < .01$

3. (a)

		D	D^2
16	18	−2	4
17	14	3	9
24	16	8	64
25	19	6	36
19	15	4	16
15	20	−5	25
28	22	6	36
25	16	9	81
22	18	4	16
22	17	5	25
$\bar{D} = 3.8$		$\Sigma D = 38$	$\Sigma D^2 = 312$

$$S_{\bar{D}} = \frac{S_D}{\sqrt{n}} = \sqrt{\frac{\Sigma D^2 - \frac{(\Sigma D)^2}{n}}{n(n-1)}} = \sqrt{\frac{312 - \frac{(38)^2}{10}}{90}} = 1.36$$

$$t = \frac{\bar{D}}{S_{\bar{D}}} = \frac{3.80}{1.36} = 2.79$$

$t = 2.79$, df $= 9$, $p < .05$

3. (b) $\Sigma x_1^2 = \Sigma X_1^2 - \frac{(\Sigma X_1)^2}{n_1} = 172.1$

$$\Sigma x_2^2 = \Sigma X_2^2 - \frac{(\Sigma X_2)^2}{n_2} = 52.5$$

$$S_{\bar{X}_1-\bar{X}_2} = \sqrt{\frac{\Sigma x_1^2 + \Sigma x_2^2}{n_1 + n_2 - 2} \times \frac{n_1 + n_2}{n_1 n_2}} = \sqrt{\frac{172.1 + 52.5}{18} \times \frac{20}{100}} = 1.58$$

$$t = \frac{\bar{X}_1 - \bar{X}_2}{S_{\bar{X}_1-\bar{X}_2}} = \frac{3.80}{1.58} = 2.41$$

$t = 2.41$, df $= 18$, $p < .05$

4. $\Sigma x_1^2 = \Sigma X_1^2 - \frac{(\Sigma X_1)^2}{n_1} = 489 - \frac{(65)^2}{10} = 66.5$

$$\Sigma x_2^2 = \Sigma X_2^2 - \frac{(\Sigma X_2)^2}{n_2} = 903 - \frac{(89)^2}{10} = 110.9$$

$$S_{\bar{X}_1-\bar{X}_2} = \sqrt{\frac{\Sigma x_1^2 + \Sigma x_2^2}{n_1 + n_2 - 2} \times \frac{n_1 + n_2}{n_1 n_2}} = \sqrt{\frac{110.9 + 66.5}{18} \times \frac{20}{100}} = 1.40$$

$$t = \frac{\bar{X}_1 - \bar{X}_2}{S_{\bar{X}_1-\bar{X}_2}} = \frac{2.40}{1.40} = 1.71$$

$t = 1.71$, df $= 18$, $p > .05$

5.

		D	D^2
16	12	4	16
4	8	−4	16
6	7	−1	1
9	7	2	4
9	9	0	0
8	6	2	4
5	5	0	0
10	4	6	36
10	4	6	36
12	3	9	81
$\bar{D} = 2.4$		$\Sigma D = 24$	$\Sigma D^2 = 194$

$$S_{\bar{D}} = \sqrt{\frac{\Sigma D^2 - \frac{(\Sigma D)^2}{n}}{n(n-1)}} = \sqrt{\frac{194 - \frac{(24)^2}{10}}{90}} = 1.23$$

$$t = \frac{\bar{D}}{S_{\bar{D}}} = \frac{2.40}{1.23} = 1.95$$

$t = 1.95$, df $= 9$, $p > .05$

6. $r_1 = .65$, $Z'_1 = .775$

   $r_2 = .35$, $Z'_2 = .365$

$$\sigma_{Z'_1-Z'_2} = \sqrt{\frac{1}{N_1 - 3} + \frac{1}{N_2 - 3}} = \sqrt{.01 + .01} = .14$$

$$z = \frac{Z'_1 - Z'_2}{\sigma_{Z'_1-Z'_2}} = \frac{.41}{.14} = 2.93,\ p < .01$$

7.

		Classical Music Dislike	Classical Music Like
Modern Art	Like	*a* 5	*b* 10
	Dislike	*c* 10	*d* 5

$$\phi = \frac{bc - ad}{\sqrt{(a + c)(b + d)(c + d)(a + b)}} = \frac{(10 \cdot 10) - (5 \cdot 5)}{\sqrt{15 \cdot 15 \cdot 15 \cdot 15}} = .33$$

$$\chi^2 = \frac{n[(bc - ad) - n/2]^2}{(a + b)(a + c)(b + d)(c + d)} = \frac{30[(100 - 25) - \frac{30}{2}]^2}{15 \cdot 15 \cdot 15 \cdot 15} = 2.13$$

$\chi^2 = 2.13$, df $= 1$, $p > .05$

8. Running time converted to speed scores

   A: 50, 33, 20, 7, 25, 8, 33, 12, 25, 4

   B: 33, 20, 25, 5, 50, 7, 50, 33, 20, 11, 17, 14

A	*B*
50	33
33	20
20	25
7	5
25	50
8	7
33	50
12	33
25	20
4	11
	17
	14

$\Sigma X_A = 217$

$\Sigma X_A^2 = 6601$

$\overline{X}_A = 21.70$

$\Sigma X_B = 285$

$\Sigma X_B^2 = 9283$

$\overline{X}_B = 23.75$

$$t = \frac{2.05}{6.35} = .32$$

$t = .32$, df $= 20$, $p > .05$

9. True. $S_{\bar{X}_1-\bar{X}_2} = \sqrt{S_{\bar{X}_1}{}^2 + S_{\bar{X}_2}{}^2}$, and $S_{\bar{X}} = \dfrac{S}{\sqrt{n}}$ shows that $S_{\bar{X}_1-\bar{X}_2}$ is directly related to the variation in the population and inversely to the square root of sample size. If we can reduce σ by more effective experimental controls, we will be able to reduce $S_{\bar{X}_1-\bar{X}_2}$ and increase the power of our statistical test. All too often experimenters are inclined to assume that sources of poor control will average out and thus, produce no bias, if they affect both experimental groups. This may well be true, but they still add to the variance within each group and reduce the power of the test. Therefore, such "averaging out" is not an effective substitute for the careful experimental control of conditions.
10. False. The sample means are normally distributed regardless of sample size *if* the population is normally distributed. The necessity for t arises from the fact that the ratios $(\bar{X} - M)/S_{\bar{X}}$ and $(\bar{X}_1 - \bar{X}_2)/S_{\bar{X}_1-\bar{X}_2}$ are not normally distributed. These ratios are not normally distributed because S is likely to underestimate σ, therefore underestimating the size of $S_{\bar{X}} = S/\sqrt{n}$ and underestimating the size of $S_{\bar{X}_1-\bar{X}_2}$. These underestimations result in too many large values of the ratios $(\bar{X} - M)/S_{\bar{X}}$ and $(\bar{X}_1 - \bar{X}_2)/S_{\bar{X}_1-\bar{X}_2}$ for appropriate evaluation by a normal curve table. The ratio is distributed as t rather than z. An appropriate table has been prepared for the evaluation of t.

CHAPTER 11

1. (a)
$$S_1{}^2 = \frac{\Sigma(X - \bar{X})^2}{(n_1 - 1)} = \frac{\Sigma X_1{}^2 - \dfrac{(\Sigma X_1)^2}{n}}{(n_1 - 1)} = \frac{4349 - \dfrac{(195)^2}{10}}{9} = 60.7$$

$$S_2{}^2 = \frac{\Sigma(X - \bar{X})^2}{(n_2 - 1)} = \frac{\Sigma X_2{}^2 - \dfrac{(\Sigma X_2)^2}{n}}{(n_2 - 1)} = \frac{3356 - \dfrac{(166)^2}{10}}{9} = 66.7$$

$$F = \frac{S_2{}^2}{S_1{}^2} = 1.10 \qquad \text{Not significant for df} = 9/9$$

(b)
$$SS_t = \sum_N X^2 - \frac{\left(\sum_N X\right)^2}{N} = 7705 - \frac{(361)^2}{20} = 1189$$

$$SS_b = \sum_k \frac{\left(\sum_n X\right)^2}{n} - \frac{\left(\sum_N X\right)^2}{N} = \frac{(195)^2}{10} + \frac{(166)^2}{10} - \frac{(361)^2}{20} = 42.1$$

$$SS_w = \sum_N X^2 - \sum_k \frac{\left(\sum_n X\right)^2}{n} = 7705.0 - 6558.1 = 1146.9$$

Source	*SS*	*df*	*MS*	*F*
Between	42.1	1	42.1	.66
Within	1146.9	18	63.7	
Total	1189.0	19		

$F = .66$, $df = 1/18$, $p > .05$

2. (a) $$SS_t = \sum_N X^2 - \frac{\left(\sum_N X\right)^2}{N} = 8055 - \frac{(413)^2}{24} = 948.0$$

$$SS_b = \sum_k \frac{\left(\sum_n X\right)^2}{n} - \frac{\left(\sum_N X\right)^2}{N} = 7862.5 - \frac{(413)^2}{24} = 755.5$$

$$SS_w = \sum_N X^2 - \sum_k \frac{\left(\sum_n X\right)^2}{n} = 8055.0 - 7862.5 = 192.5$$

Source	*SS*	*df*	*MS*	*F*
Between	755.5	3	251.8	26.2**
Within	192.5	20		
Total	948.0	23		

$F = 26.2$, df $= 3/20$, $p < .01$

Since F is significant, we can conduct Duncan's Multiple Range Test

(b)

B		*A*		*C*		*D*
9.8	(4.0)	13.8	(7.7)	21.5	(2.2)	23.7

$$S_{\bar{X}} = \sqrt{\frac{S_w^2}{n}} = \sqrt{\frac{9.6}{6}} = 1.26$$

S_w^2 has 20 df, $C_3 = 3.10$

$R_{n_3} = C_3 \cdot S_{\bar{X}} = 3.10 \times 1.26 = 3.9$

The differences among both BAC and ACD exceed the range of non-significance which is 3.9

$R_{n_2} = C_2 \cdot S_{\bar{X}} = 2.95 \times 1.26 = 3.7$

The difference between pairs BA and AC exceed R_{n_2}, but pair CD does not. This pair, CD, is underlined. Significant differences exist between any nonunderlined mean and any other mean. No significant differences exist among means having a common underline. We conclude that a significant difference exists between B and all other means, A and all other means, but not between means C and D.

3. The variation among means furnishes one estimate of σ^2, another estimate of σ^2 is provided by the variation of measures within each sample. If the estimate based on the variation among means is significantly larger than the estimate based on the variation within samples, it will be the result of variation among sample means greater than that to be expected on the basis of sampling error.

4. $$SS_t = \sum_N X^2 - \frac{\left(\sum_N X\right)^2}{N} = 4743 - \frac{(379)^2}{35} = 639$$

$$SS_b = \sum_k \frac{\left(\sum_n X\right)^2}{n} - \frac{\left(\sum_N X\right)^2}{N} = 4486.1 - \frac{(379)^2}{35} = 382.1$$

$$SS_w = \sum_N X^2 - \sum_k \frac{\left(\sum_n X\right)^2}{n} = 4743.0 - 4486.1 = 256.9$$

Source	*SS*	*df*	*MS*	*F*
Between	382.1	4	95.5	11.1**
Within	256.9	30	8.6	
Total	639.0	34		

$F = 11.1$, df $= 4/30$, $p < .01$

5. $$SS_t = \sum_N X^2 - \frac{\left(\sum_N X\right)^2}{N} = 937 - \frac{(125)^2}{22} = 226.8$$

$$SS_b = \frac{\left(\sum_{n_1} X\right)^2}{n_1} + \frac{\left(\sum_{n_2} X\right)^2}{n_2} + \frac{\left(\sum_{n_3} X\right)^2}{n_3} - \frac{\left(\sum_N X\right)^2}{N} = 794.6 - \frac{(125)^2}{22} = 84.4$$

$$SS_w = \sum_N X^2 - \left[\frac{\left(\sum_{n_1} X\right)^2}{n_1} + \frac{\left(\sum_{n_2} X\right)^2}{n_2} + \frac{\left(\sum_{n_3} X\right)^2}{n_3}\right] = 937 - 794.6 = 142.2$$

Source	*SS*	*df*	*MS*	*F*
Between	84.4	2	42.2	5.6*
Within	142.4	19	7.5	
Total	226.8	21		

$F = 5.6$, df $= 2/19$, $p < .05$

6. False. Not variances, but sums of squares.
7. True. If F is not significant none of the typical postanalysis of variance tests should be conducted.
8. True. The mean square is the sum of squares divided by degrees of freedom. This is the estimate of variance.
9. True. The total df may be partitioned just as the sums of squares are partitioned.
10. False. Only the within and between estimates of variance are independent of each other. If either of these is altered, we will also change the estimate of variance based on the variation of all measures about a grand mean.

CHAPTER 12

1. $$SS_t = \sum_N X^2 - \frac{\left(\sum_N X\right)^2}{N} = 2468 - \frac{(186)^2}{16} = 305.8$$

$$SS_{\text{Cols.}} = \sum_C \frac{\left(\sum_{n_C} X\right)^2}{n_C} - \frac{\left(\sum_N X\right)^2}{N} = 2171.2 - \frac{(186)^2}{16} = 9.0$$

$$SS_{\text{Row}} = \sum_R \frac{\left(\sum_{n_R} X\right)^2}{n_R} - \frac{\left(\sum_N X\right)^2}{N} = 2164.5 - \frac{(186)^2}{16} = 2.3$$

$$SS_{\text{within}} = \sum_N X^2 - \sum_k \frac{\left(\sum_{n_k} X\right)^2}{n_k} = 2468 - 2398.5 = 69.5$$

$$SS_{\text{interact.}} = SS_t - (SS_{\text{Col.}} + SS_{\text{Row}} + SS_{\text{within}}) = 225.0$$

Source	*SS*	*df*	*MS*	*F*
Between columns	9.0	1	9.0	1.55
Between rows	2.3	1	2.3	<1.00
Interaction	225.0	1	225.0	38.79**
Within cells	69.5	12	5.8	
Total	305.8	15		

Neither main effect is significant; however, the interaction is significant beyond the 1% level.

$$2.\ SS_t = \sum_N X^2 - \frac{\left(\sum_N X\right)^2}{N} = 15726 - \frac{(524)^2}{20} = 1997.2$$

$$SS_{A's} = \sum_{A's} \frac{\left(\sum_{n_A} X\right)^2}{n_A} - \frac{\left(\sum_N X\right)^2}{N} = 15208.0 - 13728.8 = 1479.2$$

$$SS_{B's} = \sum_{B's} \frac{\left(\sum_{n_B} X\right)^2}{n_B} - \frac{\left(\sum_N X\right)^2}{N} = 13729 - 13728.8 = .2$$

$$SS_{\text{within}} = \sum_N X^2 - \sum_k \frac{\left(\sum_{n_k} X\right)^2}{n_k} = 15726 - 15210 = 516.0$$

$$SS_{\text{interact.}} = SS_t - (SS_{A's} + SS_{B's} + SS_{\text{within}}) = 1.8$$

Source	*SS*	*df*	*MS*	*F*
A's	1479.2	1	1479.2	45.94**
B's	.2	1	.2	<1.00
interaction	1.8	1	1.8	<1.00
within	516.0	16	32.2	

Only the main effect of *A* is significant at the 1% level.

$$3.\ SS_t = \sum_N X^2 - \frac{\left(\sum_N X\right)^2}{N} = 15644 - \frac{(714)^2}{36} = 1483.0$$

$$SS_B = \sum_B \frac{\left(\sum_{n_B} X\right)^2}{n_B} - \frac{\left(\sum_N X\right)^2}{N} = \frac{(410)^2}{18} + \frac{(304)^2}{18} - \frac{(714)^2}{36} = 312.1$$

$$SS_A = \sum_A \frac{\left(\sum_{n_A} X\right)^2}{n_A} - \frac{\left(\sum_N X\right)^2}{N} = \frac{(215)^2}{12} + \frac{(233)^2}{12} + \frac{(266)^2}{12} - \frac{(714)^2}{36}$$

$$= 111.5$$

$$SS_{\text{within}} = \sum_{N} X^2 - \sum_{k} \frac{\left(\sum_{n_k} X\right)^2}{n_k} = 15644 - \left[\frac{(97)^2}{6} + \frac{(129)^2}{6} + \cdots + \frac{(82)^2}{6}\right] = 415.7$$

$$SS_{\text{interaction}} = SS_{\text{total}} - (SS_A + SS_B + SS_{\text{within}}) = 643.7$$

Source	*SS*	*df*	*MS*	*F*
A's	111.5	2	55.8	4.01*
B's	312.1	1	312.1	22.45**
Interaction	643.7	2	321.8	23.15**
Within	415.7	30	13.9	
Total	1483.0	35		

The effect of A is significant at the 5% level, the effect of B and the interaction of $A \times B$ are significant at the 1% level. At the A_1 level there is a progressive increase across levels of B. At the A_2 level this is exactly reversed and a progressive decrease is found.

4. $$SS_t = \sum_{N} X^2 - \frac{\left(\sum_{N} X\right)^2}{N} = 5471 - \frac{(397)^2}{32} = 545.7$$

$\sum_{n_{A_1}} X = 182$

$\sum_{n_{A_2}} X = 215$

$$SS_{A\text{'s}} = \frac{\left(\sum_{n_{A1}} X\right)^2}{n_A} + \frac{\left(\sum_{n_{A2}} X\right)^2}{n_A} - \frac{\left(\sum_{N} X\right)^2}{N} = \frac{(182)^2}{16} + \frac{(215)^2}{16} - \frac{(397)^2}{32} = 34.03$$

$\sum_{n_{B_1}} X = 169$

$\sum_{n_{B_2}} X = 228$

$$SS_{B\text{'s}} = \frac{\left(\sum_{n_{B1}} X\right)^2}{n_B} + \frac{\left(\sum_{n_{B2}} X\right)^2}{n_B} - \frac{\left(\sum_{N} X\right)^2}{N} = \frac{(169)^2}{16} + \frac{(228)^2}{16} - \frac{(397)^2}{32} = 108.78$$

$\sum_{n_{C_1}} X = 185$

$\sum_{n_{C_2}} X = 212$

$$SS_{C\text{'s}} = \frac{\left(\sum_{n_{C1}} X\right)^2}{n_C} + \frac{\left(\sum_{n_{C2}} X\right)^2}{n_C} - \frac{\left(\sum_{N} X\right)^2}{N} = \frac{(185)^2}{16} + \frac{(212)^2}{16} - \frac{(397)^2}{32} = 22.78$$

$$\sum_{n_{A_1B_2,A_2B_1}} X = 203 \qquad SS_{A\times B} = \frac{\left(\sum_{n_{A_1B_2,A_2B_1}} X\right)^2}{n_{A\times B}} + \frac{\left(\sum_{n_{A_1B_1,A_2B_2}} X\right)^2}{n_{A\times B}} - \frac{\left(\sum_N X\right)^2}{N}$$

$$\sum_{n_{A_1B_1,A_2B_2}} X = 194 \qquad = \frac{(203)^2}{16} + \frac{(194)^2}{16} - \frac{(397)^2}{32} = 2.53$$

$$\sum_{n_{A_1C_2,A_2C_1}} X = 191 \qquad SS_{A\times C} = \frac{\left(\sum_{n_{A_1C_2,A_2C_1}} X\right)^2}{n_{A\times C}} + \frac{\left(\sum_{n_{A_1C_1,A_2C_2}} X\right)^2}{n_{A\times C}} - \frac{\left(\sum_N X\right)^2}{N}$$

$$\sum_{n_{A_1C_1,A_2C_2}} X = 206 \qquad = \frac{(191)^2}{16} + \frac{(206)^2}{16} - \frac{(397)^2}{32} = 7.03$$

$$\sum_{n_{B_1C_2,B_2C_1}} X = 198 \qquad SS_{B\times C} = \frac{\left(\sum_{n_{B_1C_2,B_2C_1}} X\right)^2}{n_{B\times C}} + \frac{\left(\sum_{n_{B_1C_1,B_2C_2}} X\right)^2}{n_{B\times C}} - \frac{\left(\sum_N X\right)^2}{N}$$

$$\sum_{n_{B_1C_1,B_2C_2}} X = 199 \qquad = \frac{(198)^2}{16} + \frac{(199)^2}{16} - \frac{(397)^2}{32} = .03$$

$$SS_{\text{within}} = \sum_N X^2 - \frac{\left(\sum_{n_k} X\right)^2}{n_k} = 5471 - \left[\frac{(36)^2}{4} + \frac{(42)^2}{4} + \cdots + \frac{(65)^2}{4}\right]$$

$$= 370.25$$

$$SS_{A\times B\times C} = SS_t - (SS_A + SS_B + SS_C + SS_{A\times B} + SS_{A\times C} + SS_{B\times C} + SS_{\text{within}}) = .29$$

Source	*SS*	*df*	*MS*	*F*
A's	34.03	1	34.0	2.21
B's	108.78	1	108.78	7.06*
C's	22.78	1	22.78	1.48
$A \times B$	2.53	1	2.53	<1.00
$A \times C$	7.03	1	7.03	<1.00
$B \times C$	.03	1	.03	<1.00
$A \times B \times C$	.29	1	.29	<1.00
Within cells	370.25	24	15.43	
Total	545.72	31		

Only the *B* effect is significant, and that is significant only at the 5% level.

CHAPTER 13

1. True
2. True

3. $SS_{\text{total}} = (10)^2 + \cdots + (8)^2 + (10)^2 - \dfrac{(139)^2}{18} = 185.61$

$$SS_{\text{between trials}} = \frac{(56)^2}{6} + \frac{(48)^2}{6} + \frac{(35)^2}{6} - \frac{(139)^2}{18} = 37.44$$

$$SS_{\text{bet. blocks}} = \frac{(26)^2}{3} + \frac{(17)^2}{3} + \frac{(16)^2}{3} + \frac{(14)^2}{3} + \frac{(28)^2}{3} + \frac{(38)^2}{3} - \frac{(139)^2}{18} = 141.61$$

$$SS_{\text{trials}\times\text{blocks}} = SS_{\text{total}} - (SS_{\text{trials}} + SS_{\text{blocks}}) = 6.56$$

Source	*SS*	*df*	*MS*	*F*
Between trials	37.44	2	18.72	28.36**
Between blocks	141.61	5		
Blocks × trials (error)	6.56	10	.66	
Total	185.61	17		

4. $SS_{\text{total}} = (10)^2 + (8)^2 + \cdots + (18)^2 + (15)^2 - \dfrac{(163)^2}{16} = 414.44$

$$SS_{\text{between } A\text{'s}} = \frac{(86)^2}{8} + \frac{(77)^2}{8} - \frac{(163)^2}{16} = 5.07$$

$$SS_{\text{between blocks}} = \frac{(16)^2}{2} + \frac{(12)^2}{2} + \cdots + \frac{(35)^2}{2} + \frac{(29)^2}{2} - \frac{(163)^2}{16} = 379.94$$

Partitions into

$$SS_{\text{between } B\text{'s}} = \frac{(45)^2}{8} + \frac{(118)^2}{8} - \frac{(163)^2}{16} = 333.07$$

$$SS_{\substack{\text{between blocks}\\ \text{within } B\text{'s}}} = SS_{\text{between blocks}} - SS_{\text{between } B\text{'s}} = 46.87$$

$$SS_{\text{blocks}\times A\text{'s}} = SS_{\text{total}} - (\text{blocks} + A\text{'s}) = 29.43$$

Partitions into

$$SS_{\text{conditions}\times A\text{'s}} = \frac{(71)^2}{8} + \frac{(92)^2}{8} - \frac{(163)^2}{16} = 27.57$$

$$SS_{\substack{\text{blocks}\times A\text{'s}\\ \text{within } B\text{'s}}} = SS_{\text{block}\times A\text{'s}} - SS_{B\text{'s}\times\text{trials}} = 1.86$$

Source	*SS*	*df*	*MS*	*F*
Between *B*'s	333.07	1	333.07	42.7**
Between blocks within *B*'s (error)	46.87	6	7.81	
Between *A*'s	5.07	1	5.07	16.35**
B's × *A*'s	27.57	1	27.57	88.94**
Blocks × *A*'s within *B*'s (error)	1.86	6	.31	
Total	414.44	15		

5. $SS_{\text{total}} = (8)^2 + (9)^2 + \cdots + (16)^2 + (12)^2 - \frac{(225)^2}{20} = 191.75$

$$SS_{\text{between } A\text{'s}} = \frac{(95)^2}{10} + \frac{(130)^2}{10} - \frac{(225)^2}{20} = 61.25$$

$$SS_{\text{between } B\text{'s}} = \frac{(125)^2}{10} + \frac{(100)^2}{10} - \frac{(225)^2}{20} = 31.25$$

$$SS_{A\times B} = \frac{(111)^2}{10} + \frac{(114)^2}{10} - \frac{(225)^2}{20} = .45$$

$$SS_{\text{between blocks}} = \frac{(34)^2}{4} + \frac{(38)^2}{4} + \cdots + \frac{(49)^2}{4} - \frac{(225)^2}{20} = 89.00$$

$$SS_{\text{blocks}\times\text{conditions}} = SS_{\text{total}} - (A\text{'s} + B\text{'s} + A \times B + \text{between blocks}) = 9.80$$

Source	*SS*	*df*	*MS*	*F*
Between *A*'s	61.25	1	61.25	74.7**
Between *B*'s	31.25	1	31.25	38.1**
$A \times B$	.45	1	.45	
Between blocks	89.00	4		
Blocks × conditions (pooled error)	9.80	12	.82	
Total	191.75	19		

6. Summing *A* and *B* levels over the levels of *T* yields.

		A 1	*A* 2
B	1	340	430
	2	360	330

$$\frac{(790)^2}{40} + \frac{(670)^2}{40} - \frac{(1460)^2}{80} = 180$$

7. Interaction is the residual variation left between cells after *A*, *B*, and *T* sources of variance have been accounted for. Therefore, the interaction from all sources will be the difference between a cell mean and the grand mean, less the difference between *A* and the grand mean, *B* and the grand mean, and *T* and the grand mean, the quantity squared, and multiplied by *n*, the number of cases on which the cell mean is based. This complicated verbal description will be easier to follow in the formula below.

$$n\Sigma[\overline{X}_{A_1B_1T_1} - \overline{\overline{X}} - (\overline{X}_{A_1} - \overline{\overline{X}} + \overline{X}_{B_1} - \overline{\overline{X}} + \overline{X}_{T_1} - \overline{\overline{X}})]^2$$

This simplifies to

$$n\Sigma(\overline{X}_{ABT} - \overline{X}_{A_1} - \overline{X}_{B_1} - \overline{X}_{T_1} + 2\overline{\overline{X}})^2$$

For the problem at hand

$\overline{X}_{T_1} = 17.00$	$A_1B_1T_1 = 10[16 - 17.50 - 19.25 - 17.00 + 2(18.25)]^2$
$\overline{X}_{T_2} = 19.50$	$A_1B_1T_2 = 10[18 - 17.50 - 19.25 - 19.50 + 2(18.25)]^2$
$\overline{X}_{A_1} = 17.50$	$A_1B_2T_1 = 10[17 - 17.50 - 17.25 - 17.00 + 2(18.25)]^2$
$\overline{X}_{A_2} = 19.00$	$A_1B_2T_2 = 10[19 - 17.50 - 17.25 - 19.50 + 2(18.25)]^2$
$\overline{X}_{B_1} = 19.25$	$A_2B_1T_1 = 10[20 - 19.00 - 19.25 - 17.00 + 2(18.25)]^2$
$\overline{X}_{B_2} = 17.25$	$A_2B_1T_2 = 10[23 - 19.00 - 19.25 - 19.50 + 2(18.25)]^2$
$\overline{\overline{X}} = 18.25$	$A_2B_2T_1 = 10[15 - 19.00 - 17.25 - 17.00 + 2(18.25)]^2$
	$A_2B_2T_2 = 10[18 - 19.00 - 17.25 - 19.50 + 2(18.25)]^2$

$$SS_{\text{pooled interaction}} = 10[(-1.25)^2 + (-1.75)^2 + (1.75)^2 + \cdots + (-1.75)^2 + (-1.25)^2]$$

$$SS_{\text{pooled interaction}} = 185$$

CHAPTER 14

1. The sign test is appropriate for the data in Exercise 1. We shall arbitrarily designate as "+" any pair in which the "participation" member receives a superior rating. With 16 pairs our expectation under the null hypothesis of no difference between treatments would be 8 "+" and 8 "−" designations. Departures from these frequencies may be tested by chi square.

	+	−	
o	12	4	16
e	8	8	16

$$\chi^2 = \sum \frac{(|o - e| - .5)^2}{e} = \frac{(3.5)^2}{8} \times 2 = 3.06$$

$$\chi^2 = 3.06, \text{ df} = 1, p > .05$$

A very common error in problems of this type can be made if the investigator considers the lecture and participation groups as independent, *which they are not!* Such an incorrect assumption could result in the faulty analysis given below:

We shall arbitrarily designate all superior ratings in either group as "+," the other member of each pair is designated "−."

	Lecture	Participation	
+	[8] 4 (+)	[8] 12 (+)	16
−	[8] 12 (−)	[8] 4 (−)	16
	16	16	32

$$\chi^2 = \sum \frac{(|o - e| - .5)^2}{e} = \frac{(3.5)^2}{8} \times 4 = 6.12$$

$\chi^2 = 6.12$, df $= 1$, $p < .02$

2.

			Rank	Rank of sign with lowest n
A	1–7	−6	7	7
B	6–4	2	2	
C	9–4	5	5.5	
D	18–10	8	8	
E	19–21	−2	2	2
F	22–11	11	9	
G	34–19	15	10.5	
H	27–12	15	10.5	
I	16–14	2	2	
J	19–24	−5	5.5	5.5
K	41–12	29	15	
L	17–20	−3	4	4
M	49–14	35	16	
N	36–16	20	13.5	
O	28–10	18	12	
P	24–4	20	13.5	

$T = \Sigma = 18.5$

Wilcoxon's $T = 18.5$, $n = 16$. From Table W, this is significant at the 1% level of confidence. Notice the increase in power over that of the sign test conducted on the same data.

3. When the data of groups *I* and *II* are arranged in order of magnitude the median is 16.25. This yields the χ^2 table below.

	I	*II*	
Above Median	[6] 9	[6] 3	12
Below Median	[7] 4	[7] 10	14
	13	13	26

$$\chi^2 = \sum \frac{(|o - e| - .5)^2}{e} = \frac{(2.5)^2}{6} + \frac{(2.5)^2}{6} + \frac{(2.5)^2}{7} + \frac{(2.5)^2}{7} = 3.87$$

$$\chi^2 = 3.87, \text{df} = 1, p < .05$$

If the median falls exactly on the midpoint of a score interval, or if we wish to determine only the median score and several scores are found at this point, we can form the dichotomy as "scores above the median" versus "scores *not* above the median."

4. Ranking the pooled measures

I		*II*	
X_{I}	*Rank*	X_{II}	*Rank*
16	12.5	2	1
19	16.5	9	5
27	22.5	18	15
24	21	14	9
23	20	16	12.5
12	7	19	16.5
21	18.5	21	18.5
14	9	8	4
16	12.5	7	3
27	22.5	4	2
35	24	16	12.5
42	25	11	6
71	26	14	9
		$\Sigma R_{\text{II}} =$	114.0

$$U_2 = n_1 n_2 + \frac{n_2(n_2 + 1)}{2} - \Sigma R_2$$

$$= 13 \times 13 + \frac{13(13 + 1)}{2} - 114$$

$$= 146$$

To determine if U_{II} or U_{I} is the smaller: $U_{\text{I}} = n_1 n_2 - U_{\text{II}}$

$$= 169 - 146$$

$$= 23$$

Table *U* shows that a *U* of 34 is required for significance at the 1% level. Our obtained value of $U = 23.0$ is less than this and therefore significant at the 1% level.

6. Ranking the pooled measures

I		*II*	
X_I	R_I	X_{II}	R_{II}
16	11.5	18	13.5
12	8	9	6.5
14	10	6	3
13	9	3	1
9	6.5	7	4
8	5	5	2
24	16	18	13.5
27	18	26	17
16	11.5	$\Sigma R_{II} =$	60.5
21	15		
30	19		

$$U_{II} = n_1 n_2 + \frac{n_2(n_2+1)}{2} - \Sigma R_{II}$$

$$U_{II} = 11 \times 8 + \frac{8(8+1)}{2} - 60.5 = 63.5$$

To determine if U_{II} or U_I is the smaller

$$U_I = n_1 n_2 - U_{II}$$

$$= 88 - 63.5$$

$$= 24.5$$

This value of U fails to reach significance at the 5% level of confidence.

7. Ranks for the pooled data appear below.

I		II		III		IV	
X	R	X	R	X	R	X	R
47	25	40	23	17	9	1	1
41	24	50	26	14	6	8	4
29	17	21	13	16	8	4	2
36	22	26	15	19	10.5	9	5
35	21	34	20	26	15	7	3
26	15	31	18	32	19	20	12
19	10.5						
15	7						
$\Sigma R_I =$	141.5	$\Sigma R_{II} =$	115.0	$\Sigma R_{III} =$	67.5	$\Sigma R_{IV} =$	27.0

$$H = \frac{12}{N(N+1)} \cdot \sum_k \frac{(\Sigma R)^2}{n_k} - 3(N+1)$$

$$= \frac{12}{26(26+1)} \cdot \left[\frac{(141.5)^2}{8} + \frac{(115)^2}{6} + \frac{(67.5)^2}{6} + \frac{(27)^2}{6}\right] - 3(26+1)$$

$\chi^2 = 14.5$, df $= 3$, $p < .01$

8.

X_{I}	R_{I}	X_{II}	R_{II}	X_{III}	R_{III}
127	17	16	4.5	140	18
16	4.5	14	3	184	21
19	6	3	1.5	176	20
24	7	3	1.5	64	12
100	15	36	9	161	19
47	11	42	10	98	14
81	13			121	16
29	8				
	$\Sigma R_{\text{I}} = 81.5$		$\Sigma R_{\text{II}} = 29.5$		$\Sigma R_{\text{III}} = 120$

$$H = \frac{12}{N(N+1)} \cdot \sum_k \frac{(\Sigma R)^2}{n_k} - 3(N+1)$$

$$= \frac{12}{21(21+1)} \cdot \left[\frac{(81.5)^2}{8} + \frac{(29.5)^2}{6} + \frac{(120)^2}{7}\right] - 3(21+1) = 12.77$$

$\chi^2 = 12.77$, df $= 2$, $p < .01$

Table A Table of Random Numbers

Line \ Col.	(1)	(2)	(3)	(4)	(5)	(6)	(7)	(8)	(9)	(10)	(11)	(12)	(13)	(14)
1	10480	15011	01536	02011	81647	91646	69179	14194	62590	36207	20969	99570	91291	90700
2	22368	46573	25595	85393	30995	89198	27982	53402	93965	34095	52666	19174	39615	99505
3	24130	48360	22527	97265	76393	64809	15179	24830	49340	32081	30680	19655	63348	58629
4	42167	93093	06243	61680	07856	16376	39440	53537	71341	57004	00849	74917	97758	16379
5	37570	39975	81837	16656	06121	91782	60468	81305	49684	60672	14110	06927	01263	54613
6	77921	06907	11008	42751	27756	53498	18602	70659	90655	15053	21916	81825	44394	42880
7	99562	72905	56420	69994	98872	31016	71194	18738	44013	48840	63213	21069	10634	12952
8	96301	91977	05463	07972	18876	20922	94595	56869	69014	60045	18425	84903	42508	32307
9	89579	14342	63661	10281	17453	18103	57740	84378	25331	12566	58678	44947	05585	56941
10	85475	36857	53342	53988	53060	59533	38867	62300	08158	17983	16439	11458	18593	64952
11	28918	69578	88231	33276	70997	79936	56865	05859	90106	31595	01547	85590	91610	78188
12	63553	40961	48235	03427	49626	69445	18663	72695	52180	20847	12234	90511	33703	90322
13	09429	93969	52636	92737	88974	33488	36320	17617	30015	08272	84115	27156	30613	74952
14	10365	61129	87529	85689	48237	52267	67689	93394	01511	26358	85104	20285	29975	89868
15	07119	97336	71048	08178	77233	13916	47564	81056	97735	85977	29372	74461	28551	90707
16	51085	12765	51821	51259	77452	16308	60756	92144	49442	53900	70960	63990	75601	40719
17	02368	21382	52404	60268	89368	19885	55322	44819	01188	65255	64835	44919	05944	55157
18	01011	54092	33362	94904	31273	04146	18594	29852	71585	85030	51132	01915	92747	64951
19	52162	53916	46369	58586	23216	14513	83149	98736	23495	64350	94738	17752	35156	35749
20	07056	97628	33787	09998	42698	06691	76988	13602	51851	46104	88916	19509	25625	58104
21	48663	91245	85828	14346	09172	30168	90229	04734	59193	22178	30421	61666	99904	32812
22	54164	58492	22421	74103	47070	25306	76468	26384	58151	06646	21524	15227	96909	44592
23	32639	32363	05597	24200	13363	38005	94342	28728	35806	06912	17012	64161	18296	22851
24	29334	27001	87637	87308	58731	00256	45834	15398	46557	41135	10367	07684	36188	18510
25	[illegible]	[illegible]	[illegible]	[illegible]	[illegible]	[illegible]	[illegible]	[illegible]	[illegible]	[illegible]	[illegible]	[illegible]	[illegible]	[illegible]

26	81525	72295	04839	96423	24878	82651	66566	14778	76797	14780	13300	87074	79666	95725
27	29676	20591	68086	26432	46901	20849	89768	81536	86645	12659	92259	57102	80428	25280
28	00742	57392	39064	66432	84673	40027	32832	61362	98947	96067	64760	64584	96096	98253
29	05366	04213	25669	26422	44407	44048	37937	63904	45766	66134	75470	66520	34693	90449
30	91921	26418	64117	94305	26766	25940	39972	22209	71500	64568	91402	42416	07844	69618
31	00582	04711	87917	77341	42206	35126	74087	99547	81817	42607	43808	76655	62028	76630
32	00725	69884	62797	56170	86324	88072	76222	36086	84637	93161	76038	65855	77919	88006
33	69011	65795	95876	55293	18988	27354	26575	08625	40801	59920	29841	80150	12777	48501
34	25976	57948	29888	88604	67917	48708	18912	82271	65424	69774	33611	54262	85963	03547
35	09763	83473	73577	12908	30883	18317	28290	35797	05998	41688	34952	37888	38917	88050
36	91567	42595	27958	30134	04024	86385	29880	99730	55536	84855	29080	09250	79656	73211
37	17955	56349	90999	49127	20044	59931	06115	20542	18059	02008	73708	83517	36103	42791
38	46503	18584	18845	49618	02304	51038	20655	58727	28168	15475	56942	53389	20562	87338
39	92157	89634	94824	78171	84610	82834	09922	25417	44137	48413	25555	21246	35509	20468
40	14577	62765	35605	81263	39667	47358	56873	56307	61607	49518	89686	20103	77490	18062
41	98427	07523	33362	64270	01638	92477	66969	98420	04880	45585	46565	04102	46880	45709
42	34914	63976	88720	82765	34476	17032	87589	40836	32427	70002	70663	88863	77775	69348
43	70060	28277	39475	46473	23219	53416	94970	25832	69975	94884	19661	72828	00102	66794
44	53976	54914	06990	67245	68350	82948	11398	42878	80287	88267	47363	46634	06541	97809
45	76072	29515	40980	07391	58745	25774	22987	80059	39911	96189	41151	14222	60697	59583
46	90725	52210	83974	29992	65831	38857	50490	83765	55657	14361	31720	57375	56228	41546
47	64364	67412	33339	31926	14883	24413	59744	92351	97473	89286	35931	04110	23726	51900
48	08962	00358	31662	25388	61642	34072	81249	35648	56891	69352	48373	45578	78547	81788
49	95012	68379	93526	70765	10592	04542	76463	54328	02349	17247	28865	14777	62730	92277
50	15664	10493	20492	38391	91132	21999	59516	81652	27195	48223	46751	22923	32261	85653

Source: Table of 105,000 Random Decimal Digits, Statement No. 4914, File No. 261-A-1, Interstate Commerce Commission, Washington, D.C., May 1949, Page 1.

TABLE B Coefficients of Terms for the Binomial Expansion

n/terms	1	2	3	4	5	6	7	8	9	10	11	12	13	14	15	16
1	1	1														
2	1	2	1													
3	1	3	3	1												
4	1	4	6	4	1											
5	1	5	10	10	5	1										
6	1	6	15	20	15	6	1									
7	1	7	21	35	35	21	7	1								
8	1	8	28	56	70	56	28	8	1							
9	1	9	36	84	126	126	84	36	9	1						
10	1	10	45	120	210	252	210	120	45	10	1					
11	1	11	55	165	330	462	462	330	165	55	11	1				
12	1	12	66	220	495	792	924	792	495	220	66	12	1			
13	1	13	78	286	715	1287	1716	1716	1287	715	286	78	13	1		
14	1	14	91	364	1001	2002	3003	3432	3003	2002	1001	364	91	14	1	
15	1	15	105	455	1365	3003	5005	6435	6435	5005	3003	1365	455	105	15	1

Powers of	2	3	4	5
2	4	9	16	25
3	8	27	64	125
4	16	81	256	625
5	32	243	1,024	3,125
6	64	729	4,096	15,625
7	128	2,187	16,384	78,125
8	256	6,561	65,536	390,625
9	512	19,683	262,144	1,953,125
10	1,024	59,049	1,048,576	9,765,625
11	2,048	177,147	4,194,304	48,828,125
12	4,096	531,441	16,777,216	244,140,625
13	8,192	1,594,323	67,109,264	1,220,703,125
14	16,384	4,782,969	268,437,056	6,103,515,625
15	32,768	14,348,909	1,073,748,224	30,517,578,125

TABLE C Table of χ^2

Degrees of Freedom df	P = .99	.98	.95	.90	.80	.70	.50	.30	.20	.10	.05	.02	.01
1	.000157	.000628	.00393	.0158	.0642	.148	.455	1.074	1.642	2.706	3.841	5.412	6.635
2	.0201	.0404	.103	.211	.446	.713	1.386	2.408	3.219	4.605	5.991	7.824	9.210
3	.115	.185	.352	.584	1.005	1.424	2.366	3.665	4.642	6.251	7.815	9.837	11.341
4	.297	.429	.711	1.064	1.649	2.195	3.357	4.878	5.989	7.779	9.488	11.668	13.277
5	.554	.752	1.145	1.610	2.343	3.000	4.351	6.064	7.289	9.236	11.070	13.388	15.086
6	.872	1.134	1.635	2.204	3.070	3.828	5.348	7.231	8.558	10.645	12.592	15.033	16.812
7	1.239	1.564	2.167	2.833	3.822	4.671	6.346	8.383	9.803	12.017	14.067	16.622	18.475
8	1.646	2.032	2.733	3.490	4.594	5.527	7.344	9.524	11.030	13.362	15.507	18.168	20.090
9	2.088	2.532	3.325	4.168	5.380	6.393	8.343	10.656	12.242	14.684	16.919	19.679	21.666
10	2.558	3.059	3.940	4.865	6.179	7.267	9.342	11.781	13.442	15.987	18.307	21.161	23.209
11	3.053	3.609	4.575	5.578	6.989	8.148	10.341	12.899	14.631	17.275	19.675	22.618	24.725
12	3.571	4.178	5.226	6.304	7.807	9.034	11.340	14.011	15.812	18.549	21.026	24.054	26.217
13	4.107	4.765	5.892	7.042	8.634	9.926	12.340	15.119	16.985	19.812	22.362	25.472	27.688
14	4.660	5.368	6.571	7.790	9.467	10.821	13.339	16.222	18.151	21.064	23.685	26.873	29.141
15	5.229	5.985	7.261	8.547	10.307	11.721	14.339	17.322	19.311	22.307	24.996	28.259	30.578
16	5.812	6.614	7.962	9.312	11.152	12.624	15.338	18.418	20.465	23.542	26.296	29.633	32.000
17	6.408	7.255	8.672	10.085	12.002	13.531	16.338	19.511	21.615	24.769	27.587	30.995	33.409
18	7.015	7.906	9.390	10.865	12.857	14.440	17.338	20.601	22.760	25.989	28.869	32.346	34.805
19	7.633	8.567	10.117	11.651	13.716	15.352	18.338	21.689	23.900	27.204	30.144	33.687	36.191
20	8.260	9.237	10.851	12.443	14.578	16.266	19.337	22.775	25.038	28.412	31.410	35.020	37.566
21	8.897	9.915	11.591	13.240	15.445	17.182	20.337	23.858	26.171	29.615	32.671	36.343	38.932
22	9.542	10.600	12.338	14.041	16.314	18.101	21.337	24.939	27.301	30.813	33.924	37.659	40.289
23	10.196	11.293	13.091	14.848	17.187	19.021	22.337	26.018	28.429	32.007	35.172	38.968	41.638
24	10.856	11.992	13.848	15.659	18.062	19.943	23.337	27.096	29.553	33.196	36.415	40.270	42.980
25	11.524	12.697	14.611	16.473	18.940	20.867	24.337	28.172	30.675	34.382	37.652	41.566	44.314
26	12.198	13.409	15.379	17.292	19.820	21.792	25.336	29.246	31.795	35.563	38.885	42.856	45.642
27	12.879	14.125	16.151	18.114	20.703	22.719	26.336	30.319	32.912	36.741	40.113	44.140	46.963
28	13.565	14.847	16.928	18.939	21.588	23.647	27.336	31.391	34.027	37.916	41.337	45.419	48.278
29	14.256	15.574	17.708	19.768	22.475	24.577	28.336	32.461	35.139	39.087	42.557	46.693	49.588
30	14.953	16.306	18.493	20.599	23.364	25.508	29.336	33.530	36.250	40.256	43.773	47.962	50.892

Table C is reprinted from Table III of Fisher: Statistical Methods for Research Workers, Oliver & Boyd Ltd., Edinburgh, by permission of the author and publishers.

For larger values of df, the expression $\sqrt{2x^3} - \sqrt{2(df) - 1}$ may be used as a normal deviate with unit standard error.

TABLE D Critical Values for Duncan's New Multiple Range Test $\alpha = .01$

df/K	2	3	4	5	6	7	8	9	10	12	14	16
2	14.04											
3	8.201	8.321										
4	6.512	6.677	6.740									
5	5.702	5.893	5.989	6.040								
6	5.243	5.439	5.549	5.614	5.655							
7	4.949	5.145	5.260	5.334	5.383	5.416						
8	4.746	4.939	5.057	5.135	5.189	5.227	5.256					
9	4.596	4.787	4.906	4.986	5.043	5.086	5.118	5.142				
10	4.482	4.671	4.790	4.871	4.931	4.975	5.010	5.037	5.058			
11	4.392	4.579	4.697	4.780	4.841	4.887	4.924	4.952	4.975			
12	4.320	4.504	4.622	4.706	4.767	4.815	4.852	4.883	4.907	4.944		
13	4.260	4.442	4.560	4.644	4.706	4.755	4.793	4.824	4.850	4.889		
14	4.210	4.391	4.508	4.591	4.654	4.704	4.743	4.775	4.802	4.843	4.872	
15	4.168	4.347	4.463	4.547	4.610	4.660	4.700	4.733	4.760	4.803	4.834	
16	4.131	4.309	4.425	4.509	4.572	4.622	4.663	4.696	4.724	4.768	4.800	4.825
17	4.099	4.275	4.391	4.475	4.539	4.589	4.630	4.664	4.693	4.738	4.771	4.797
18	4.071	4.246	4.362	4.445	4.509	4.560	4.601	4.635	4.664	4.711	4.745	4.772
19	4.046	4.220	4.335	4.419	4.483	4.534	4.575	4.610	4.639	4.686	4.722	4.749
20	4.024	4.197	4.312	4.395	4.459	4.510	4.552	4.587	4.617	4.664	4.701	4.729
24	3.956	4.126	4.239	4.322	4.386	4.437	4.480	4.516	4.546	4.596	4.634	4.665
30	3.889	4.056	4.168	4.250	4.314	4.366	4.409	4.445	4.477	4.528	4.569	4.601
40	3.825	3.988	4.098	4.180	4.244	4.296	4.339	4.376	4.408	4.461	4.503	4.537
60	3.762	3.922	4.031	4.111	4.174	4.226	4.270	4.307	4.340	4.394	4.438	4.474
120	3.702	3.858	3.965	4.044	4.107	4.158	4.202	4.239	4.272	4.327	4.372	3.410
∞	3.643	3.796	3.900	3.978	4.040	4.091	4.135	4.172	4.205	4.261	4.307	4.345

TABLE D **Critical Values for Duncan's New Multiple Range Test $\alpha = .05$ (Continued)**

df/K	2	3	4	5	6	7	8	9	10	12	14	16
2	6.085											
3	4.501	4.516										
4	3.927	4.013	4.033									
5	3.635	3.749	3.797	3.814								
6	3.461	3.587	3.649	3.680	3.694							
7	3.344	3.477	3.548	3.588	3.611	3.622						
8	3.261	3.399	3.475	3.521	3.549	3.566	3.575					
9	3.199	3.339	3.420	3.470	3.502	3.523	3.536	3.544				
10	3.151	3.293	3.376	3.430	3.465	3.489	3.505	3.516	3.522			
11	3.113	3.256	3.342	3.397	3.435	3.462	3.480	3.493	3.501			
12	3.082	3.225	3.313	3.370	3.410	3.439	3.459	3.474	3.484	3.496		
13	3.055	3.200	3.289	3.348	3.389	3.419	3.442	3.458	3.470	3.484		
14	3.033	3.178	3.268	3.329	3.372	3.403	3.426	3.444	3.457	3.474	3.482	
15	3.014	3.160	3.250	3.312	3.356	3.389	3.413	3.432	3.446	3.465	3.476	
16	2.998	3.144	3.235	3.298	3.343	3.376	3.402	3.422	3.437	3.458	3.470	3.477
17	2.984	3.130	3.222	3.285	3.331	3.366	3.392	3.412	3.429	3.451	3.465	3.473
18	2.971	3.118	3.210	3.274	3.321	3.356	3.383	3.405	3.421	3.445	3.460	3.470
19	2.960	3.107	3.199	3.264	3.311	3.347	3.375	3.397	3.415	3.440	3.456	3.467
20	2.950	3.097	3.190	3.255	3.303	3.339	3.368	3.391	3.409	3.436	3.453	3.464
24	2.919	3.066	3.160	3.226	3.276	3.315	3.345	3.370	3.390	3.420	3.441	3.456
30	2.888	3.035	3.131	3.199	3.250	3.290	3.322	3.349	3.371	3.405	3.430	3.447
40	2.858	3.006	3.102	3.171	3.224	3.266	3.300	3.328	3.352	3.390	3.418	3.439
60	2.829	2.976	3.073	3.143	3.198	3.241	3.277	3.307	3.333	3.374	3.406	3.431
120	2.800	2.947	3.045	3.116	3.172	3.217	3.254	3.287	3.314	3.359	3.394	3.423
∞	2.772	2.918	2.017	3.089	3.146	3.193	3.232	3.265	3.294	3.343	3.382	3.414

Table D is reprinted from values reported by H. Leon Harter, Critical Values for Duncan's New Multiple Range Test. *Biometrics*, 1960, **16,** 671–685 with the kind permission of the author and the editor of *Biometrics*.

TABLE F Critical Values of F. Values Marking 5% (Roman Type) and 1% (Bold Face Type) of the Right Tail of the F Distribution

Degrees of freedom for lesser mean square	Degrees of freedom for greater mean square																							
	1	2	3	4	5	6	7	8	9	10	11	12	14	16	20	24	30	40	50	75	100	200	500	∞
1	161	200	216	225	230	234	237	239	241	242	243	244	245	246	248	249	250	251	252	253	253	254	254	254
	4,052	**4,999**	**5,403**	**5,625**	**5,764**	**5,859**	**5,928**	**5,981**	**6,022**	**6,056**	**6,082**	**6,106**	**6,142**	**6,169**	**6,208**	**6,234**	**6,258**	**6,286**	**6,302**	**6,323**	**6,334**	**6,352**	**6,361**	**6,366**
2	18.51	19.00	19.16	19.25	19.30	19.33	19.36	19.37	19.38	19.39	19.40	19.41	19.42	19.43	19.44	19.45	19.46	19.47	19.47	19.48	19.49	19.49	19.50	19.50
	98.49	**99.00**	**99.17**	**99.25**	**99.30**	**99.33**	**99.34**	**99.36**	**99.38**	**99.40**	**99.41**	**99.42**	**99.43**	**99.44**	**99.45**	**99.46**	**99.47**	**99.48**	**99.48**	**99.49**	**99.49**	**99.49**	**99.50**	**99.50**
3	10.13	9.55	9.28	9.12	9.01	8.94	8.88	8.84	8.81	8.78	8.76	8.74	8.71	8.69	8.66	8.64	8.62	8.60	8.58	8.57	8.56	8.54	8.54	8.53
	34.12	**30.82**	**29.46**	**28.71**	**28.24**	**27.91**	**27.67**	**27.49**	**27.34**	**27.23**	**27.13**	**27.05**	**26.92**	**26.83**	**26.69**	**26.60**	**26.50**	**26.41**	**26.35**	**26.27**	**26.23**	**26.18**	**26.14**	**26.12**
4	7.71	6.94	6.59	6.39	6.26	6.16	6.09	6.04	6.00	5.96	5.93	5.91	5.87	5.84	5.80	5.77	5.74	5.71	5.70	5.68	5.66	5.65	5.64	5.63
	21.20	**18.00**	**16.69**	**15.98**	**15.52**	**15.21**	**14.98**	**14.80**	**14.66**	**14.54**	**14.45**	**14.37**	**14.24**	**14.15**	**14.02**	**13.93**	**13.83**	**13.74**	**13.69**	**13.61**	**13.57**	**13.52**	**13.48**	**13.46**
5	6.61	5.79	5.41	5.19	5.05	4.95	4.88	4.82	4.78	4.74	4.70	4.68	4.64	4.60	4.54	4.53	4.50	4.46	4.44	4.42	4.40	4.38	4.73	4.36
	16.26	**13.27**	**12.06**	**11.39**	**10.97**	**10.67**	**10.45**	**10.27**	**10.15**	**10.05**	**9.96**	**9.89**	**9.77**	**9.68**	**9.55**	**9.47**	**9.38**	**9.29**	**9.24**	**9.17**	**9.13**	**9.07**	**9.04**	**9.02**
6	5.99	5.14	4.76	4.53	4.39	4.28	4.21	4.15	4.10	4.06	4.03	4.00	3.96	3.92	3.87	3.84	3.81	3.77	3.75	3.72	3.71	3.69	3.68	3.67
	13.74	**10.92**	**9.78**	**9.15**	**8.75**	**8.47**	**8.26**	**8.10**	**7.98**	**7.87**	**7.79**	**7.72**	**7.60**	**7.52**	**7.39**	**7.31**	**7.23**	**7.14**	**7.09**	**7.02**	**6.99**	**6.94**	**6.90**	**6.88**
7	5.59	4.74	4.35	4.12	3.97	3.87	3.79	3.73	3.68	3.63	3.60	3.57	3.52	3.49	3.44	3.41	3.38	3.34	3.32	3.29	3.28	3.25	3.24	3.23
	12.25	**9.55**	**8.45**	**7.85**	**7.46**	**7.19**	**7.00**	**6.84**	**6.71**	**6.62**	**6.54**	**6.47**	**6.35**	**6.27**	**6.15**	**6.07**	**5.98**	**5.90**	**5.85**	**5.78**	**5.75**	**5.70**	**5.67**	**5.65**
8	5.32	4.46	4.07	3.84	3.69	3.58	3.50	3.44	3.39	3.34	3.31	3.28	3.23	3.20	3.15	3.12	3.08	3.05	3.03	3.00	2.98	2.96	2.94	2.93
	11.26	**8.65**	**7.59**	**7.01**	**6.63**	**6.37**	**6.19**	**6.03**	**5.91**	**5.82**	**5.74**	**5.67**	**5.56**	**5.48**	**5.36**	**5.28**	**5.20**	**5.11**	**5.06**	**5.00**	**4.96**	**4.91**	**4.88**	**4.86**
9	5.12	4.26	3.86	3.63	3.48	3.37	3.29	3.23	3.18	3.13	3.10	3.07	3.02	2.98	2.93	2.90	2.86	2.82	2.80	2.77	2.76	2.73	2.72	2.71
	10.56	**8.02**	**6.99**	**6.42**	**6.06**	**5.80**	**5.62**	**5.47**	**5.35**	**5.26**	**5.18**	**5.11**	**5.00**	**4.92**	**4.80**	**4.73**	**4.64**	**4.56**	**4.51**	**4.45**	**4.41**	**4.36**	**4.33**	**4.31**
10	4.96	4.10	3.71	3.48	3.33	3.22	3.14	3.07	3.02	2.97	2.94	2.91	2.86	2.82	2.77	2.74	2.70	2.67	2.64	2.61	2.59	2.56	2.55	2.54
	10.04	**7.56**	**6.55**	**5.99**	**5.64**	**5.39**	**5.21**	**5.06**	**4.95**	**4.85**	**4.78**	**4.71**	**4.60**	**4.52**	**4.41**	**4.33**	**4.25**	**4.17**	**4.12**	**4.05**	**4.01**	**3.96**	**3.93**	**3.91**

11	4.84	3.98	3.59	3.36	3.20	3.09	3.01	2.95	2.90	2.86	2.82	2.79	2.74	2.70	2.65	2.61	2.57	2.53	2.50	2.47	2.45	2.42	2.41	2.40
	9.65	**7.20**	**6.22**	**5.67**	**5.32**	**5.07**	**4.88**	**4.74**	**4.63**	**4.54**	**4.46**	**4.40**	**4.29**	**4.21**	**4.10**	**4.02**	**3.94**	**3.86**	**3.80**	**3.74**	**3.70**	**3.66**	**3.62**	**3.60**
12	4.75	3.88	3.49	3.26	3.11	3.00	2.92	2.85	2.80	2.76	2.72	2.69	2.64	2.60	2.54	2.50	2.46	2.42	2.40	2.36	2.35	2.32	2.31	2.30
	9.33	**6.93**	**5.95**	**5.41**	**5.06**	**4.82**	**4.65**	**4.50**	**4.39**	**4.30**	**4.22**	**4.16**	**4.05**	**3.98**	**3.86**	**3.78**	**3.70**	**3.61**	**3.56**	**3.49**	**3.46**	**3.41**	**3.38**	**3.36**
13	4.67	3.80	3.41	3.18	3.02	2.92	2.84	2.77	2.72	2.67	2.63	2.60	2.55	2.51	2.46	2.42	2.38	2.34	2.32	2.28	2.26	2.24	2.22	2.21
	9.07	**6.70**	**5.74**	**5.20**	**4.86**	**4.62**	**4.44**	**4.30**	**4.19**	**4.10**	**4.02**	**3.96**	**3.85**	**3.78**	**3.67**	**3.59**	**3.51**	**3.42**	**3.37**	**3.30**	**3.27**	**3.21**	**3.18**	**3.16**
14	4.60	3.74	3.34	3.11	2.96	2.85	2.77	2.70	2.65	2.60	2.56	2.53	2.48	2.44	2.39	2.35	2.31	2.27	2.24	2.21	2.19	2.16	2.14	2.13
	8.86	**6.51**	**5.56**	**5.03**	**4.69**	**4.46**	**4.28**	**4.14**	**4.03**	**3.94**	**3.86**	**3.80**	**3.70**	**3.62**	**3.51**	**3.43**	**3.34**	**3.26**	**3.21**	**3.14**	**3.11**	**3.06**	**3.02**	**3.00**
15	4.54	3.68	3.29	3.06	2.90	2.79	2.70	2.64	2.59	2.55	2.51	2.48	2.43	2.39	2.33	2.29	2.25	2.21	2.18	2.15	2.12	2.10	2.08	2.07
	8.68	**6.36**	**5.42**	**4.89**	**4.56**	**4.32**	**4.14**	**4.00**	**3.89**	**3.80**	**3.73**	**3.67**	**3.56**	**3.48**	**3.36**	**3.29**	**3.20**	**3.12**	**3.07**	**3.00**	**2.97**	**2.92**	**2.89**	**2.87**
16	4.49	3.63	3.24	3.01	2.85	2.74	2.66	2.59	2.54	2.49	2.45	2.42	2.37	2.33	2.28	2.24	2.20	2.16	2.13	2.09	2.07	2.04	2.02	2.01
	8.53	**6.23**	**5.29**	**4.77**	**4.44**	**4.20**	**4.03**	**3.89**	**3.78**	**3.69**	**3.61**	**3.55**	**3.45**	**3.37**	**3.25**	**3.18**	**3.10**	**3.01**	**2.96**	**2.89**	**2.86**	**2.80**	**2.77**	**2.75**
17	4.45	3.59	3.20	2.96	2.81	2.70	2.62	2.55	2.50	2.45	2.41	2.38	2.33	2.29	2.23	2.19	2.15	2.11	2.08	2.04	2.02	1.99	1.97	1.96
	8.40	**6.11**	**5.18**	**4.67**	**4.34**	**4.10**	**3.93**	**3.79**	**3.68**	**3.59**	**3.52**	**3.45**	**3.35**	**3.27**	**3.16**	**3.08**	**3.00**	**2.92**	**2.86**	**2.79**	**2.76**	**2.70**	**2.67**	**2.65**
18	4.41	3.55	3.16	2.93	2.77	2.66	2.58	2.51	2.46	2.41	2.37	2.34	2.29	2.25	2.19	2.15	2.11	2.07	2.04	2.00	1.98	1.95	1.93	1.92
	8.28	**6.01**	**5.09**	**4.58**	**4.25**	**4.01**	**3.85**	**3.71**	**3.60**	**3.51**	**3.44**	**3.37**	**3.27**	**3.19**	**3.07**	**3.00**	**2.91**	**2.83**	**2.78**	**2.71**	**2.68**	**2.62**	**2.59**	**2.57**
19	4.38	3.52	3.13	2.90	2.74	2.63	2.55	2.48	2.43	2.38	2.34	2.31	2.26	2.21	2.15	2.11	2.07	2.02	2.00	1.96	1.94	1.91	1.90	1.88
	8.18	**5.93**	**5.01**	**4.50**	**4.17**	**3.94**	**3.77**	**3.63**	**3.52**	**3.43**	**3.36**	**3.30**	**3.19**	**3.12**	**3.00**	**2.92**	**2.84**	**2.76**	**2.70**	**2.63**	**2.60**	**2.54**	**2.51**	**2.49**
20	4.35	3.49	3.10	2.87	2.71	2.60	2.52	2.45	2.40	2.35	2.31	2.28	2.23	2.18	2.12	2.08	2.04	1.99	1.96	1.92	1.90	1.87	1.85	1.84
	8.10	**5.85**	**4.94**	**4.43**	**4.10**	**3.87**	**3.71**	**3.56**	**3.45**	**3.37**	**3.30**	**3.23**	**3.13**	**3.05**	**2.94**	**2.86**	**2.77**	**2.69**	**2.63**	**2.56**	**2.53**	**2.47**	**2.44**	**2.42**
21	4.32	3.47	3.07	2.84	2.68	2.57	2.49	2.42	2.37	2.32	2.28	2.25	2.20	2.15	2.09	2.05	2.00	1.96	1.93	1.89	1.87	1.84	1.82	1.81
	8.02	**5.78**	**4.87**	**4.37**	**4.04**	**3.81**	**3.65**	**3.51**	**3.40**	**3.31**	**3.24**	**3.17**	**3.07**	**2.99**	**2.88**	**2.80**	**2.72**	**2.63**	**2.58**	**2.51**	**2.47**	**2.42**	**2.38**	**2.36**
22	4.30	3.44	3.05	2.82	2.66	2.55	2.47	2.40	2.35	2.30	2.26	2.23	2.18	2.13	2.07	2.03	1.98	1.93	1.91	1.87	1.84	1.81	1.80	1.78
	7.94	**5.72**	**4.82**	**4.31**	**3.99**	**3.76**	**3.59**	**3.45**	**3.35**	**3.26**	**3.18**	**3.12**	**3.02**	**2.94**	**2.83**	**2.75**	**2.67**	**2.58**	**2.53**	**2.46**	**2.42**	**2.37**	**2.33**	**2.31**
23	4.28	3.42	3.03	2.80	2.64	2.53	2.45	2.38	2.32	2.28	2.24	2.20	2.14	2.10	2.04	2.00	1.96	1.91	1.88	1.84	1.82	1.79	1.77	1.76
	7.88	**5.66**	**4.76**	**4.26**	**3.94**	**3.71**	**3.54**	**3.41**	**3.30**	**3.21**	**3.14**	**3.07**	**2.97**	**2.89**	**2.78**	**2.70**	**2.62**	**2.53**	**2.48**	**2.41**	**2.37**	**2.32**	**2.28**	**2.26**
24	4.26	3.40	3.01	2.78	2.62	2.51	2.43	2.36	2.30	2.26	2.22	2.18	2.13	2.09	2.02	1.98	1.94	1.89	1.86	1.82	1.80	1.76	1.74	1.73
	7.82	**5.61**	**4.72**	**4.22**	**3.90**	**3.67**	**3.50**	**3.36**	**3.25**	**3.17**	**3.09**	**3.03**	**2.93**	**2.85**	**2.74**	**2.66**	**2.58**	**2.49**	**2.44**	**2.36**	**2.33**	**2.27**	**2.23**	**2.21**
25	4.24	3.38	2.99	2.76	2.60	2.49	2.41	2.34	2.28	2.24	2.20	2.16	2.11	2.06	2.00	1.96	1.92	1.87	1.84	1.80	1.77	1.74	1.72	1.71
	7.77	**5.57**	**4.68**	**4.18**	**3.86**	**3.63**	**3.46**	**3.32**	**3.21**	**3.13**	**3.05**	**2.99**	**2.89**	**2.81**	**2.70**	**2.62**	**2.54**	**2.45**	**2.40**	**2.32**	**2.29**	**2.23**	**2.19**	**2.17**

Table F Critical Values of F. Values Marking 5% (Roman Type) and 1% (Bold Face Type) of the Right Tail of the F Distribution (Continued)

Degrees of freedom for lesser mean square	Degrees of freedom for greater mean square																							
	1	2	3	4	5	6	7	8	9	10	11	12	14	16	20	20	30	40	50	75	100	200	500	∞
26	4.22	3.37	2.98	2.74	2.59	2.47	2.39	2.32	2.27	2.22	2.18	2.15	2.10	2.05	1.99	1.95	1.90	1.85	1.82	1.78	1.76	1.72	1.70	1.69
	7.72	**5.53**	**4.64**	**4.14**	**3.82**	**3.59**	**3.42**	**3.29**	**3.17**	**3.09**	**3.02**	**2.96**	**2.86**	**2.77**	**2.66**	**2.58**	**2.50**	**2.41**	**2.36**	**2.28**	**2.25**	**2.19**	**2.15**	**2.13**
27	4.21	3.35	2.96	2.73	2.57	2.46	2.37	2.30	2.25	2.20	2.16	2.13	2.08	2.03	1.97	1.93	1.88	1.84	1.80	1.76	1.74	1.71	1.68	1.67
	7.68	**5.49**	**4.60**	**4.11**	**3.79**	**3.56**	**3.39**	**3.26**	**3.14**	**3.06**	**2.98**	**2.93**	**2.83**	**2.74**	**2.63**	**2.55**	**2.47**	**2.38**	**2.33**	**2.25**	**2.21**	**2.16**	**2.12**	**2.10**
28	4.20	3.34	2.95	2.71	2.56	2.44	2.36	2.29	2.24	2.19	2.15	2.12	2.06	2.02	1.96	1.91	1.87	1.81	1.78	1.75	1.72	1.69	1.67	1.65
	7.64	**5.45**	**4.57**	**4.07**	**3.76**	**3.53**	**3.36**	**3.23**	**3.11**	**3.03**	**2.95**	**2.90**	**2.80**	**2.71**	**2.60**	**2.52**	**2.44**	**2.35**	**2.30**	**2.22**	**2.18**	**2.13**	**2.09**	**2.06**
29	4.18	3.33	2.93	2.70	2.54	2.43	2.35	2.28	2.22	2.18	2.14	2.10	2.05	2.00	1.94	1.90	1.85	1.80	1.77	1.73	1.71	1.68	1.65	1.64
	7.60	**5.42**	**4.54**	**4.04**	**3.73**	**3.50**	**3.33**	**3.20**	**3.08**	**3.00**	**2.92**	**2.87**	**2.77**	**2.68**	**2.57**	**2.49**	**2.41**	**2.32**	**2.27**	**2.19**	**2.15**	**2.10**	**2.06**	**2.03**
30	4.17	3.32	2.92	2.69	2153	2.42	2.34	2.27	2.21	2.16	2.12	2.09	2.04	1.99	1.93	1.89	1.84	1.79	1.76	1.72	1.69	1.66	1.64	1.62
	7.56	**5.39**	**4.51**	**4.02**	**3.70**	**3.47**	**3.30**	**3.17**	**3.06**	**2.98**	**2.90**	**2.84**	**2.74**	**2.66**	**2.55**	**2.47**	**2.38**	**2.29**	**2.24**	**2.16**	**2.13**	**2.07**	**2.03**	**2.01**
32	4.15	3.30	2.90	2.67	2.51	2.40	2.32	2.25	2.19	2.14	2.10	2.07	2.02	1.97	1.91	1.86	1.82	1.76	1.74	1.69	1.67	1.64	1.61	1.59
	7.50	**5.34**	**4.46**	**3.97**	**3.66**	**3.42**	**3.25**	**3.12**	**3.01**	**2.94**	**2.86**	**2.80**	**2.70**	**2.62**	**2.51**	**2.42**	**2.34**	**2.25**	**2.20**	**2.12**	**2.08**	**2.02**	**1.98**	**1.96**
34	4.13	3.28	2.88	2.65	2.49	2.38	2.30	2.23	2.17	2.12	2.08	2.05	2.00	1.95	1.89	1.84	1.80	1.74	1.71	1.67	1.64	1.61	1.59	1.57
	7.44	**5.29**	**4.42**	**3.93**	**3.61**	**3.38**	**3.21**	**3.08**	**2.97**	**2.89**	**2.82**	**2.76**	**2.66**	**2.58**	**2.47**	**2.38**	**2.30**	**2.21**	**2.15**	**2.08**	**2.04**	**1.98**	**1.94**	**1.91**
36	4.11	3.26	2.86	2.63	2.48	2.36	2.28	2.21	2.15	2.10	2.06	2.03	1.98	1.93	1.87	1.82	1.78	1.72	1.69	1.65	1.62	1.59	1.56	1.55
	7.39	**5.25**	**4.38**	**3.89**	**3.58**	**3.35**	**3.18**	**3.04**	**2.94**	**2.86**	**2.78**	**2.72**	**2.62**	**2.54**	**2.43**	**2.35**	**2.26**	**2.17**	**2.12**	**2.04**	**2.00**	**1.94**	**1.90**	**1.87**
38	4.10	3.25	2.85	2.62	2.46	2.35	2.26	2.19	2.14	2.09	2.05	2.02	1.96	1.92	1.85	1.80	1.76	1.71	1.67	1.63	1.60	1.57	1.54	1.53
	7.35	**5.21**	**4.34**	**3.86**	**3.54**	**3.32**	**3.15**	**3.02**	**2.91**	**2.82**	**2.75**	**2.69**	**2.59**	**2.51**	**2.40**	**2.32**	**2.22**	**2.14**	**2.08**	**2.00**	**1.97**	**1.90**	**1.86**	**1.84**
40	4.08	3.23	2.84	2.61	2.45	2.34	2.25	2.18	2.12	2.07	2.04	2.00	1.95	1.90	1.84	1.79	1.74	1.69	1.66	1.61	1.59	1.55	1.53	1.51
	7.31	**5.18**	**4.31**	**3.83**	**3.51**	**3.29**	**3.12**	**2.99**	**2.88**	**2.80**	**2.73**	**2.66**	**2.56**	**2.49**	**2.37**	**2.29**	**2.20**	**2.11**	**2.05**	**1.97**	**1.94**	**1.88**	**1.84**	**1.81**
42	4.07	3.22	2.83	2.59	2.44	2.32	2.24	2.17	2.11	2.06	2.02	1.99	1.94	1.89	1.82	1.78	1.73	1.68	1.64	1.60	1.57	1.54	1.51	1.49
	7.27	**5.15**	**4.29**	**3.80**	**3.49**	**3.26**	**3.10**	**2.96**	**2.86**	**2.77**	**2.70**	**2.64**	**2.54**	**2.46**	**2.35**	**2.26**	**2.17**	**2.08**	**2.02**	**1.94**	**1.91**	**1.85**	**1.80**	**1.78**
44	4.06	3.21	2.82	2.58	2.43	2.31	2.23	2.16	2.10	2.05	2.01	1.98	1.92	1.88	1.81	1.76	1.72	1.66	1.63	1.58	1.56	1.52	1.50	1.48
	7.24	**5.12**	**4.26**	**3.78**	**3.46**	**3.24**	**3.07**	**2.94**	**2.84**	**2.75**	**2.68**	**2.62**	**2.52**	**2.44**	**2.32**	**2.24**	**2.15**	**2.06**	**2.00**	**1.92**	**1.88**	**1.82**	**1.78**	**1.75**

46	4.05	3.20	2.81	2.57	2.42	2.30	2.22	2.14	2.09	2.04	2.00	1.97	1.91	1.87	1.80	1.75	1.71	1.65	1.62	1.57	1.54	1.51	1.48	1.46
	7.21	**5.10**	**4.24**	**3.76**	**3.44**	**3.22**	**3.05**	**2.92**	**2.82**	**2.73**	**2.66**	**2.60**	**2.50**	**2.42**	**2.30**	**2.22**	**2.13**	**2.04**	**1.98**	**1.90**	**1.86**	**1.80**	**1.76**	**1.72**
48	4.04	3.19	2.80	2.56	2.41	2.30	2.21	2.14	2.08	2.03	1.99	1.96	1.90	1.86	1.79	1.74	1.70	1.64	1.61	1.56	1.53	1.50	1.47	1.45
	7.19	**5.08**	**4.22**	**3.74**	**3.42**	**3.20**	**3.04**	**2.90**	**2.80**	**2.71**	**2.64**	**2.58**	**2.48**	**2.40**	**2.28**	**2.20**	**2.11**	**2.02**	**1.96**	**1.88**	**1.84**	**1.78**	**1.73**	**1.70**
50	4.03	3.18	2.79	2.56	2.40	2.29	2.20	2.13	2.07	2.02	1.98	1.95	1.90	1.85	1.78	1.74	1.69	1.63	1.60	1.55	1.52	1.48	1.46	1.44
	7.17	**5.06**	**4.20**	**3.72**	**3.41**	**3.18**	**3.02**	**2.88**	**2.78**	**2.70**	**2.62**	**2.56**	**2.46**	**2.39**	**2.26**	**2.18**	**2.10**	**2.00**	**1.94**	**1.86**	**1.82**	**1.76**	**1.71**	**1.68**
55	4.02	3.17	2.78	2.54	2.38	2.27	2.18	2.11	2.05	2.00	1.97	1.93	1.88	1.83	1.76	1.72	1.67	1.61	1.58	1.52	1.50	1.46	1.43	1.41
	7.12	**5.01**	**4.16**	**3.68**	**3.37**	**3.15**	**2.98**	**2.85**	**2.75**	**2.66**	**2.59**	**2.53**	**2.43**	**2.35**	**2.23**	**2.15**	**2.06**	**1.96**	**1.90**	**1.82**	**1.78**	**1.71**	**1.66**	**1.64**
60	4.00	3.15	2.76	2.52	2.37	2.25	2.17	2.10	2.04	1.99	1.95	1.92	1.86	1.81	1.75	1.70	1.65	1.59	1.56	1.50	1.48	1.44	1.41	1.39
	7.08	**4.98**	**4.13**	**3.65**	**3.34**	**3.12**	**2.95**	**2.82**	**2.72**	**2.63**	**2.56**	**2.50**	**2.40**	**2.32**	**2.20**	**2.12**	**2.03**	**1.93**	**1.87**	**1.79**	**1.74**	**1.68**	**1.63**	**1.60**
65	3.99	3.14	2.75	2.51	2.36	2.24	2.15	2.08	2.02	1.98	1.94	1.90	1.85	1.80	1.73	1.68	1.63	1.57	1.54	1.49	1.46	1.42	1.39	1.37
	7.04	**4.95**	**4.10**	**3.62**	**3.31**	**3.09**	**2.93**	**2.79**	**2.70**	**2.61**	**2.54**	**2.47**	**2.37**	**2.30**	**2.18**	**2.09**	**2.00**	**1.90**	**1.84**	**1.76**	**1.71**	**1.64**	**1.60**	**1.56**
70	3.98	3.13	2.74	2.50	2.35	2.23	2.14	2.07	2.01	1.97	1.93	1.89	1.84	1.79	1.72	1.67	1.62	1.56	1.53	1.47	1.45	1.40	1.37	1.35
	7.01	**4.92**	**4.08**	**3.60**	**3.29**	**3.07**	**2.91**	**2.77**	**2.67**	**2.59**	**2.51**	**2.45**	**2.35**	**2.28**	**2.15**	**2.07**	**1.98**	**1.88**	**1.82**	**1.74**	**1.69**	**1.62**	**1.56**	**1.53**
80	3.96	3.11	2.72	2.48	2.33	2.21	2.12	2.05	1.99	1.95	1.91	1.88	1.82	1.77	1.70	1.65	1.60	1.54	1.51	1.45	1.42	1.38	1.35	1.32
	6.96	**4.88**	**4.04**	**3.56**	**3.25**	**3.04**	**2.87**	**2.74**	**2.64**	**2.55**	**2.46**	**2.41**	**2.32**	**2.24**	**2.11**	**20.3**	**1.94**	**1.84**	**1.78**	**1.70**	**1.65**	**1.57**	**1.52**	**1.49**
100	3.94	3.09	2.70	2.46	2.30	2.19	2.10	2.03	1.97	1.92	1.88	1.85	1.79	1.75	1.68	1.63	1.57	1.51	1.48	1.42	1.39	1.34	1.30	1.28
	6.90	**4.82**	**3.98**	**3.51**	**3.20**	**2.99**	**2.82**	**2.69**	**2.59**	**2.51**	**2.43**	**2.86**	**2.26**	**2.19**	**2.06**	**1.98**	**1.89**	**1.79**	**1.73**	**1.64**	**1.59**	**1.51**	**1.46**	**1.43**
125	3.92	3.07	2.68	2.44	2.29	2.17	2.08	2.01	1.95	1.90	1.86	1.83	1.77	1.72	1.65	1.60	1.55	1.49	1.45	1.39	1.36	1.31	1.27	1.25
	6.84	**4.78**	**3.94**	**3.47**	**3.17**	**2.95**	**2.79**	**2.65**	**2.56**	**2.47**	**2.40**	**2.33**	**2.23**	**2.15**	**2.03**	**1.94**	**1.85**	**1.75**	**1.68**	**1.59**	**1.54**	**1.46**	**1.40**	**1.37**
150	3.9.	3.06	2.67	2.43	2.27	2.16	2.07	2.00	1.94	1.89	1.85	1.82	1.76	1.71	1.64	1.59	1.54	1.47	1.44	1.37	1.34	1.29	1.25	1.22
	6.81	**4.75**	**3.91**	**3.44**	**3.14**	**2.92**	**2.76**	**2.62**	**2.53**	**2.44**	**2.37**	**2.80**	**2.20**	**2.12**	**2.00**	**1.91**	**1.83**	**1.72**	**1.66**	**1.56**	**1.51**	**1.43**	**1.37**	**1.33**
200	3.89	3.04	2.65	2.41	2.26	2.14	2.05	1.98	1.92	1.87	1.83	1.80	1.74	1.69	1.62	1.57	1.52	1.45	1.42	1.35	1.32	1.26	1.22	1.19
	6.76	**4.71**	**3.88**	**3.41**	**3.11**	**2.90**	**2.73**	**2.60**	**2.50**	**2.41**	**2.34**	**2.28**	**2.17**	**2.09**	**1.97**	**1.88**	**1.79**	**1.69**	**1.62**	**1.55**	**1.48**	**1.39**	**1.33**	**1.28**
400	3.86	3.02	2.62	2.39	2.23	2.12	2.03	1.96	1.90	1.85	1.81	1.78	1.72	1.67	1.60	1.54	1.49	1.42	1.38	1.32	1.28	1.22	1.16	1.13
	6.70	**4.66**	**3.82**	**3.36**	**3.06**	**3.85**	**2.69**	**2.55**	**2.46**	**2.37**	**2.29**	**2.23**	**2.12**	**2.04**	**1.92**	**1.84**	**1.74**	**1.64**	**1.57**	**1.47**	**1.42**	**1.32**	**1.24**	**1.19**
1000	3.85	3.00	2.61	2.38	2.22	2.10	2.02	1.95	1.89	1.84	1.80	1.76	1.70	1.65	1.58	1.53	1.47	1.41	1.36	1.30	1.26	1.19	1.13	1.08
	6.66	**4.62**	**3.80**	**3.34**	**3.04**	**2.82**	**2.66**	**2.53**	**2.43**	**2.34**	**2.26**	**2.20**	**2.09**	**2.01**	**1.89**	**1.81**	**1.71**	**1.61**	**1.54**	**1.44**	**1.38**	**1.28**	**1.19**	**1.11**
∞	3.84	2.99	2.60	2.37	2.21	2.09	2.01	1.94	1.88	1.83	1.79	1.75	1.69	1.64	1.57	1.52	1.46	1.40	1.35	1.28	1.24	1.17	1.11	1.00
	6.64	**4.60**	**3.78**	**3.32**	**3.02**	**2.80**	**2.64**	**2.51**	**2.41**	**2.32**	**2.24**	**2.18**	**2.07**	**1.99**	**1.87**	**1.79**	**1.69**	**1.59**	**1.52**	**1.41**	**1.36**	**1.25**	**1.15**	**1.00**

Reprinted, by permission, from G. W. Snedecor, *Statistical methods*, 5th ed., pp. 246–249, Iowa State College Press, Ames, Iowa, 1956

Table H Critical Values for Hartley's Maximum F-Ratio Significance Test for Homogeneity of Variance

[*Alpha* = .05 and *.01* (in italics)]

d.f. \ k	2	3	4	5	6	7	8	9	10	11	12	k / d.f.
2	39.0 *199.*	87.5 *448.*	142. *729.*	202. *1036.*	266. *1362.*	333. *1705.*	403. *2063.*	475. *2432.*	550. *2813.*	626. *3204.*	704. *3605.*	2
3	15.4 *47.5*	27.8 *85.*	39.2 *120.*	50.7 *151.*	62.0 *184.*	72.9 *216.* *	83.5 *249.* *	93.9 *281.* *	104. *310.* *	114. *337.* *	124. *361.* *	3
4	9.60 *23.2*	15.5 *37.*	20.6 *49.*	25.2 *59.*	29.5 *69.*	33.6 *79.*	37.5 *89.*	41.1 *97.*	44.6 *106.*	48.0 *113.*	51.4 *120.*	4
5	7.15 *14.9*	10.8 *22.*	13.7 *28.*	16.3 *33.*	18.7 *38.*	20.8 *42.*	22.9 *46.*	24.7 *50.*	26.5 *54.*	28.2 *57.*	29.9 *60.*	5
6	5.82 *11.1*	8.38 *15.5*	10.4 *19.1*	12.1 *22.*	13.7 *25.*	15.0 *27.*	16.3 *30.*	17.5 *32.*	18.6 *34.*	19.7 *36.*	20.7 *37.*	6
7	4.99 *8.89*	6.94 *12.1*	8.44 *14.5*	9.70 *16.5*	10.8 *18.4*	11.8 *20.*	12.7 *22.*	13.5 *23.*	14.3 *24.*	15.1 *26.*	15.8 *27.*	7
8	4.43 *7.50*	6.00 *9.9*	7.18 *11.7*	8.12 *13.2*	9.03 *14.5*	9.78 *15.8*	10.5 *16.9*	11.1 *17.9*	11.7 *18.9*	12.2 *19.8*	12.7 *21.*	8
9	4.03 *6.54*	5.34 *8.5*	6.31 *9.9*	7.11 *11.1*	7.80 *12.1*	8.41 *13.1*	8.95 *13.9*	9.45 *14.7*	9.91 *15.3*	10.3 *16.0*	10.7 *16.6*	9

10	3.72	4.85	5.67	6.34	6.92	7.42	7.87	8.28	8.66	9.01	9.34	10
	5.85	*7.4*	*8.6*	*9.6*	*10.4*	*11.1*	*11.8*	*12.4*	*12.9*	*13.4*	*13.9*	
12	3.28	4.16	4.79	5.30	5.72	6.09	6.42	6.72	7.00	7.25	7.48	12
	4.91	*6.1*	*6.9*	*7.6*	*8.2*	*8.7*	*9.1*	*9.5*	*9.9*	*10.2*	*10.6*	
15	2.86	3.54	4.01	4.37	4.68	4.95	5.19	5.40	5.59	5.77	5.93	15
	4.07	*4.9*	*5.5*	*6.0*	*6.4*	*6.7*	*7.1*	*7.3*	*7.5*	*7.8*	*8.0*	
20	2.46	2.95	3.29	3.54	3.76	3.94	4.10	4.24	4.37	4.49	4.59	20
	3.32	*3.8*	*4.3*	*4.6*	*4.9*	*5.1*	*5.3*	*5.5*	*5.6*	*5.8*	*5.9*	
30	2.07	2.40	2.61	2.78	2.91	3.02	3.12	3.21	3.29	3.36	3.39	30
	2.63	*3.0*	*3.3*	*3.4*	*3.6*	*3.7*	*3.8*	*3.9*	*4.0*	*4.1*	*4.2*	
60	1.67	1.85	1.96	2.04	2.11	2.17	2.22	2.26	2.30	2.33	2.36	60
	1.96	*2.2*	*2.3*	*2.4*	*2.4*	*2.5*	*2.5*	*2.6*	*2.6*	*2.7*	*2.7*	
∞	1.00	1.00	1.00	1.00	1.00	1.00	1.00	1.00	1.00	1.00	1.00	∞
	1.00	*1.00*	*1.00*	*1.00*	*1.00*	*1.00*	*1.00*	*1.00*	*1.00*	*1.00*	*1.00*	

Source: Reprinted from Table 31 of E. S. Pearson and H. Ö. Hartley, *Biometrika Tables for Statisticians*, vol. 1, 2nd ed., 1958, published by the Syndics of the Cambridge University Press, London; reproduced by permission of the authors and publishers.

* Values in the column $k = 2$ and in the rows d.f. = 2 and ∞ are exact. Elsewhere the third digit may be in error by a few units for $F_{.95}$ and several units for $F_{.99}$. The third digit figures of values marked by an asterisk are the most uncertain.

TABLE N Percent of Total Area under the Normal Curve between Mean Ordinate And Ordinate at Any Given Sigma-Distance from the Mean

$\frac{x}{\sigma}$	.00	.01	.02	.03	.04	.05	.06	.07	.08	.09
0.0	00.00	00.40	00.80	01.20	01.60	01.99	02.39	02.79	03.19	03.59
0.1	03.98	04.38	04.78	05.17	05.57	05.96	06.36	06.75	07.14	07.53
0.2	07.93	08.32	08.71	09.10	09.48	09.87	10.26	10.64	11.03	11.41
0.3	11.79	12.17	12.55	12.93	13.31	13.68	14.06	14.43	14.80	15.17
0.4	15.54	15.91	16.28	16.64	17.00	17.36	17.72	18.08	18.44	18.79
0.5	19.15	19.50	19.85	20.19	20.54	20.88	21.23	21.57	21.90	22.24
0.6	22.57	22.91	23.24	23.57	23.89	24.22	24.54	24.86	25.17	25.49
0.7	25.80	26.11	26.42	26.73	27.04	27.34	27.64	27.94	28.23	28.52
0.8	28.81	29.10	29.39	29.67	29.95	30.23	30.51	30.78	31.06	31.33
0.9	31.59	31.86	32.12	32.38	32.64	32.90	33.15	33.40	33.65	33.89
1.0	34.13	34.38	34.61	34.85	35.08	35.31	35.54	35.77	35.99	36.21
1.1	36.43	36.65	36.86	37.08	37.29	37.49	37.70	37.90	38.10	38.30
1.2	38.49	38.69	38.88	39.07	39.25	39.44	39.62	39.80	39.97	40.15
1.3	40.32	40.49	40.66	40.82	40.99	41.15	41.31	41.47	41.62	41.77
1.4	41.92	42.07	42.22	42.36	42.51	42.65	42.79	42.92	43.06	43.19
1.5	43.32	43.45	43.57	43.70	43.83	43.94	44.06	44.18	44.29	44.41
1.6	44.52	44.63	44.74	44.84	44.95	45.05	45.15	45.25	45.35	45.45
1.7	45.54	45.64	45.73	45.82	45.91	45.99	46.08	46.16	46.25	46.33
1.8	46.41	46.49	46.56	46.64	46.71	46.78	46.86	46.93	46.99	47.06
1.9	47.13	47.19	47.26	47.32	47.38	47.44	47.50	47.56	47.61	47.67
2.0	47.72	47.78	47.83	47.88	47.93	47.98	48.03	48.08	48.12	48.17
2.1	48.21	48.26	48.30	48.34	48.38	48.42	48.46	48.50	48.54	48.57
2.2	48.61	48.64	48.68	48.71	48.75	48.78	48.81	48.84	48.87	48.90
2.3	48.93	48.96	48.98	49.01	49.04	49.06	49.09	49.11	49.13	49.16
2.4	49.18	49.20	49.22	49.25	49.27	49.29	49.31	49.32	49.34	49.36
2.5	49.38	49.40	49.41	49.43	49.45	49.46	49.48	49.49	49.51	49.52
2.6	49.53	49.55	49.56	49.57	49.59	49.60	49.61	49.62	49.63	49.64
2.7	49.65	49.66	49.67	49.68	49.69	49.70	49.71	49.72	49.73	49.74
2.8	49.74	49.75	49.76	49.77	49.77	49.78	49.79	49.79	49.80	49.81
2.9	49.81	49.82	49.82	49.83	49.84	49.84	49.85	49.85	49.86	49.86
3.0	49.87									
3.5	49.98									
4.0	49.997									
5.0	49.99997									

Table R Critical Values of the Correlation Coefficient

	Level of Significance for One-Tailed Test			
	.05	.025	.01	.005
	Level of Significance for Two-Tailed Test			
df	.10	.05	.02	.01
1	.9877	.9969	.9995	.9999
2	.9000	.9500	.9800	.9900
3	.8054	.8783	.9343	.9587
4	.7293	.8114	.8822	.9172
5	.6694	.7545	.8329	.8745
6	.6215	.7067	.7887	.8343
7	.5822	.6664	.7498	.7977
8	.5494	.6319	.7155	.7646
9	.5214	.6021	.6851	.7348
10	.4973	.5760	.6581	.7079
11	.4762	.5529	.6339	.6835
12	.4575	.5324	.6120	.6614
13	.4409	.5139	.5923	.6411
14	.4259	.4973	.5742	.6226
15	.4124	.4821	.5577	.6055
16	.4000	.4683	.5425	.5897
17	.3887	.4555	.5285	.5751
18	.3783	.4438	.5155	.5614
19	.3687	.4329	.5034	.5487
20	.3598	.4227	.4921	.5368
25	.3233	.3809	.4451	.4869
30	.2960	.3494	.4093	.4487
35	.2746	.3246	.3810	.4182
40	.2573	.3044	.3578	.3932
45	.2428	.2875	.3384	.3721
50	.2306	.2732	.3218	.3541
60	.2108	.2500	.2948	.3248
70	.1954	.2319	.2737	.3017
80	.1829	.2172	.2565	.2830
90	.1726	.2050	.2422	.2675
100	.1638	.1946	.2301	.2540

Table R is abridged from Table VI of Fisher and Yates: "Statistical Tables for Biological, Agricultural and Medical Research," published by Oliver and Boyd Ltd., Edinburgh, and by permission of the authors and publishers.

Table S **Squares and Square Roots**

N	N^2	$\sqrt{N}$	$\sqrt{10N}$
1.00	1.0000	1.00000	3.16228
1.01	1.0201	1.00499	3.17805
1.02	1.0404	1.00995	3.19374
1.03	1.0609	1.01489	3.20936
1.04	1.0816	1.01980	3.22490
1.05	1.1025	1.02470	3.24037
1.06	1.1236	1.02956	3.25576
1.07	1.1449	1.03441	3.27109
1.08	1.1664	1.03923	3.28634
1.09	1.1881	1.04403	3.30151
1.10	1.2100	1.04881	3.31662
1.11	1.2321	1.05357	3.33167
1.12	1.2544	1.05830	3.34664
1.13	1.2769	1.06301	3.36155
1.14	1.2996	1.06771	3.37639
1.15	1.3225	1.07238	3.39116
1.16	1.3456	1.07703	3.40588
1.17	1.3689	1.08167	3.42053
1.18	1.3924	1.08628	3.43511
1.19	1.4161	1.09087	3.44964
1.20	1.4400	1.09545	3.46410
1.21	1.4641	1.10000	3.47851
1.22	1.4884	1.10454	3.49285
1.23	1.5129	1.10905	3.50714
1.24	1.5376	1.11355	3.52136
1.25	1.5625	1.11803	3.53553
1.26	1.5876	1.12250	3.54965
1.27	1.6129	1.12694	3.56371
1.28	1.6384	1.13137	3.57771
1.29	1.6641	1.13578	3.59166
1.30	1.6900	1.14018	3.60555
1.31	1.7161	1.14455	3.61939
1.32	1.7424	1.14891	3.63318
1.33	1.7689	1.15326	3.64692
1.34	1.7956	1.15758	3.66060
1.35	1.8225	1.16190	3.67423
1.36	1.8496	1.16619	3.68782
1.37	1.8769	1.17047	3.70135
1.38	1.9044	1.17473	3.71484
1.39	1.9321	1.17898	3.72827
1.40	1.9600	1.18322	3.74166
1.41	1.9881	1.18743	3.75500
1.42	2.0164	1.19164	3.76829
1.43	2.0449	1.19583	3.78153
1.44	2.0736	1.20000	3.79473
1.45	2.1025	1.20416	3.80789
1.46	2.1316	1.20830	3.82099
1.47	2.1609	1.21244	3.83406
1.48	2.1904	1.21655	3.84708
1.49	2.2201	1.22066	3.86005
1.50	2.2500	1.22474	3.87298
N	N^2	$\sqrt{N}$	$\sqrt{10N}$

N	N^2	$\sqrt{N}$	$\sqrt{10N}$
1.50	2.2500	1.22474	3.87298
1.51	2.2801	1.22882	3.88587
1.52	2.3104	1.23288	3.89872
1.53	2.3409	1.23693	3.91152
1.54	2.3716	1.24097	3.92428
1.55	2.4025	1.24499	3.93700
1.56	2.4336	1.24900	3.94968
1.57	2.4649	1.25300	3.96232
1.58	2.4964	1.25698	3.97492
1.59	2.5281	1.26095	3.98748
1.60	2.5600	1.26491	4.00000
1.61	2.5921	1.26886	4.01248
1.62	2.6244	1.27279	4.02492
1.63	2.6569	1.27671	4.03733
1.64	2.6896	1.28062	4.04969
1.65	2.7225	1.28452	4.06202
1.66	2.7556	1.28841	4.07431
1.67	2.7889	1.29228	4.08656
1.68	2.8224	1.29615	4.09878
1.69	2.8561	1.30000	4.11096
1.70	2.8900	1.30384	4.12311
1.71	2.9241	1.30767	4.13521
1.72	2.9584	1.31149	4.14729
1.73	2.9929	1.31529	4.15933
1.74	3.0276	1.31909	4.17133
1.75	3.0625	1.32288	4.18330
1.76	3.0976	1.32665	4.19524
1.77	3.1329	1.33041	4.20714
1.78	3.1684	1.33417	4.21900
1.79	3.2041	1.33791	4.23084
1.80	3.2400	1.34164	4.24264
1.81	3.2761	1.34536	4.25441
1.82	3.3124	1.34907	4.26615
1.83	3.3489	1.35277	4.27785
1.84	3.3856	1.35647	4.28952
1.85	3.4225	1.36015	4.30116
1.86	3.4596	1.36382	4.31277
1.87	3.4969	1.36748	4.32435
1.88	3.5344	1.37113	4.33590
1.89	3.5721	1.37477	4.34741
1.90	3.6100	1.37840	4.35890
1.91	3.6481	1.38203	4.37035
1.92	3.6864	1.38564	4.38178
1.93	3.7249	1.38924	4.39318
1.94	3.7636	1.39284	4.40454
1.95	3.8025	1.39642	4.41588
1.96	3.8416	1.40000	4.42719
1.97	3.8809	1.40357	4.43847
1.98	3.9204	1.40712	4.44972
1.99	3.9601	1.41067	4.46094
2.00	4.0000	1.41421	4.47214
N	N^2	$\sqrt{N}$	$\sqrt{10N}$

N	N^2	$\sqrt{N}$	$\sqrt{10N}$
2.00	4.0000	1.41421	4.47214
2.01	4.0401	1.41774	4.48330
2.02	4.0804	1.42127	4.49444
2.03	4.1209	1.42478	4.50555
2.04	4.1616	1.42829	4.51664
2.05	4.2025	1.43178	4.52769
2.06	4.2436	1.43527	4.53872
2.07	4.2849	1.43875	4.54973
2.08	4.3264	1.44222	4.56070
2.09	4.3681	1.44568	4.57165
2.10	4.4100	1.44914	4.58258
2.11	4.4521	1.45258	4.59347
2.12	4.4944	1.45602	4.60435
2.13	4.5369	1.45945	4.61519
2.14	4 5796	1.46287	4.62601
2.15	4.6225	1.46629	4.63681
2.16	4.6656	1.46969	4.64758
2.17	4.7089	1.47309	4.65833
2.18	4.7524	1.47648	4.66905
2.19	4.7961	1.47986	4.67974
2.20	4.8400	1.48324	4.69042
2.21	4.8841	1.48861	4.70106
2.22	4.9284	1.48997	4.71169
2.23	4.9729	1.49332	4.72229
2.24	5.0176	1.49666	4.73286
2.25	5.0625	1.50000	4.74342
2.26	5.1076	1.50333	4.75395
2.27	5.1529	1.50665	4.76445
2.28	5.1984	1.50997	4.77493
2.29	5.2441	1.51327	4.78539
2.30	5.2900	1.51658	4.79583
2.31	5.3361	1.51987	4.80625
2.32	5.3824	1.52315	4.81664
2.33	5.4289	1.52643	4.82701
2.34	5.4756	1.52971	4.83735
2.35	5.5225	1.53297	4.84768
2.36	5.5696	1.53623	4.85798
2.37	5.6169	1.53948	4.86826
2.38	5.6644	1.54272	4.87852
2.39	5.7121	1.54596	4.88876
2.40	5.7600	1.54919	4.89898
2.41	5.8081	1.55242	4.90918
2.42	5.8564	1.55563	4.91935
2.43	5.9049	1.55885	4.92950
2.44	5.9536	1.56205	4.93964
2.45	6.0025	1.56525	4.94975
2.46	6.0516	1.56844	4.95984
2.47	6.1009	1.57162	4.96991
2.48	6.1504	1.57480	4.97996
2.49	6.2001	1.57797	4.98999
2.50	6 2500	1.58114	5.00000
N	N^2	$\sqrt{N}$	$\sqrt{10N}$

N	N^2	$\sqrt{N}$	$\sqrt{10N}$
2.50	6.2500	1.58114	5.00000
2.51	6.3001	1.58430	5.00999
2.52	6.3504	1.58745	5.01996
2.53	6.4009	1.59060	5.02991
2.54	6.4516	1.59374	5.03984
2.55	6.5025	1.59687	5.04975
2.56	6.5536	1.60000	5.05964
2.57	6.6049	1.60312	5.06952
2.58	6.6564	1.60624	5.07937
2.59	6.7081	1.60935	5.08920
2.60	6.7600	1.61245	5.09902
2.61	6.8121	1.61555	5.10882
2.62	6.8644	1.61864	5.11859
2.63	6.9169	1.62173	5.12835
2.64	6 9696	1.62481	5.13809
2.65	7.0225	1.62788	5.14782
2.66	7.0756	1.63095	5.15752
2.67	7.1289	1.63401	5.16720
2.68	7.1824	1.63707	5.17687
2.69	7.2361	1.64012	5.18652
2.70	7.2900	1.64317	5.19615
2.71	7.3441	1.64621	5.20577
2.72	7.3984	1.64924	5.21536
2.73	7.4529	1.65227	5.22494
2.74	7.5076	1.65529	5.23450
2.75	7.5625	1.65831	5.24404
2.76	7.6176	1.66132	5.25357
2.77	7.6729	1.66433	5.26308
2.78	7.7284	1.66733	5.27257
2.79	7.7841	1.67033	5.28205
2.80	7.8400	1.67332	5.29150
2.81	7.8961	1.67631	5.30094
2.82	7.9524	1.67929	5.31037
2.83	8 0089	1.68226	5.31977
2.84	8.0656	1.68523	5.32917
2.85	8.1225	1.68819	5.33854
2.86	8.1796	1.69115	5.34790
2.87	8.2369	1.69411	5.35724
2.88	8.2944	1.69706	5.36656
2.89	8.3521	1.70000	5.37587
2.90	8.4100	1.70294	5.38516
2.91	8.4681	1.70587	5.39444
2.92	8.5264	1.70880	5.40370
2.93	8.5849	1.71172	5.41295
2.94	8.6436	1.71464	5.42218
2.95	8.7025	1.71756	5.43139
2.96	8.7616	1.72047	5.44059
2.97	8.8209	1.72337	5.44977
2.98	8.8804	1.72627	5.45894
2.99	8.9401	1.72916	5.46809
3.00	9.0000	1.73205	5.47723
N	N^2	$\sqrt{N}$	$\sqrt{10N}$

N	N^2	$\sqrt{N}$	$\sqrt{10N}$
3.00	9.0000	1.73205	5.47723
3.01	9.0601	1.73494	5.48635
3.02	9.1204	1.73781	5.49545
3.03	9.1809	1.74069	5.50454
3.04	9.2416	1.74356	5.51362
3.05	9.3025	1.74642	5.52268
3.06	9.3636	1.74929	5.53173
3.07	9.4249	1.75214	5.54076
3.08	9.4864	1.75499	5.54977
3.09	9.5481	1.75784	5.55878
3.10	9.6100	1.76068	5.56776
3.11	9.6721	1.76352	5.57674
3.12	9.7344	1.76635	5.58570
3.13	9.7969	1.76918	5.59464
3.14	9.8596	1.77200	5.60357
3.15	9.9225	1.77482	5.61249
3.16	9.9856	1.77764	5.62139
3.17	10.0489	1.78045	5.63028
3.18	10.1124	1.78326	5.63915
3.19	10.1761	1.78606	5.64801
3.20	10.2400	1.78885	5.65685
3.21	10.3041	1.79165	5.66569
3.22	10.3684	1.79444	5.67450
3.23	10.4329	1.79722	5.68331
3.24	10.4976	1.80000	5.69210
3.25	10.5625	1.80278	5.70088
3.26	10.6276	1.80555	5.70964
3.27	10.6929	1.80831	5.71839
3.28	10.7584	1.81108	5.72713
3.29	10.8241	1.81384	5.73585
3.30	10.8900	1.81659	5.74456
3.31	10.9561	1.81934	5.75326
3.32	11.0224	1.82209	5.76194
3.33	11.0889	1.82483	5.77062
3.34	11.1556	1.82757	5.77927
3.35	11.2225	1.83030	5.78792
3.36	11.2896	1.83303	5.79655
3.37	11.3569	1.83576	5.80517
3.38	11.4244	1.83848	5.81378
3.39	11.4921	1.84120	5.82237
3.40	11.5600	1.84391	5.83095
3.41	11.6281	1.84662	5.83952
3.42	11.6964	1.84932	5 84808
3.43	11.7649	1.85203	5.85662
3.44	11.8336	1.85472	5.86515
3.45	11.9025	1.85742	5.87367
3.46	11.9716	1.86011	5.88218
3.47	12.0409	1.86279	5.89067
3.48	12.1104	1.86548	5.89915
3.49	12.1801	1.86815	5.90762
3.50	12.2500	1.87083	5.91608
N	N^2	$\sqrt{N}$	$\sqrt{10N}$

N	N^2	$\sqrt{N}$	$\sqrt{10N}$
3.50	12.2500	1.87083	5.91608
3.51	12.3201	1.87350	5.92453
3.52	12.3904	1.87617	5.93296
3.53	12.4609	1.87883	5.94138
3.54	12.5316	1.88149	5.94979
3.55	12.6025	1.88414	5.95819
3.56	12.6736	1.88680	5.96657
3.57	12.7449	1.88944	5.97495
3.58	12.8164	1.89209	5.98331
3.59	12.8881	1.89473	5.99166
3.60	12.9600	1.89737	6.00000
3.61	13.0321	1.90000	6.00833
3.62	13.1044	1.90263	6.01664
3.63	13.1769	1.90526	6.02495
3.64	13.2496	1.90788	6.03324
3.65	13.3225	1.91050	6.04152
3.66	13.3956	1.91311	6.04979
3.67	13.4689	1.91572	6.05805
3.68	13.5424	1.91833	6.06630
3.69	13.6161	1.92094	6.07454
3.70	13.6900	1.92354	6.08276
3.71	13.7641	1.92614	6.09098
3.72	13.8384	1.92873	6.09918
3.73	13.9129	1.93132	6.10737
3.74	13.9876	1.93391	6.11555
3.75	14.0625	1.93649	6.12372
3.76	14.1376	1.93907	6.13188
3.77	14.2129	1.94165	6.14003
3.78	14.2884	1.94422	6.14817
3.79	14.3641	1.94679	6.15630
3.80	14.4400	1.94936	6.16441
3.81	14.5161	1.95192	6.17252
3.82	14.5924	1.95448	6.18061
3.83	14.6689	1.95704	6.18870
3.84	14.7456	1.95959	6.19677
3.85	14.8225	1.96214	6.20484
3.86	14.8996	1.96469	6.21289
3.87	14.9769	1.96723	6.22093
3.88	15.0544	1.96977	6.22896
3.89	15.1321	1.97231	6.23699
3.90	15.2100	1.97484	6.24500
3.91	15.2881	1.97737	6.25300
3.92	15.3664	1.97990	6.26099
3.93	15.4449	1.98242	6.26897
3.94	15.5236	1.98494	6.27694
3.95	15.6025	1.98746	6.28490
3.96	15.6816	1.98997	6.29285
3.97	15.7609	1.99249	6.30079
3.98	15.8404	1.99499	6.30872
3.99	15.9201	1.99750	6.31664
4.00	16.0000	2.00000	6.32456
N	N^2	$\sqrt{N}$	$\sqrt{10N}$

N	N^2	$\sqrt{N}$	$\sqrt{10N}$
4.00	16.0000	2.00000	6.32456
4.01	16.0801	2.00250	6.33246
4.02	16.1604	2.00499	6.34035
4.03	16.2409	2.00749	6.34823
4.04	16.3216	2.00998	6.35610
4.05	16.4025	2.01246	6.36396
4.06	16.4836	2.01494	6.37181
4.07	16.5649	2.01742	6.37966
4.08	16.6464	2.01990	6.38749
4.09	16.7281	2.02237	6.39531
4.10	16.8100	2.02485	6.40312
4.11	16.8921	2.02731	6.41093
4.12	16.9744	2.02978	6.41872
4.13	17.0569	2.03224	6.42651
4.14	17.1396	2.03470	6.43428
4.15	17.2225	2.03715	6.44205
4.16	17.3056	2.03961	6.44981
4.17	17.3889	2.04206	6.45755
4.18	17.4724	2.04450	6.46529
4.19	17.5561	2.04695	6.47302
4.20	17.6400	2.04939	6.48074
4.21	17.7241	2.05183	6.48845
4.22	17.8084	2.05426	6.49615
4.23	17.8929	2.05670	6.50384
4.24	17.9776	2.05913	6.51153
4.25	18.0625	2.06155	6.51920
4.26	18.1476	2.06398	6.52687
4.27	18.2329	2.06640	6.53452
4.28	18.3184	2.06882	6.54217
4.29	18.4041	2.07123	6.54981
4.30	18.4900	2.07364	6.55744
4.31	18.5761	2.07605	6.56506
4.32	18.6624	2.07846	6.57267
4.33	18.7489	2.08087	6.58027
4.34	18.8356	2.08327	6.58787
4.35	18.9225	2.08567	6.59545
4.36	19.0096	2.08806	6.60303
4.37	19.0969	2.09045	6.61060
4.38	19.1844	2.09284	6.61816
4.39	19.2721	2.09523	6.62571
4.40	19.3600	2.09762	6.63325
4.41	19.4481	2.10000	6.64078
4.42	19.5364	2.10238	6.64831
4.43	19.6249	2.10476	6.65582
4.44	19.7136	2.10713	6.66333
4.45	19.8025	2.10950	6.67083
4.46	19.8916	2.11187	6.67832
4.47	19.9809	2.11424	6.68581
4.48	20.0704	2.11660	6.69328
4.49	20.1601	2.11896	6.70075
4.50	20.2500	2.12132	6.70820
N	N^2	$\sqrt{N}$	$\sqrt{10N}$

N	N^2	$\sqrt{N}$	$\sqrt{10N}$
4.50	20.2500	2.12132	6.70820
4.51	20.3401	2.12368	6.71565
4.52	20.4304	2.12603	6.72309
4.53	20.5209	2.12838	6.73053
4.54	20.6116	2.13073	6.73795
4.55	20.7025	2.13307	6.74537
4.56	20.7936	2.13542	6.75278
4.57	20.8849	2.13776	6.76018
4.58	20.9764	2.14009	6.76757
4.59	21.0681	2.14243	6.77495
4.60	21.1600	2.14476	6.78233
4.61	21.2521	2.14709	6.78970
4.62	21.3444	2.14942	6.79706
4.63	21.4369	2.15174	6.80441
4.64	21.5296	2.15407	6.81175
4.65	21.6225	2.15639	6.81909
4.66	21.7156	2.15870	6.82642
4.67	21.8089	2.16102	6.83374
4.68	21.9024	2.16333	6.84105
4.69	21.9961	2.16564	6.84836
4.70	22.0900	2.16795	6.85565
4.71	22.1841	2.17025	6.86294
4.72	22.2784	2.17256	6.87023
4.73	22.3729	2.17486	6.87750
4.74	22.4676	2.17715	6.88477
4.75	22.5625	2.17945	6.89202
4.76	22.6576	2.18174	6.89928
4.77	22.7529	2.18403	6.90652
4.78	22.8484	2.18632	6.91375
4.79	22.9441	2.18861	6.92098
4.80	23.0400	2.19089	6.92820
4.81	23.1361	2.19317	6.93542
4.82	23.2324	2.19545	6.94262
4.83	23.3289	2.19773	6.94982
4.84	23.4256	2.20000	6.95701
4.85	23.5225	2.20227	6.96419
4.86	23.6196	2.20454	6.97137
4.87	23.7169	2.20681	6.97854
4.88	23.8144	2.20907	6.98570
4.89	23.9121	2.21133	6.99285
4.90	24.0100	2.21359	7.00000
4.91	24.1081	2.21585	7.00714
4.92	24.2064	2.21811	7.01427
4.93	24.3049	2.22036	7.02140
4.94	24.4036	2.22261	7.02851
4.95	24.5025	2.22486	7.03562
4.96	24.6016	2.22711	7.04273
4.97	24.7009	2.22935	7.04982
4.98	24.8004	2.23159	7.05691
4.99	24.9001	2.23383	7.06399
5.00	25.0000	2.23607	7.07107
N	N^2	$\sqrt{N}$	$\sqrt{10N}$

N	N^2	$\sqrt{N}$	$\sqrt{10N}$
5.00	25.0000	2.23607	7.07107
5.01	25.1001	2.23830	7.07814
5.02	25.2004	2.24054	7.08520
5.03	25.3009	2.24277	7.09225
5.04	25.4016	2.24499	7.09930
5.05	25.5025	2.24722	7.10634
5.06	25.6036	2.24944	7.11337
5.07	25.7049	2.25167	7.12039
5.08	25.8064	2.25389	7.12741
5.09	25.9081	2.25610	7.13442
5.10	26.0100	2.25832	7.14143
5.11	26.1121	2.26053	7.14843
5.12	26.2144	2.26274	7.15542
5.13	26.3169	2.26495	7.16240
5.14	26.4196	2.26716	7.16938
5.15	26.5225	2.26936	7.17635
5.16	26.6256	2.27156	7.18331
5.17	26.7289	2.27376	7.19027
5.18	26.8324	2.27596	7.19722
5.19	26.9361	2.27816	7.20417
5.20	27.0400	2.28035	7.21110
5.21	27.1441	2.28254	7.21803
5.22	27.2484	2.28473	7.22496
5.23	27.3529	2.28692	7.23187
5.24	27.4576	2.28910	7.23878
5.25	27.5625	2.29129	7.24569
5.26	27.6676	2.29347	7.25259
5.27	27.7729	2.29565	7.25948
5.28	27.8784	2.29783	7.26636
5.29	27.9841	2.30000	7.27324
5.30	28.0900	2.30217	7.28011
5.31	28.1961	2.30434	7.28697
5.32	28.3024	2.30651	7.29383
5.33	28.4089	2.30868	7.30068
5.34	28.5156	2.31084	7.30753
5.35	28.6225	2.31301	7.31437
5.36	28.7296	2.31517	7.32120
5.37	28.8369	2.31733	7.32803
5.38	28.9444	2.31948	7.33485
5.39	29.0521	2.32164	7.34166
5.40	29.1600	2.32379	7.34847
5.41	29.2681	2.32594	7.35527
5.42	29.3764	2.32809	7.36206
5.43	29.4849	2.33024	7.36885
5.44	29.5936	2.33238	7.37564
5.45	29.7025	2.33452	7.38241
5.46	29.8116	2.33666	7.38918
5.47	29.9209	2.33880	7.39594
5.48	30.0304	2.34094	7.40270
5.49	30.1401	2.34307	7.40945
5.50	30.2500	2.34521	7.41620
N	N^2	$\sqrt{N}$	$\sqrt{10N}$

N	N^2	$\sqrt{N}$	$\sqrt{10N}$
5.50	30.2500	2.34521	7.41620
5.51	30.3601	2.34734	7.42294
5.52	30.4704	2.34947	7.42967
5.53	30.5809	2.35160	7.43640
5.54	30.6916	2.35372	7.44312
5.55	30.8025	2.35584	7.44983
5.56	30.9136	2.35797	7.45654
5.57	31.0249	2.36008	7.46324
5.58	31.1364	2.36220	7.46994
5.59	31.2481	2.36432	7.47663
5.60	31.3600	2.36643	7.48331
5.61	31.4721	2.36854	7.48999
5.62	31.5844	2.37065	7.49667
5.63	31.6969	2.37276	7.50333
5.64	31.8096	2.37487	7.50999
5.65	31.9225	2.37697	7.51665
5.66	32.0356	2.37908	7.52330
5.67	32.1489	2.38118	7.52994
5.68	32.2624	2.38328	7.53658
5.69	32.3761	2.38537	7.54321
5.70	32.4900	2.38747	7.54983
5.71	32.6041	2.38956	7.55645
5.72	32.7184	2.39165	7.56307
5.73	32.8329	2.39374	7.56968
5.74	32.9476	2.39583	7.57628
5.75	33.0625	2.39792	7.58288
5.76	33.1776	2.40000	7.58947
5.77	33.2929	2.40208	7.59605
5.78	33.4084	2.40416	7.60263
5.79	33.5241	2.40624	7.60920
5.80	33.6400	2.40832	7.61577
5.81	33.7561	2.41039	7.62234
5.82	33.8724	2.41247	7.62889
5.83	33.9889	2.41454	7.63544
5.84	34.1056	2.41661	7.64199
5.85	34.2225	2.41868	7.64853
5.86	34.3396	2.42074	7.65506
5.87	34.4569	2.42281	7.66159
5.88	34.5744	2.42487	7.66812
5.89	34.6921	2.42693	7.67463
5.90	34.8100	2.42899	7.68115
5.91	34.9281	2.43105	7.68765
5.92	35.0464	2.43311	7.69415
5.93	35.1649	2.43516	7.70065
5.94	35.2836	2.43721	7.70714
5.95	35.4025	2.43926	7.71362
5.96	35.5216	2.44131	7.72010
5.97	35.6409	2.44336	7.72658
5.98	35.7604	2.44540	7.73305
5.99	35.8801	2.44745	7.73951
6.00	36.0000	2.44949	7.74597
N	N^2	$\sqrt{N}$	$\sqrt{10N}$

N	N^2	$\sqrt{N}$	$\sqrt{10N}$	N	N^2	$\sqrt{N}$	$\sqrt{10N}$
6.00	36.0000	2.44949	7.74597	**6.50**	42.2500	2.54951	8.06226
6.01	36.1201	2.45153	7.75242	6.51	42.3801	2.55147	8.06846
6.02	36.2404	2.45357	7.75887	6.52	42.5104	2.55343	8.07465
6.03	36.3609	2.45561	7.76531	6.53	42.6409	2.55539	8.08084
6.04	36.4816	2.45764	7.77174	6.54	42.7716	2.55734	8.08703
6.05	36.6025	2.45967	7.77817	6.55	42.9025	2.55930	8.09321
6.06	36.7236	2.46171	7.78460	6.56	43.0336	2.56125	8.09938
6.07	36.8449	2.46374	7.79102	6.57	43.1649	2.56320	8.10555
6.08	36.9664	2.46577	7.79744	6.58	43.2964	2.56515	8.11172
6.09	37.0881	2.46779	7.80385	6.59	43.4281	2.56710	8.11788
6.10	37.2100	2.46982	7.81025	**6.60**	43.5600	2.56905	8.12404
6.11	37.3321	2.47184	7.81665	6.61	43.6921	2.57099	8.13019
6.12	37.4544	2.47386	7.82304	6.62	43.8244	2.57294	8.13634
6.13	37.5769	2.47588	7.82943	6.63	43.9569	2.57488	8.14248
6.14	37.6996	2.47790	7.83582	6.64	44.0896	2.57682	8.14862
6.15	37.8225	2.47992	7.84219	6.65	44.2225	2.57876	8.15475
6.16	37.9456	2.48193	7.84857	6.66	44.3556	2.58070	8.16088
6.17	38.0689	2.48395	7.85493	6.67	44.4889	2.58263	8.16701
6.18	38.1924	2.48596	7.86130	6.68	44.6224	2.58457	8.17313
6.19	38.3161	2.48797	7.86766	6.69	44.7561	2.58650	8.17924
6.20	38.4400	2.48998	7.87401	**6.70**	44.8900	2.58844	8.18535
6.21	38.5641	2.49199	7.88036	6.71	45.0241	2.59037	8.19146
6.22	38.6884	2.49399	7.88670	6.72	45.1584	2.59230	8.19756
6.23	38.8129	2.49600	7.89303	6.73	45.2929	2.59422	8.20366
6.24	38.9376	2.49800	7.89937	6.74	45.4276	2.59615	8.20975
6.25	39.0625	2.50000	7.90569	6.75	45.5625	2.59808	8.21584
6.26	39.1876	2.50200	7.91202	6.76	45.6976	2.60000	8.22192
6.27	39.3129	2.50400	7.91833	6.77	45.8329	2.60192	8.22800
6.28	39.4384	2.50599	7.92465	6.78	45.9684	2.60384	8.23408
6.29	39.5641	2.50799	7.93095	6.79	46.1041	2.60576	8.24015
6.30	39.6900	2.50998	7.93725	**6.80**	46.2400	2.60768	8.24621
6.31	39.8161	2.51197	7.94355	6.81	46.3761	2.60960	8.25227
6.32	39.9424	2.51396	7.94984	6.82	46.5124	2.61151	8.25833
6.33	40.0689	2.51595	7.95613	6.83	46.6489	2.61343	8.26438
6.34	40.1956	2.51794	7.96241	6.84	46.7856	2.61534	8.27043
6.35	40.3225	2.51992	7.96869	6.85	46.9225	2.61725	8.27647
6.36	40.4496	2.52190	7.97496	6.86	47.0596	2.61916	8.28251
6.37	40.5769	2.52389	7.98123	6.87	47.1969	2.62107	8.28855
6.38	40.7044	2.52587	7.98749	6.88	47.3344	2.62298	8.29458
6.39	40.8321	2.52784	7.99375	6.89	47.4721	2.62488	8.30060
6.40	40.9600	2.52982	8.00000	**6.90**	47.6100	2.62679	8.30662
6.41	41.0881	2.53180	8.00625	6.91	47.7481	2.62869	8.31264
6.42	41.2164	2.53377	8.01249	6.92	47.8864	2.63059	8.31865
6.43	41.3449	2.53574	8.01873	6.93	48.0249	2.63249	8.32466
6.44	41.4736	2.53772	8.02496	6.94	48.1636	2.63439	8.33067
6.45	41.6025	2.53969	8.03119	6.95	48.3025	2.63629	8.33667
6.46	41.7316	2.54165	8.03741	6.96	48.4416	2.63818	8.34266
6.47	41.8609	2.54362	8.04363	6.97	48.5809	2.64008	8.34865
6.48	41.9904	2.54558	8.04984	6.98	48.7204	2.64197	8.35464
6.49	42.1201	2.54755	8.05605	6.99	48.8601	2.64386	8.36062
6.50	42.2500	2.54951	8.06226	**7.00**	49.0000	2.64575	8.36660
N	N^2	$\sqrt{N}$	$\sqrt{10N}$	N	N^2	$\sqrt{N}$	$\sqrt{10N}$

N	N^2	$\sqrt{N}$	$\sqrt{10N}$
7.00	49.0000	2.64575	8.36660
7.01	49.1401	2.64764	8.37257
7.02	49.2804	2.64953	8.37854
7.03	49.4209	2.65141	8.38451
7.04	49.5616	2.65330	8.39047
7.05	49.7025	2.65518	8.39643
7.06	49.8436	2.65707	8.40238
7.07	49.9849	2.65895	8.40833
7.08	50.1264	2.66083	8.41427
7.09	50.2681	2.66271	8.42021
7.10	50.4100	2.66458	8.42615
7.11	50.5521	2.66646	8.43208
7.12	50.6944	2.66833	8.43801
7.13	50.8369	2.67021	8.44393
7.14	50.9796	2.67208	8.44985
7.15	51.1225	2.67395	8.45577
7.16	51.2656	2.67582	8.46168
7.17	51.4089	2.67769	8.46759
7.18	51.5524	2.67955	8.47349
7.19	51.6961	2.68142	8.47939
7.20	51.8400	2.68328	8.48528
7.21	51.9841	2.68514	8.49117
7.22	52.1284	2.68701	8.49706
7.23	52.2729	2.68887	8.50294
7.24	52.4176	2.69072	8.50882
7.25	52.5625	2.69258	8.51469
7.26	52.7076	2.69444	8.52056
7.27	52.8529	2.69629	8.52643
7.28	52.9984	2.69815	8.53229
7.29	53.1441	2.70000	8.53815
7.30	53.2900	2.70185	8.54400
7.31	53.4361	2.70370	8.54985
7.32	53.5824	2.70555	8.55570
7.33	53.7289	2.70740	8.56154
7.34	53.8756	2.70924	8.56738
7.35	54.0225	2.71109	8.57321
7.36	54.1696	2.71293	8.57904
7.37	54.3169	2.71477	8.58487
7.38	54.4644	2.71662	8.59069
7.39	54.6121	2.71846	8.59651
7.40	54.7600	2.72029	8.60233
7.41	54.9081	2.72213	8.60814
7.42	55.0564	2.72397	8.61394
7.43	55.2049	2.72580	8.61974
7.44	55.3536	2.72764	8.62554
7.45	55.5025	2.72947	8.63134
7.46	55.6516	2.73130	8.63713
7.47	55.8009	2.73313	8.64292
7.48	55.9504	2.73496	8.64870
7.49	56.1001	2.73679	8.65448
7.50	56.2500	2.73861	8.66025
N	N^2	$\sqrt{N}$	$\sqrt{10N}$

N	N^2	$\sqrt{N}$	$\sqrt{10N}$
7.50	56.2500	2.73861	8.66025
7.51	56.4001	2.74044	8.66603
7.52	56.5504	2.74226	8.67179
7.53	56.7009	2.74408	8.67756
7.54	56.8516	2.74591	8.68332
7.55	57.0025	2.74773	8.68907
7.56	57.1536	2.74955	8.69483
7.57	57.3049	2.75136	8.70057
7.58	57.4564	2.75318	8.70632
7.59	57.6081	2.75500	8.71206
7.60	57.7600	2.75681	8.71780
7.61	57.9121	2.75862	8.72353
7.62	58.0644	2.76043	8.72926
7.63	58.2169	2.76225	8.73499
7.64	58.3696	2.76405	8.74071
7.65	58.5225	2.76586	8.74643
7.66	58.6756	2.76767	8.75214
7.67	58.8289	2.76948	8.75785
7.68	58.9824	2.77128	8.76356
7.69	59.1361	2.77308	8.76926
7.70	59.2900	2.77489	8.77496
7.71	59.4441	2.77669	8.78066
7.72	59.5984	2.77849	8.78635
7.73	59.7529	2.78029	8.79204
7.74	59.9076	2.78209	8.79773
7.75	60.0625	2.78388	8.80341
7.76	60.2176	2.78568	8.80909
7.77	60.3729	2.78747	8.81476
7.78	60.5284	2.78927	8.82043
7.79	60.6841	2.79106	8.82610
7.80	60.8400	2.79285	8.83176
7.81	60.9961	2.79464	8.83742
7.82	61.1524	2.79643	8.84308
7.83	61.3089	2.79821	8.84873
7.84	61.4656	2.80000	8.85438
7.85	61.6225	2.80179	8.86002
7.86	61.7796	2.80357	8.86566
7.87	61.9369	2.80535	8.87130
7.88	62.0944	2.80713	8.87694
7.89	62.2521	2.80891	8.88257
7.90	62.4100	2.81069	8.88819
7.91	62.5681	2.81247	8.89382
7.92	62.7264	2.81425	8.89944
7.93	62.8849	2.81603	8.90505
7.94	63.0436	2.81780	8.91067
7.95	63.2025	2.81957	8.91628
7.96	63.3616	2.82135	8.92188
7.97	63.5209	2.82312	8.92749
7.98	63.6804	2.82489	8.93308
7.99	63.8401	2.82666	8.93868
8.00	64.0000	2.82843	8.94427
N	N^2	$\sqrt{N}$	$\sqrt{10N}$

N	N^2	$\sqrt{N}$	$\sqrt{10N}$	N	N^2	$\sqrt{N}$	$\sqrt{10N}$
8.00	64.0000	2.82843	8.94427	**8.50**	72.2500	2.91548	9.21954
8.01	64.1601	2.83019	8.94986	8.51	72.4201	2.91719	9.22497
8.02	64.3204	2.83196	8.95545	8.52	72.5904	2.91890	9.23038
8.03	64.4809	2.83373	8.96103	8.53	72.7609	2.92062	9.23580
8.04	64.6416	2.83549	8.96660	8.54	72.9316	2.92233	9.24121
8.05	64.8025	2.83725	8.97218	8.55	73.1025	2.92404	9.24662
8.06	64.9636	2.83901	8.97775	8.56	73.2736	2.92575	9.25203
8.07	65.1249	2.84077	8.98332	8.57	73.4449	2.92746	9.25743
8.08	65.2864	2.84253	8.98888	8.58	73.6164	2.92916	9.26283
8.09	65.4481	2.84429	8.99444	8.59	73.7881	2.93087	9.26823
8.10	65.6100	2.84605	9.00000	**8.60**	73.9600	2.93258	9.27362
8.11	65.7721	2.84781	9.00555	8.61	74.1321	2.93428	9.27901
8.12	65.9344	2.84956	9.01110	8.62	74.3044	2.93598	9.28440
8.13	66.0969	2.85132	9.01665	8.63	74.4769	2.93769	9.28978
8.14	66.2596	2.85307	9.02219	8.64	74.6496	2.93939	9.29516
8.15	66.4225	2.85482	9.02774	8.65	74.8225	2.94109	9.30054
8.16	66.5856	2.85657	9.03327	8.66	74.9956	2.94279	9.30591
8.17	66.7489	2.85832	9.03881	8.67	75.1689	2.94449	9.31128
8.18	66.9124	2.86007	9.04434	8.68	75.3424	2.94618	9.31665
8.19	67.0761	2.86182	9.04986	8.69	75.5161	2.94788	9.32202
8.20	67.2400	2.86356	9.05539	**8.70**	75.6900	2.94958	9.32738
8.21	67.4041	2.86531	9.06091	8.71	75.8641	2.95127	9.33274
8.22	67.5684	2.86705	9.06642	8.72	76.0384	2.95296	9.33809
8.23	67.7329	2.86880	9.07193	8.73	76.2129	2.95466	9.34345
8.24	67.8976	2.87054	9.07744	8.74	76.3876	2.95635	9.34880
8.25	68.0625	2.87228	9.08295	8.75	76.5625	2.95804	9.35414
8.26	68.2276	2.87402	9.08845	8.76	76.7376	2.95973	9.35949
8.27	68.3929	2.87576	9.09395	8.77	76.9129	2.96142	9.36483
8.28	68.5584	2.87750	9.09945	8.78	77.0884	2.96311	9.37017
8.29	68.7241	2.87924	9.10494	8.79	77.2641	2.96479	9.37550
8.30	68.8900	2.88097	9.11043	**8.80**	77.4400	2.96648	9.38083
8.31	69.0561	2.88271	9.11592	8.81	77.6161	2.96816	9.38616
8.32	69.2224	2.88444	9.12140	8.82	77.7924	2.96985	9.39149
8.33	69.3889	2.88617	9.12688	8.83	77.9689	2.97153	9.39681
8.34	69.5556	2.88791	9.13236	8.84	78.1456	2.97321	9.40213
8.35	69.7225	2.88964	9.13783	8.85	78.3225	2.97489	9.40744
8.36	69.8896	2.89137	9.14330	8.86	78.4996	2.97658	9.41276
8.37	70.0569	2.89310	9.14877	8.87	78.6769	2.97825	9.41807
8.38	70.2244	2.89482	9.15423	8.88	78.8544	2.97993	9.42338
8.39	70.3921	2.89655	9.15969	8.89	79.0321	2.98161	9.42868
8.40	70.5600	2.89828	9.16515	**8.90**	79.2100	2.98329	9.43398
8.41	70.7281	2.90000	9.17061	8.91	79.3881	2.98496	9.43928
8.42	70.8964	2.90172	9.17606	8.92	79.5664	2.98664	9.44458
8.43	71.0649	2.90345	9.18150	8.93	79.7449	2.98831	9.44987
8.44	71.2336	2.90517	9.18695	8.94	79.9236	2.98998	9.45516
8.45	71.4025	2.90689	9.19239	8.95	80.1025	2.99166	9.46044
8.46	71.5716	2.90861	9.19783	8.96	80.2816	2.99333	9.46573
8.47	71.7409	2.91033	9.20326	8.97	80.4609	2.99500	9.47101
8.48	71.9104	2.91204	9.20869	8.98	80.6404	2.99666	9.47629
8.49	72.0801	2.91376	9.21412	8.99	80.8201	2.99833	9.48156
8.50	72.2500	2.91548	9.21954	**9.00**	81.0000	3.00000	9.48683
N	N^2	$\sqrt{N}$	$\sqrt{10N}$	N	N^2	$\sqrt{N}$	$\sqrt{10N}$

N	N^2	$\sqrt{N}$	$\sqrt{10N}$
9.00	81.0000	3.00000	9.48683
9.01	81.1801	3.00167	9.49210
9.02	81.3604	3.00333	9.49737
9.03	81.5409	3.00500	9.50263
9.04	81.7216	3.00666	9.50789
9.05	81.9025	3.00832	9.51315
9.06	82.0836	3.00998	9.51840
9.07	82.2649	3.01164	9.52365
9.08	82.4464	3.01330	9.52890
9.09	82.6281	3.01496	9.53415
9.10	82.8100	3.01662	9.53939
9.11	82.9921	3.01828	9.54463
9.12	83.1744	3.01993	9.54987
9.13	83.3569	3.02159	9.55510
9.14	83.5396	3.02324	9.56033
9.15	83.7225	3.02490	9.56556
9.16	83.9056	3.02655	9.57079
9.17	84.0889	3.02820	9.57601
9.18	84.2724	3.02985	9.58123
9.19	84.4561	3.03150	9.58645
9.20	84.6400	3.03315	9.59166
9.21	84.8241	3.03480	9.59687
9.22	85.0084	3.03645	9.60208
9.23	85.1929	3.03809	9.60729
9.24	85.3776	3.03974	9.61249
9.25	85.5625	3.04138	9.61769
9.26	85.7476	3.04302	9.62289
9.27	85.9329	3.04467	9.62808
9.28	86.1184	3.04631	9.63328
9.29	86.3041	3.04795	9.63846
9.30	86.4900	3.04959	9.64365
9.31	86.6761	3.05123	9.64883
9.32	86.8624	3.05287	9.65401
9.33	87.0489	3.05450	9.65919
9.34	87.2356	3.05614	9.66437
9.35	87.4225	3.05778	9.66954
9.36	87.6096	3.05941	9.67471
9.37	87.7969	3.06105	9.67988
9.38	87.9844	3.06268	9.68504
9.39	88.1721	3.06431	9.69020
9.40	88.3600	3.06594	9.69536
9.41	88.5481	3.06757	9.70052
9.42	88.7364	3.06920	9.70567
9.43	88.9249	3.07083	9.71082
9.44	89.1136	3.07246	9.71597
9.45	89.3025	3.07409	9.72111
9.46	89.4916	3.07571	9.72625
9.47	89.6809	3.07734	9.73139
9.48	89.8704	3.07896	9.73653
9.49	90.0601	3.08058	9.74166
9.50	90.2500	3.08221	9.74679
N	N^2	$\sqrt{N}$	$\sqrt{10N}$

N	N^2	$\sqrt{N}$	$\sqrt{10N}$
9.50	90.2500	3.08221	9.74679
9.51	90.4401	3.08383	9.75192
9.52	90.6304	3.08545	9.75705
9.53	90.8209	3.08707	9.76217
9.54	91.0116	3.08869	9.76729
9.55	91.2025	3.09031	9.77241
9.56	91.3936	3.09192	9.77753
9.57	91.5849	3.09354	9.78264
9.58	91.7764	3.09516	9.78775
9.59	91.9681	3.09677	9.79285
9.60	92.1600	3.09839	9.79796
9.61	92.3521	3.10000	9.80306
9.62	92.5444	3.10161	9.80816
9.63	92.7369	3.10322	9.81326
9.64	92.9296	3.10483	9.81835
9.65	93.1225	3.10644	9.82344
9.66	93.3156	3.10805	9.82853
9.67	93.5089	3.10966	9.83362
9.68	93.7024	3.11127	9.83870
9.69	93.8961	3.11288	9.84378
9.70	94.0900	3.11448	9.84886
9.71	94.2841	3.11609	9.85393
9.72	94.4784	3.11769	9.85901
9.73	94.6729	3.11929	9.86408
9.74	94.8676	3.12090	9.86914
9.75	95.0625	3.12250	9.87421
9.76	95.2576	3.12410	9.87927
9.77	95.4529	3.12570	9.88433
9.78	95.6484	3.12730	9.88939
9.79	95.8441	3.12890	9.89444
9.80	96.0400	3.13050	9.89949
9.81	96.2361	3.13209	9.90454
9.82	96.4324	3.13369	9.90959
9.83	96.6289	3.13528	9.91464
9.84	96.8256	3.13688	9.91968
9.85	97.0225	3.13847	9.92472
9.86	97.2196	3.14006	9.92975
9.87	97.4169	3.14166	9.93479
9.88	97.6144	3.14325	9.93982
9.89	97.8121	3.14484	9.94485
9.90	98.0100	3.14643	9.94987
9.91	98.2081	3.14802	9.95490
9.92	98.4064	3.14960	9.95992
9.93	98.6049	3.15119	9.96494
9.94	98.8036	3.15278	9.96995
9.95	99.0025	3.15436	9.97497
9.96	99.2016	3.15595	9.97998
9.97	99.4009	3.15753	9.98499
9.98	99.6004	3.15911	9.98999
9.99	99.8001	3.16070	9.99500
10.00	100.000	3.16228	10.0000
N	N^2	$\sqrt{N}$	$\sqrt{10N}$

TABLE T Critical Values of t for a Two-Tailed Test

df	.90	.80	.70	.60	.50	.40	.30	.20	.10	.05	.02	.01
1	.158	.325	.510	.727	1.000	1.376	1.963	3.078	6.314	12.706	31.821	63.657
2	.142	.289	.445	.617	.816	1.061	1.386	1.886	2.920	4.303	6.965	9.925
3	.137	.277	.424	.584	.765	.978	1.250	1.638	2.353	3.182	4.541	5.841
4	.134	.271	.414	.569	.741	.941	1.190	1.533	2.132	2.776	3.747	4.604
5	.132	.267	.408	.559	.727	.920	1.156	1.476	2.015	2.571	3.365	4.032
6	.131	.265	.404	.553	.718	.906	1.134	1.440	1.943	2.447	3.143	3.707
7	.130	.263	.402	.549	.711	.896	1.119	1.415	1.895	2.365	2.998	3.499
8	.130	.262	.399	.546	.706	.889	1.108	1.397	1.860	2.306	2.896	3.355
9	.129	.261	.398	.543	.703	.883	1.100	1.383	1.833	2.262	2.821	3.250
10	.129	.260	.397	.542	.700	.879	1.093	1.372	1.812	2.228	2.764	3.169
11	.129	.260	.396	.540	.697	.876	1.088	1.363	1.796	2.201	2.718	3.106
12	.128	.259	.395	.539	.695	.873	1.083	1.356	1.782	2.179	2.681	3.055
13	.128	.259	.394	.538	.694	.870	1.079	1.350	1.771	2.160	2.650	3.012
14	.128	.258	.393	.537	.692	.868	1.076	1.345	1.761	2.145	2.624	2.977
15	.128	.258	.393	.536	.691	.866	1.074	1.341	1.753	2.131	2.602	2.947
16	.128	.258	.392	.535	.690	.865	1.071	1.337	1.746	2.120	2.583	2.921
17	.128	.257	.392	.534	.689	.863	1.069	1.333	1.740	2.110	2.567	2.898
18	.127	.257	.392	.534	.688	.862	1.067	1.330	1.734	2.101	2.552	2.878
19	.127	.257	.391	.533	.688	.861	1.066	1.328	1.729	2.093	2.539	2.861
20	.127	.257	.391	.533	.687	.860	1.064	1.325	1.725	2.086	2.528	2.845
21	.127	.257	.391	.532	.686	.859	1.063	1.323	1.721	2.080	2.518	2.831
22	.127	.256	.390	.532	.686	.858	1.061	1.321	1.717	2.074	2.508	2.819
23	.127	.256	.390	.532	.685	.858	1.060	1.319	1.714	2.069	2.500	2.807
24	.127	.256	.390	.531	.685	.857	1.059	1.318	1.711	2.064	2.492	2.797
25	.127	.256	.390	.531	.684	.856	1.058	1.316	1.708	2.060	2.485	2.787
26	.127	.256	.390	.531	.684	.856	1.058	1.315	1.706	2.056	2.479	2.779
27	.127	.256	.389	.531	.684	.855	1.057	1.314	1.703	2.052	2.473	2.771
28	.127	.256	.389	.530	.683	.855	1.056	1.313	1.701	2.048	2.467	2.763
29	.127	.256	.389	.530	.683	.854	1.055	1.311	1.699	2.045	2.462	2.756
30	.127	.256	.389	.530	.683	.854	1.055	1.310	1.697	2.042	2.457	2.750
∞	.12566	.25335	.38532	.52440	.67449	.84162	1.03643	1.28155	1.64485	1.95996	2.32634	2.57582

Table T is reprinted from Table IV of Fisher: Statistical Methods for Research Workers, Oliver & Boyd Ltd., Edinburgh, by permission of the author and publishers.

Table U Significant Values of U for the Mann-Whitney Test

Two-Tailed Test, $\alpha = .05$

N_1 \ N_2	3	4	5	6	7	8	9	10	11	12	13	14	15	16	17	18	19	20
1	—	—	—	—	—	—	—	—	—	—	—	—	—	—	—	—	—	—
2	—	—	—	—	—	0	0	0	0	1	1	1	1	1	2	2	2	2
3	—	—	0	1	1	2	2	3	3	4	4	5	5	6	6	7	7	8
4		0	1	2	3	4	4	5	6	7	8	9	10	11	11	12	13	13
5			2	3	5	6	7	8	9	11	12	13	14	15	17	18	19	20
6				5	6	8	10	11	13	14	16	17	19	21	22	24	25	27
7					8	10	12	14	16	18	20	22	24	26	28	30	32	34
8						13	15	17	19	22	24	26	29	31	34	36	38	41
9							17	20	23	26	28	31	34	37	39	42	45	48
10								23	26	29	33	36	39	42	45	48	52	55
11									30	33	37	40	44	47	51	55	58	62
12										37	41	45	49	53	57	61	65	69
13											45	50	54	59	63	67	72	76
14												55	59	64	67	74	78	83
15													64	70	75	80	85	90
16														75	81	86	92	98
17															87	93	99	105
18																99	106	112
19																	113	119
20																		127

$z = 1.960$

Two-Tailed Test, $\alpha = .01$

N_1 \ N_2	3	4	5	6	7	8	9	10	11	12	13	14	15	16	17	18	19	20
1	—	—	—	—	—	—	—	—	—	—	—	—	—	—	—	—	—	—
2	—	—	—	—	—	—	—	—	—	—	—	—	—	—	—	—	0	0
3	—	—	—	—	—	—	0	0	0	1	1	1	2	2	2	2	3	3
4		—	—	0	0	1	1	2	2	3	4	4	5	5	6	6	7	8
5			0	1	2	3	3	4	5	6	7	7	8	9	10	11	12	13
6				2	3	4	5	6	7	9	10	11	12	13	15	16	17	18
7					4	6	7	9	10	12	13	15	16	18	19	21	22	24
8						8	10	12	14	16	18	19	21	23	25	27	29	31
9							11	13	16	18	20	22	24	27	29	31	33	36
10								16	18	21	24	26	29	31	34	37	39	42
11									21	24	27	30	33	36	39	42	45	48
12										27	31	34	37	41	44	47	51	54
13											34	38	42	45	49	53	56	60
14												42	46	50	54	58	63	67
15													51	55	60	64	69	73
16														60	65	70	74	79
17															70	75	81	86
18																81	87	92
19																	93	99
20																		105

$z = 2.576$

Prepared from Mann, H. B. & Whitney, D. R. On a test of whether one of two random variables is statistically larger than the other. *Annuals of Math. Statistics, 18*, 1947, 50–60, and from Auble, D. Extended tables for the Mann-Whitney Statistic, *Bulletin of the Institute of Educational Research at Indiana University* 1, 1953, by permission of the authors and the editors.

TABLE W The Wilcoxon Signed Rank Test. Minimum Significant Values of T for a Two-Tailed Test

N	5%	1%
6	1	—
7	2	—
8	4	0
9	6	2
10	8	3
11	11	5
12	14	7
13	17	10
14	21	13
15	25	16
16	30	19
17	35	23
18	40	28
19	46	32
20	52	37
21	59	43
22	66	49
23	73	55
24	81	61
25	90	68

The entries in Table W are reprinted from Frank Wilcoxon and Roberta A. Wilcox, *Some rapid approximate statistical procedures,* The American Cynamid Company, New Jersey, 1964, by permission.

Table Z Transformation of r to Z'

r	Z'	r	Z'	r	Z'	r	Z'	r	Z'
.000	.000	.200	.203	.400	.424	.600	.693	.800	1.099
.005	.005	.205	.208	.405	.430	.605	.701	.805	1.113
.010	.010	.210	.213	.410	.436	.610	.709	.810	1.127
.015	.015	.215	.218	.415	.442	.615	.717	.815	1.142
.020	.020	.220	.224	.420	.448	.620	.725	.820	1.157
.025	.025	.225	.229	.425	.454	.625	.733	.825	1.172
.030	.030	.230	.234	.430	.460	.630	.741	.830	1.188
.035	.035	.235	.239	.435	.466	.635	.750	.835	1.204
.040	.040	.240	.245	.440	.472	.640	.758	.840	1.221
.045	.045	.245	.250	.445	.478	.645	.767	.845	1.238
.050	.050	.250	.255	.450	.485	.650	.775	.850	1.256
.055	.055	.255	.261	.455	.491	.655	.784	.855	1.274
.060	.060	.260	.266	.460	.497	.660	.793	.860	1.293
.065	.065	.265	.271	.465	.504	.665	.802	.865	1.313
.070	.070	.270	.277	.470	.510	.670	.811	.870	1.333
.075	.075	.275	.282	.475	.517	.675	.820	.875	1.354
.080	.080	.280	.288	.480	.523	.680	.829	.880	1.376
.085	.085	.285	.293	.485	.530	.685	.838	.885	1.398
.090	.090	.290	.299	.490	.536	.690	.848	.890	1.422
.095	.095	.295	.304	.495	.543	.695	.858	.895	1.447
.100	.100	.300	.310	.500	.549	.700	.867	.900	1.472
.105	.105	.305	.315	.505	.556	.705	.877	.905	1.499
.110	.110	.310	.321	.510	.563	.710	.887	.910	1.528
.115	.116	.315	.326	.515	.570	.715	.897	.915	1.557
.120	.121	.320	.332	.520	.576	.720	.908	.920	1.589
.125	.126	.325	.337	.525	.583	.725	.918	.925	1.623
.130	.131	.330	.343	.530	.590	.730	.929	.930	1.658
.135	.136	.335	.348	.535	.597	.735	.940	.935	1.697
.140	.141	.340	.354	.540	.604	.740	.950	.940	1.738
.145	.146	.345	.360	.545	.611	.745	.962	.945	1.783
.150	.151	.350	.365	.550	.618	.750	.973	.950	1.832
.155	.156	.355	.371	.555	.626	.755	.984	.955	1.886
.160	.161	.360	.377	.560	.633	.760	.996	.960	1.946
.165	.167	.365	.383	.565	.640	.765	1.008	.965	2.014
.170	.172	.370	.388	.570	.648	.770	1.020	.970	2.092
.175	.177	.375	.394	.575	.655	.775	1.033	.975	2.185
.180	.182	.380	.400	.580	.662	.780	1.045	.980	2.298
.185	.187	.385	.406	.585	.670	.785	1.058	.985	2.443
.190	.192	.390	.412	.590	.678	.790	1.071	.990	2.647
.195	.198	.395	.418	.595	.685	.795	1.085	.995	2.994

Reprinted, by permission, from Allen L. Edwards, *Statistical methods for the behavioral sciences*, Rinehart & Company, Inc., New York.

Index